P9-BZA-359

REFRIGERATION AND AIR-CONDITIONING

Refrigeration and Air-Conditioning

A. R. Trott

McGRAW-HILL Book Company (UK) Limited

London · New York · St Louis · San Francisco · Auckland
Bogotá · Guatemala · Hamburg · Johannesburg · Lisbon
Madrid · Mexico · Montreal · New Delhi · Panama · Paris
San Juan · São Paulo · Singapore · Sydney · Tokyo · Toronto

Published by
McGraw-Hill Book Company (UK) Limited
Maidenhead · Berkshire · England

British Library Cataloguing in Publication Data

Trott, A R
Refrigeration and air-conditioning.
1. Refrigeration and refrigerating machinery
2. Air conditioning
I. Title
621.5′6 TP492 80-41474

ISBN 0-07-084543-3

1 2 3 4 5 8 3 2 1

Typeset by Santype International Ltd., Salisbury
PRINTED AND BOUND
Robert Hartnoll Limited Bodmin Cornwall

CONTENTS

REFRIGERATION: The process of removing heat.

AIR-CONDITIONING: A form of air treatment whereby temperature, humidity, ventilation, and air cleanliness are all controlled within limits determined by the requirements of the air conditioned enclosure.

BS 5643 : 1979

CHAPTER
ONE

UNITS OF MEASUREMENT. BASIC PHYSICS

1-1 INTERNATIONAL SYSTEM (SI)

The *International System of Units* (SI) provides a coherent system of measurement units, and all the physical quantities required for refrigeration and air-conditioning can be derived from the basic standards:

Length	metre	m
Mass	kilogram	kg
Time	second	s
Electric current	ampere	A
Temperature	kelvin	K
Electric potential	volt	V

From these basic units are derived

Area	square metre	m^2
Volume	cubic metre	m^3
Liquid volume	litre	$m^3 \times 10^{-3}$
Power	watt	W (ampere volt)

Force	newton	N (kg m/s^2)
Energy (Work)	joule	J (N m or W s)
Pressure	pascal	Pa (N/m^2)
also	bar	bar (Pa × 10^5)
Temperature	degree Celsius	°C (K − 273.15)

From these, in turn, can be derived other units for use in the calculation of refrigeration and air-conditioning loads:

Specific heat capacity	J/kg K or kJ/kg K
Specific enthalpy	J/kg or kJ/kg
Thermal conductivity	W/m K (W m/m^2 K)
Thermal conductance	W/m^2 K

In addition to SI, there are a number of expressions which remain in common use, since much available data is still recorded in these units, and practising engineers should be familiar with them:

Thermal energy	British thermal unit	Btu	= 1.055 kJ
	therm (Btu × 10^5)	therm	= 105.5 MJ
	kilocalorie	kcal	= 4.187 kJ
Thermal work	British thermal units per hour	Btu/h	= 0.293 W
	kilocalories per hour	kcal/h	= 1.163 W
	ton refrigeration	TR or t.r.	= 3.517 kW
Electrical energy	'unit of electricity'	kW h	= 3.6 MJ
Volume	Imperial gallon	Imp gal	= 4.546 litre
	US gallon	US gal	= 3.785 litre
Mass	pound	lb	= 0.4537 kg
	Imperial ton (2240 lb)	ton	= 1016 kg
	US ton (2000 lb)	US ton	= 907 kg
Length	foot	ft	= 0.305 m
Temperature	degree Fahrenheit	°F	= (1.8 × °C) + 32

Force	pound-force	lbf	= 4.448 N
Pressure	pound-force per square inch	lbf/in^2	= 6.895 kPa
	kilogram-force per square centimetre	kgf/cm^2	= 98.07 kPa
	inch water gauge	in w.g.	= 249 Pa

Other terms not given here may be encountered from time to time and will be found in standard reference works.[1, 2, 3, 4]

1-2 BASIC PHYSICS—HEAT

This book does not deal with basic physical processes other than to remind the reader of those properties which will be met in the subject matter.

Heat is a form of energy and the heat of a body is the molecular energy within it.

If a change of heat content can be sensed as a change of temperature, it is called sensible heat. If it results in a change of state (solid to liquid, liquid to gas, or vice versa), without any change of sensed temperature, it is called latent heat. The total heat of a body is expressed as *enthalpy* (see Fig. 1-1), measured from a base temperature, usually 0 K or −40°C (also −40°F).

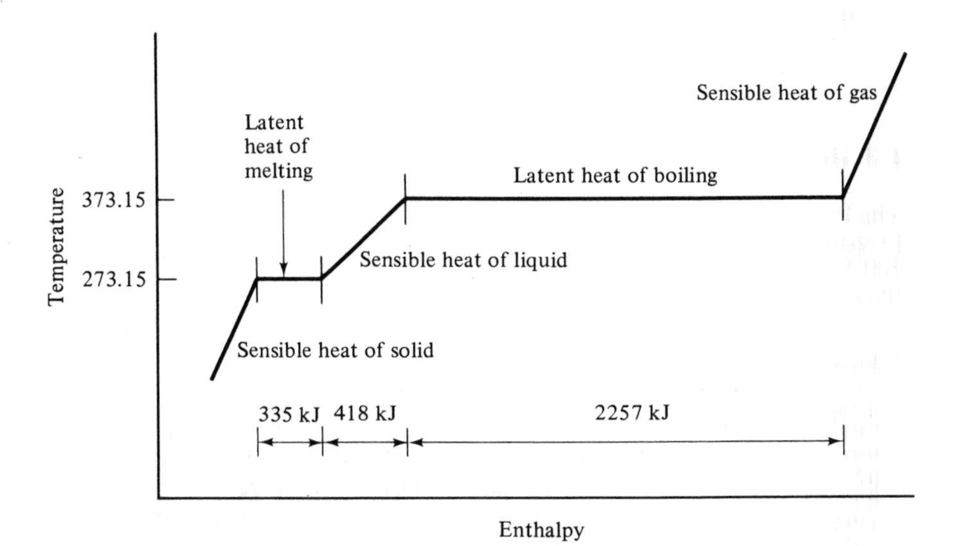

Figure 1-1 Change of temperature and state with enthalpy.

Example 1-1 For water, the latent heat of freezing is 335 kJ/kg and the specific heat capacity of the liquid averages 4.187 kJ/kg K. The quantity of heat to be removed from 1 kg of water at 30°C in order to turn it into ice at 0°C is

$$4.187(30 - 0) + 335 = 460.6 \text{ kJ/kg}$$

Example 1-2 If the latent heat of boiling of water at 1.013 bar is 2257 kJ/kg, the quantity of heat which must be added to 1 kg of water at 30°C in order to boil it is

$$4.187(100 - 30) + 2257 = 2550 \text{ kJ/kg}$$

Example 1-3 The specific enthalpy of water, taken from 0°C base, is 334.91 kg/kg at 80°C. What is the average specific heat capacity through the range 0 to 80°C?

$$\frac{334.91}{80} = 4.186 \text{ kJ/kg K}$$

1-3 BOILING POINT

The temperature at which a liquid boils is not constant, but varies with the pressure. Thus, while the *boiling point* of water is commonly taken as 100°C, this is only true at a pressure of one standard atmosphere (1.013 bar) and, by varying the pressure, the boiling point can be changed, viz.

Pressure, bar	Boiling point, °C
0.006	0
0.04	29
0.08	41.5
0.2	60.1
0.5	81.4
1.013	100.0

This pressure–temperature property can be shown graphically (see Fig. 1-2).

The boiling point is limited by the *critical temperature* at the upper end, beyond which it cannot exist as a liquid, and by the *triple point* at the lower end, which is at the freezing temperature. Between these two limits, if the fluid is at a pressure higher than its boiling pressure it will remain a liquid, while if the temperature is higher than its boiling temperature it will exist as a gas. If both liquid and vapour are present in the same enclosure, and no other volatile substance is present, the condition must lie on the p–T line.

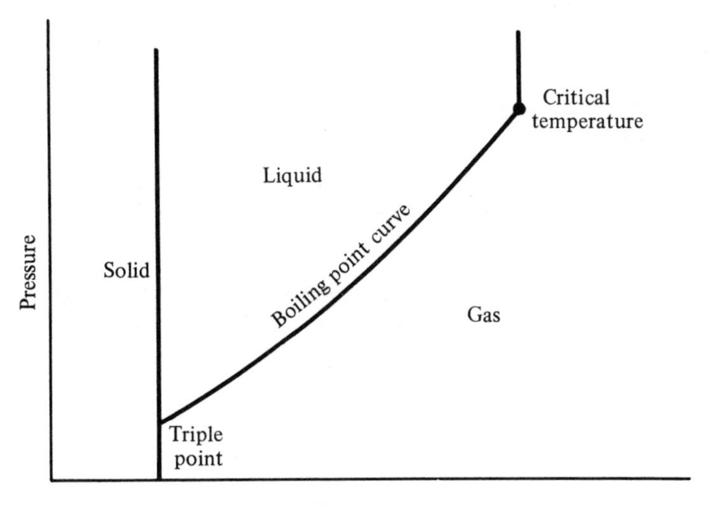

Figure 1-2 Change of state with pressure and temperature.

At a pressure below the triple point pressure, the solid can change directly to a gas (sublimation) and the reverse, as in the formation of carbon dioxide snow from the released gas.

The liquid zone to the left of the boiling point line is sub-cooled liquid. The gas under this line is superheated gas.

1-4 GENERAL GAS LAWS. BOYLE'S LAW

Boyle's Law states that, for an ideal gas, the product of pressure and volume at constant temperature is a constant:

$$pV = \text{constant}$$

No gas obeys this law exactly, so there is no 'ideal' gas, but the agreement is close enough for general engineering purposes.

Example 1-4 A volume of an ideal gas in a cylinder and at atmospheric pressure is compressed to half the volume at constant temperature. What is the new pressure?

$$p_1 V_1 = \text{constant}$$

$$= p_2 V_2$$

$$\frac{V_1}{V_2} = 2$$

$$\text{so} \quad p_2 = 2 \times p_1$$

$$= 2 \times 1.013\,25 \text{ bar } (101\,325 \text{ Pa})$$

$$= 2.026\,5 \text{ bar}$$

1-5 CHARLES' LAW

Charles' Law states that, for an ideal gas, the volume at constant pressure is proportional to the absolute temperature:

$$\frac{V}{T} = \text{constant}$$

Example 1-5 A mass of an ideal gas occupies 0.75 m³ at 20°C and is heated at constant pressure to 90°C. What is the final volume?

$$V_2 = V_1 \times \frac{T_2}{T_1}$$

$$= 0.75 \times \frac{273.15 + 90}{273.15 + 20}$$

$$= 0.93 \text{ m}^3$$

1-6 GENERAL GAS LAW

These two are combined in the *General Gas Law*, which is expressed as

$$pV = (\text{a constant}) \times T$$

The constant is mass × R, where R is the *gas constant*, so

$$pV = mRT$$

Example 1-6 What is the volume of 5 kg of a gas, having a gas constant of 287 J/kg K, at a pressure of one standard atmosphere and at 25°C?

$$pV = mRT$$

$$V = \frac{mRT}{p}$$

$$= \frac{5 \times 287(273.15 + 25)}{101\,325}$$

$$= 4.22 \text{ m}^3$$

1-7 DALTON'S LAW

Dalton's Law of partial pressures considers a mixture of two or more gases, and states that the total pressure of the mixture is equal to the sum of the individual pressures, if each gas separately occupied the space.

Example 1-7 A cubic metre of air contains 0.906 kg nitrogen of gas constant 297, 0.384 kg oxygen (gas constant 260), and 0.015 kg argon (gas constant 208). What will be the total pressure at 20°C?

$$pV = mRT$$

$$V = 1$$

$$\text{so} \quad p = mRT$$

For the nitrogen $p_N = 0.906 \times 297 \times 293.15 = 78\,881$ Pa

For the oxygen $p_O = 0.278 \times 260 \times 293.15 = 21\,189$ Pa

For the argon $p_A = 0.015 \times 208 \times 293.15 = 915$ Pa

Total pressure $= 100\,985$ Pa

(1.009 85 bar)

1-8 HEAT TRANSFER

Heat will move from a hot body to a colder one, and can do so by three methods:

Conduction. Direct from one body touching the other, or through a continuous solid or liquid mass

Convection. By means of a heat-carrying fluid moving between one and the other, the bodies not being in contact

Radiation. Mainly by infrared waves, which are independent of contact or an intermediate fluid

Conduction through a continuous material is expressed directly by its area, thickness, and a conduction coefficient. For the plane surface,

$$\text{Conductivity} = \frac{\text{area} \times \text{specific conductance}}{\text{thickness}}$$

$$= \frac{A \times k}{L}$$

and the heat conducted is

$$Q_f = \text{conductivity} \times (T_1 - T_2)$$

Example 1-8 A brick wall, 225 mm thick and having a specific conductance of 0.60 W/m K, measures 10 m long by 3 m high, and has a temperature difference between the inside and outside faces of 25 K. What is the rate of heat conduction?

$$Q_f = \frac{10 \times 3 \times 0.60 \times 25}{0.225}$$

$$= 2000 \text{ W (or 2 kW)}$$

Thermal conductivities, in watts per metre kelvin, for various common materials are as follows:

Copper	200	Cork	0.040
Mild steel	50	Expanded polystyrene	0.034
Concrete	1.5	Polyurethane foam	0.026
Water	0.62	Still air	0.026

Conductivities for other materials can be found from standard reference works.[1, 2, 5]

Convection requires a fluid, liquid or gaseous, which is free to move between the hot and cold bodies. This mode of heat transfer is very complex and depends firstly on whether the flow of fluid is 'natural', i.e., caused by thermal currents set up in the fluid as it expands, or 'forced' by fans or pumps. Other parameters are the density, specific heat capacity, and viscosity of the fluid and the shape of the interacting surface.

With so many factors, expressions for convective heat flow cannot be as simple as those for conduction. The interpretation of observed data has been made possible by the use of a number of groups which combine the variables and which can then be used to estimate convective heat flow.

The main groups used in such estimates are as follows:

Number	Sign	Parameters
Reynolds	Re	Velocity of fluid Density of fluid Viscosity of fluid Dimension of surface
Grashof	Gr	Coefficient of expansion of fluid Density of fluid Viscosity of fluid Force of gravity Temperature difference Dimension of surface
Nusselt	Nu	Thermal conductivity of fluid Dimension of surface Rate of heat transfer
Prandtl	Pr	Specific heat capacity of fluid Viscosity of fluid Thermal conductivity of fluid

The calculation of every heat transfer coefficient for a refrigeration or air-conditioning system would be a very time-consuming process, even with modern methods of calculation. Formulas based on these factors will be found in standard reference works, expressed in terms of heat transfer coefficients under different conditions of fluid flow.[1, 4, 6, 7, 8, 9]

Example 1-9 A formula for heat transfer between air and a vertical plane surface [see Ref. 1, sec. 6(19)] gives

$$h' = 0.99 + 0.21V$$

where h' is the heat transfer in British thermal units per hour degree Fahrenheit square foot and V is the velocity of air in feet per second.

What is the thermal conductance for an air velocity of 3 m/s?

$$V = 3/0.305$$

$$= 9.84 \text{ ft/s}$$

$$h' = 0.99 + (0.21 \times 9.84)$$

$$= 3.056 \text{ Btu/h °F ft}^2$$

$$= 17.3 \text{ W/m}^2 \text{ K}$$

In most cases of heat transfer by conduction, there will be some convective effect at the surfaces. The overall heat transfer is calculated as an overall resistance, this being the sum of all the resistances to heat flow:

$$R_t = R_i + R_c + R_o$$

$$U = \frac{1}{R_t}$$

where R_t = total thermal resistance
R_i = inside convective resistance
R_c = conductive resistance
R_o = outside convective resistance
U = overall transmittance

Example 1-10 A brick wall, plastered on one face, has a thermal conductivity of 2.8 W/m² K, an inside surface resistance of 0.3 m² K/W, and an outside surface resistance of 0.05 m² K/W. What is the overall transmittance?

$$R_t = R_i + R_c + R_o$$

$$= 0.3 + \frac{1}{2.8} + 0.05$$

$$= 0.707$$

$$U = 1.414 \text{ W/m}^2 \text{ K}$$

Typical overall thermal transmittances are as follows:

Cavity brick wall 260 mm thick, sheltered exposure on outside	1.5 $W/m^2 K$
Chilled water inside copper tube, forced draught air flow outside	28 $W/m^2 K$
Condensing ammonia gas inside steel tube, thin film of water outside	650 $W/m^2 K$

Special note should be taken of the influence of geometrical shape, where other than plain surfaces are involved.

The overall thermal transmittance, U, is used to calculate the total heat flow. For a plane surface of area A and a steady temperature difference ΔT, it is

$$Q_f = A \times U \times \Delta T$$

If a fluid is being heated or cooled it will undergo a change of sensible heat, changing its temperature, so that ΔT across the heat exchanger wall will not be constant. Since the rate of temperature change (heat flow) will be proportional to the ΔT at any one point on its surface, the space–temperature curve will approximate to a parabola (see Fig. 1-3).

The average temperature difference over the length of the curve is expressed as

$$\Delta T = \frac{\Delta T_{max} - \Delta T_{min}}{\ln (\Delta T_{max}/\Delta T_{min})}$$

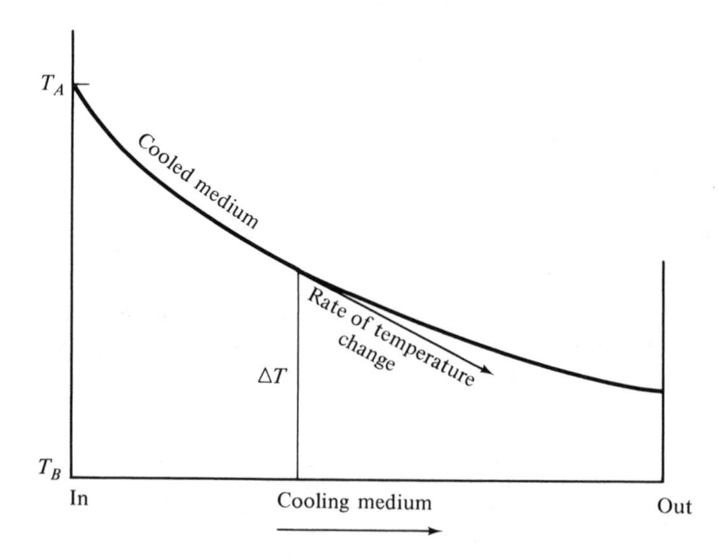

Figure 1-3 Changing temperature difference of a cooled fluid.

This is applicable to any heat transfer where either or both the media change in temperature (see Fig. 1-4). This derived term is the logarithmic mean temperature difference (ln MTD) and can be used as ΔT in the general equation, providing U is constant throughout the cooling range, or an average figure is known, giving

$$Q_f = A \times U \times \ln \text{MTD}$$

Example 1-11 A liquid chiller evaporates at 3°C and cools water from 11.5 to 6.4°C. What is the logarithmic mean temperature difference and what is the heat transfer if it has a surface of 420 m^2 and the thermal transmittance is 110 W/m^2 K?

$$\Delta T_{\max} = 11.5 - 3 = 8.5 \text{ K}$$

$$\Delta T_{\min} = 6.4 - 3 = 3.4 \text{ K}$$

$$\ln \text{MTD} = \frac{8.5 - 3.4}{\ln (8.5/3.4)}$$

$$= 5.566 \text{ K}$$

$$Q_f = 420 \times 110 \times 5.566$$

$$= 257\,000 \text{ W or } 257 \text{ kW}$$

If the heat exchanger was of infinite size, the space–temperature curves would eventually meet and no further heat could be transferred. The chiller in Example 1-11 would cool the water down to 3°C. The *effectiveness* of a heat exchanger can be expressed as the ratio of heat actually transferred to the ideal maximum:

$$\Sigma = \frac{T_{A\,\text{in}} - T_{A\,\text{out}}}{T_{A\,\text{in}} - T_{B\,\text{in}}}$$

Taking the heat exchanger in Example 1-11,

$$\Sigma = \frac{11.5 - 6.4}{11.5 - 3.0}$$

$$= 0.6 \text{ or } 60\%$$

Example 1-12 A finned coil cools air from 24 to 13°C, using evaporating refrigerant at 8.6°C. What is the coil effectiveness?

$$\Sigma = \frac{24 - 13}{24 - 8.6}$$

$$= 0.71 \text{ or } 71\%$$

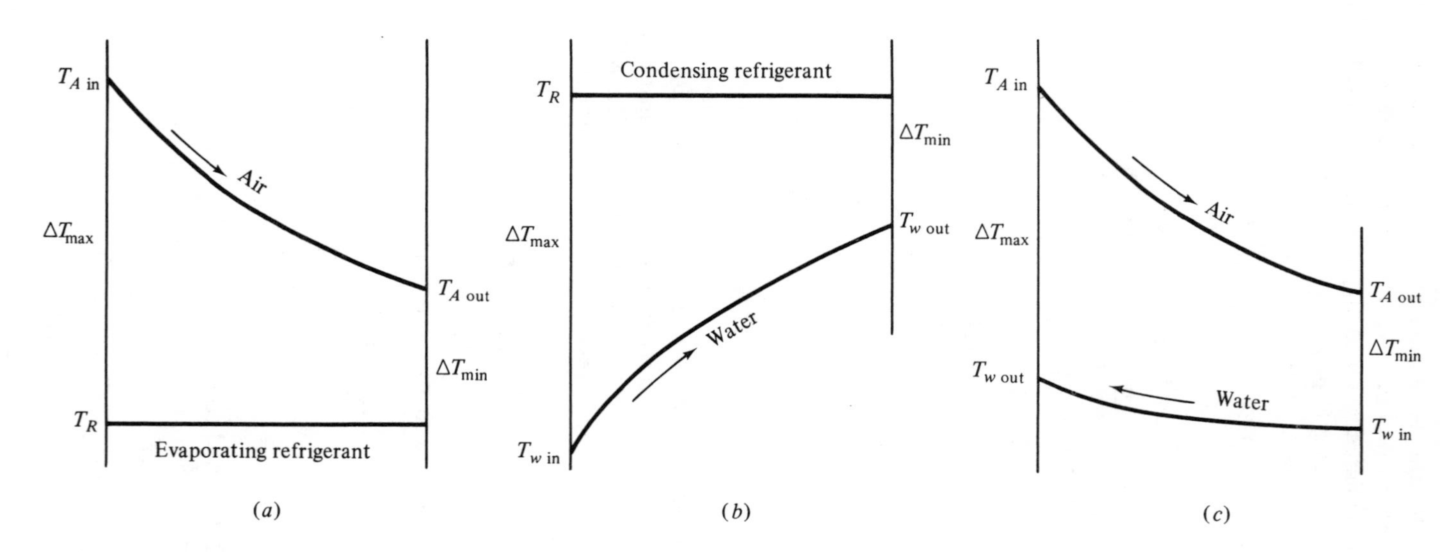

Figure 1-4 Temperature change: (*a*) refrigerant cooling fluid, (*b*) fluid cooling refrigerant, (*c*) two fluids.

Radiation of heat, mainly by infrared waves, was shown by Boltzman and Stefan to be proportional to the fourth power of the absolute temperature and to depend on the colour, material, and texture of the surface:

$$Q_f = \sigma \varepsilon T^4$$

where σ is Stefan's constant ($= 5.67 \times 10^{-8}$ W/m^2 K^4) and ε is the surface emissivity.

Emissivity figures for common materials have been determined, and are expressed as the ratio to the radiation by a perfectly black body, viz.

Rough surfaces such as brick, concrete, or tile, regardless of colour	0.85–0.95
Metallic paints	0.40–0.60
Unpolished metals	0.20–0.30
Polished metals	0.02–0.28

Where one body is losing heat to another by radiation, it must be borne in mind that both are radiating heat according to their respective temperatures and the net loss by the hotter body will be the difference of these. For two plane surfaces facing each other, the net loss becomes

$$Q_f = 5.67 \times 10^{-8} \times A(\varepsilon_1 T_1^4 - \varepsilon_2 T_2^4)$$

Example 1-13 What is the radiant heat transfer between two unpolished aluminium surfaces, of area 4 m^2, one being at 6°C and the other at 45°C, where ε is 0.27 for both surfaces?

$$Q_f = 5.67 \times 10^{-8} \times 4 \times 0.27(318.15^4 - 279.15^4)$$
$$= 256 \text{ W}$$

(This example indicates the amount of radiant heat transfer which might occur between a cooling coil and a re-heat coil if placed close together in the same duct. It serves to illustrate that radiation heat transfer can usually be ignored in refrigeration and air-conditioning calculations, with the exception of solar heat gain.)

1-9 MASS TRANSFER

Heat may be transferred as sensible or latent heat. In the latter case boiling or condensation will take place at the heat exchanger surface.

Where there is a mixture of two gases, notably water vapour and air, the vapour will diffuse through the air to condense on a cold surface. The rate at which it may condense will therefore depend on the rate of heat transfer available and the concentration of vapour in the mixture. The rate of change of this concentration is considered in the same way as the temperature difference in heat flow, to give mass transfer coefficients.[1, 6, 10]

CHAPTER

TWO

THE REFRIGERATION CYCLE

2-1 BASIC CYCLE

A liquid boils and condenses—the change between the liquid and gaseous states—at a temperature which depends on its pressure, within the limits of its freezing point and critical temperature. In boiling it must obtain the latent heat of evaporation and in condensing the latent heat must be given up again.

The basic refrigeration cycle (Fig. 2-1) makes use of the boiling and condensing of a working fluid at different temperatures and, therefore, at different pressures. In boiling at the lower temperature and pressure, the fluid takes in latent heat and is changed to a dry gas. The gas is raised in pressure within a mechanical device to the higher pressure corresponding to the condensing temperature and gives up its latent heat at the higher pressure to change back to a liquid.

The total refrigeration effect will be the heat transferred to the working fluid in the boiling or evaporating vessel, i.e., the change in enthalpy between condensed liquid and the gas leaving the evaporator:

Enthalpy of fluid leaving condenser	= 91.4 kJ/kg
Enthalpy of fluid entering evaporator	= 91.4 kJ/kg
Enthalpy of gas leaving evaporator	= 249.9 kJ/kg
Refrigerating effect	= 249.9 − 91.4 = 158.5 kJ/kg

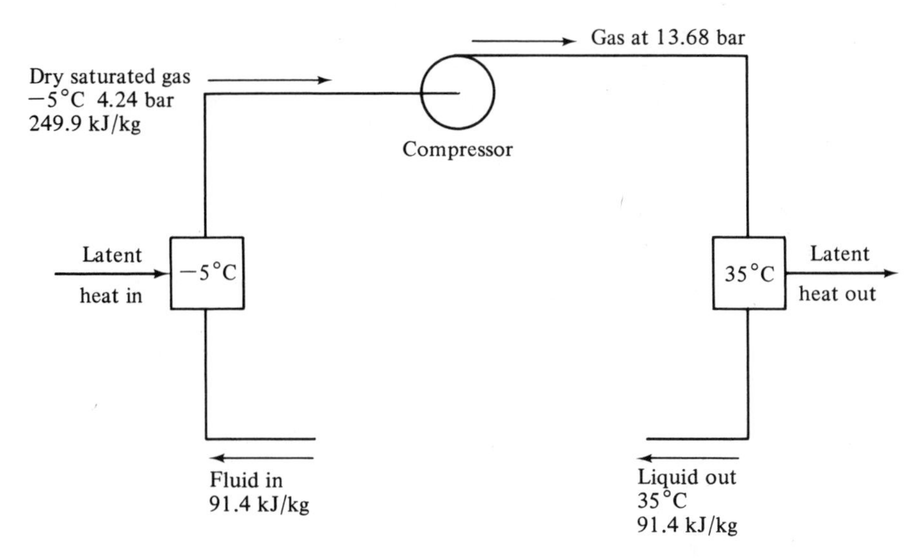

Figure 2-1 Basic refrigeration cycle.

The complete circuit (Fig. 2-2) will require a connection between the condenser and the inlet to the evaporator, having a pressure-reducing valve. The reduction in pressure at this valve will cause a drop in temperature of the working fluid, and some of it will flash off to vapour to remove the energy for this cooling. The volume of the working fluid therefore increases at the valve by this amount of flash gas and gives rise to its name, the expansion valve.

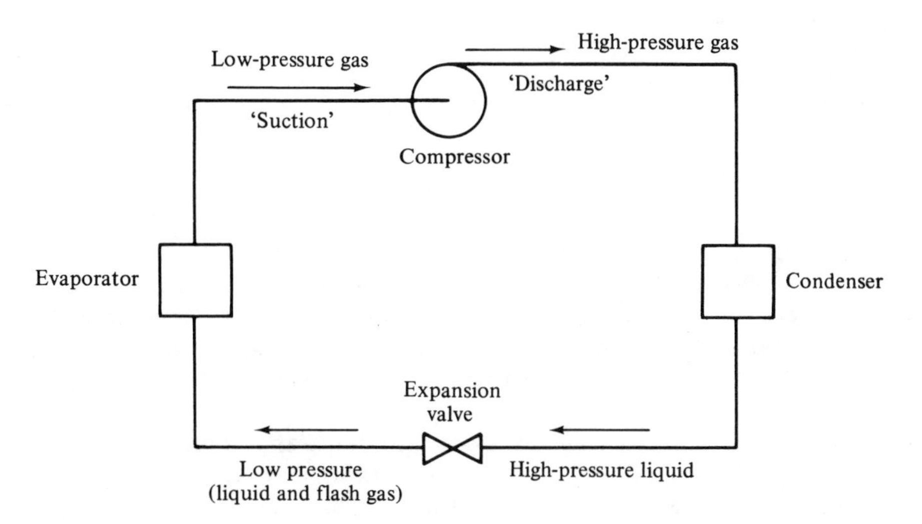

Figure 2-2 Complete basic cycle.

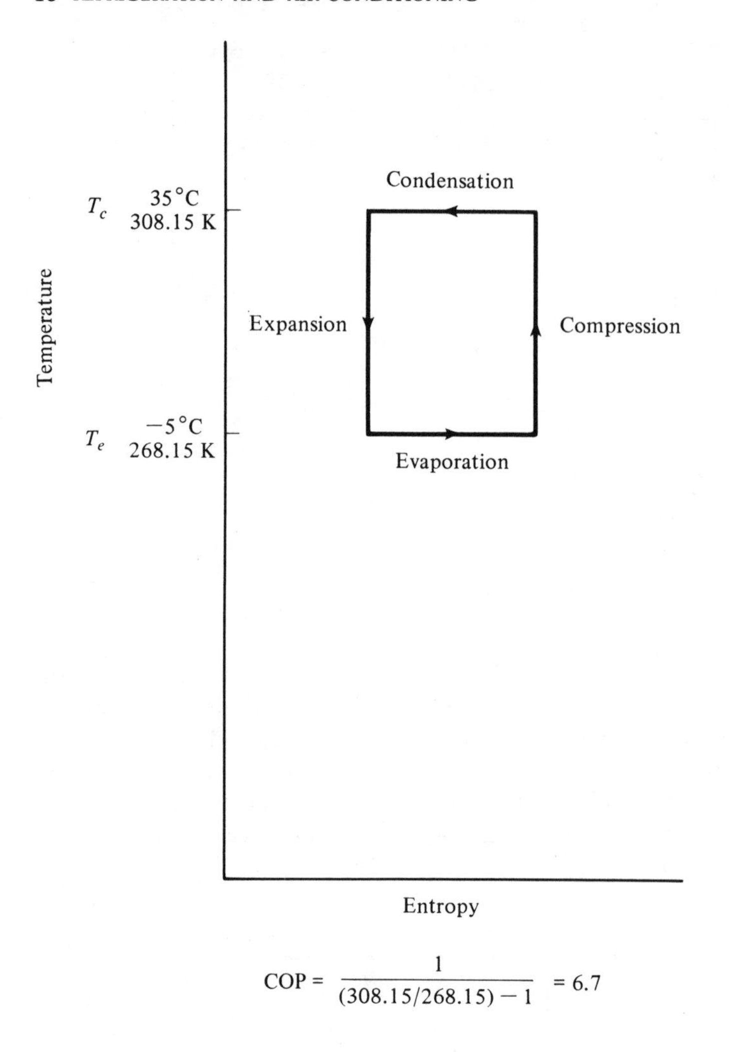

$$\text{COP} = \frac{1}{(308.15/268.15) - 1} = 6.7$$

Figure 2-3 Ideal reversed Carnot cycle.

The cycle can be represented in terms of the *reversed Carnot* cycle (Fig. 2-3) which supposes no temperature difference at the two heat exchangers—all heat passing at constant temperature. Since some temperature difference is required for heat to flow into the evaporator and out of the condenser, the *modified* reversed Carnot cycle (Fig. 2-4) is more accurate.

A more informative diagram is based on the graph of pressure against enthalpy, and gives a direct measure of heat transferred in the processes. In practice, the gas leaving the evaporator is slightly superheated, while the liquid leaving the condenser will be slightly sub-cooled, points A_1 and C_1

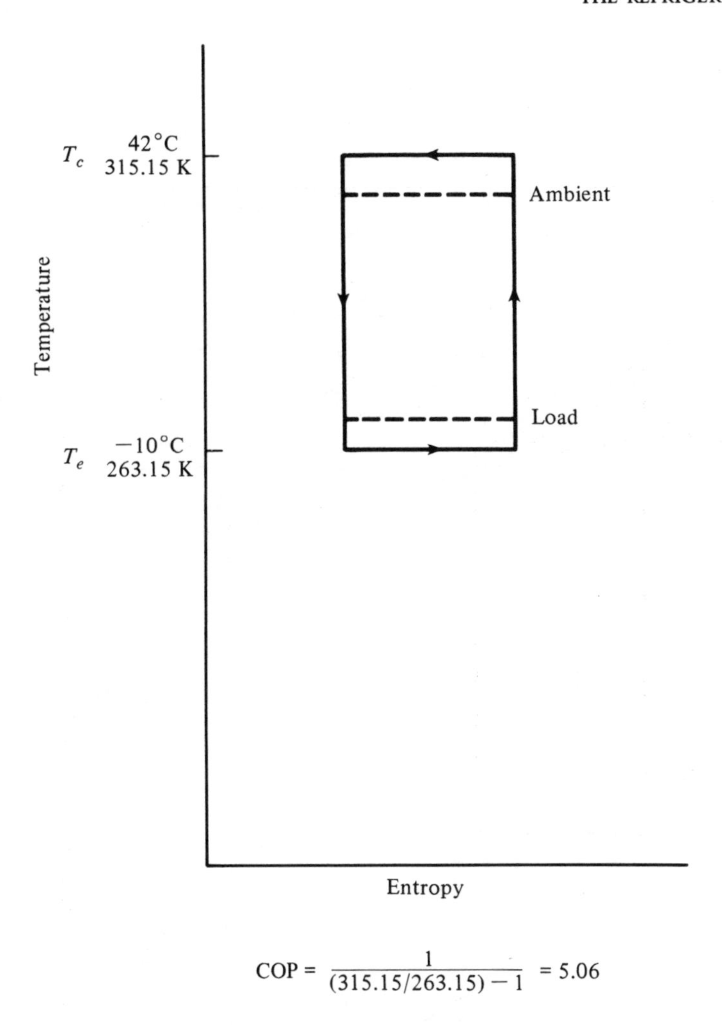

$$\text{COP} = \frac{1}{(315.15/263.15) - 1} = 5.06$$

Figure 2-4 Modified reversed Carnot cycle.

in Fig. 2-5. The refrigerating effect and the energy put in by the compressor may be read off directly in terms of the enthalpy of the fluid. Compression is assumed to be adiabatic, but may vary according to the type of compressor. Pressure losses will occur across the gas inlets and outlets, and there will be pressure losses through the evaporator and the condenser. The final temperature at the end of compression will depend on the working conditions and on the refrigerant.

It will be seen from the pressure–enthalpy diagram (also known as the *Mollier chart*) that the states of the working fluid are indicated by the sector on the diagram. The liquid leaving the condenser, slightly subcooled, at C_1 enters the expansion valve and loses pressure to point D_1.

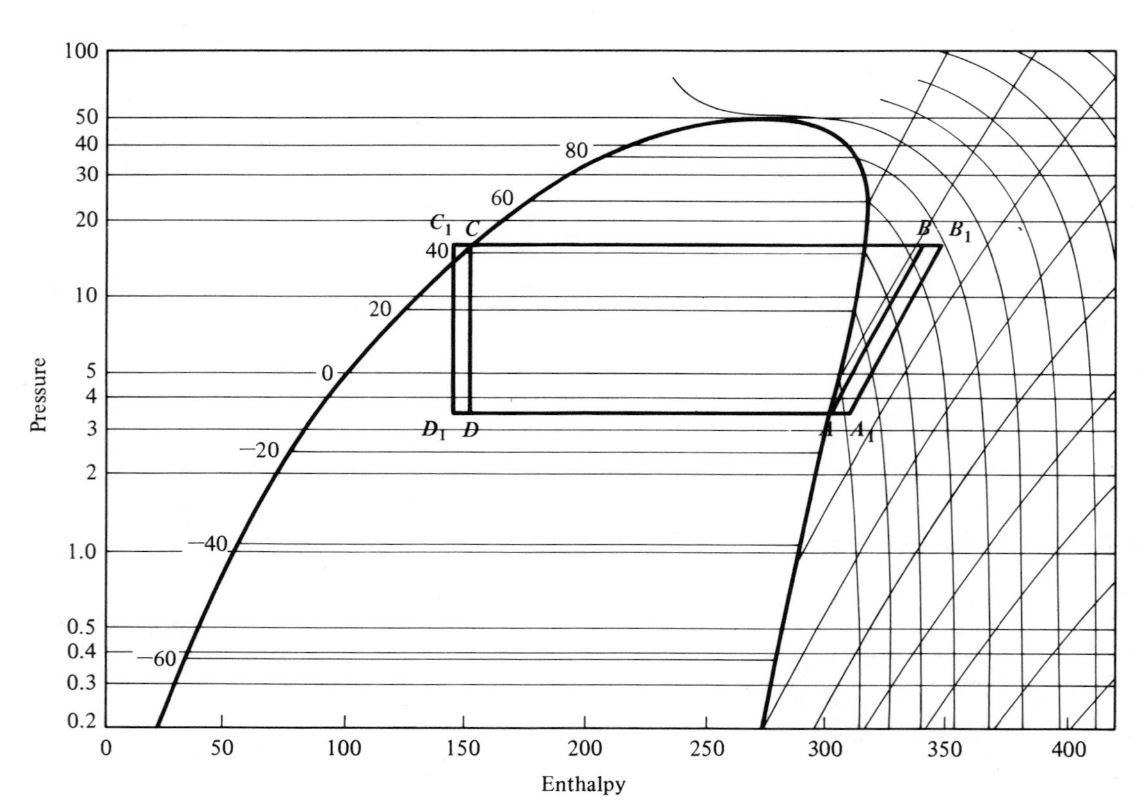

Figure 2-5 Pressure–enthalpy or Mollier diagram. *(From Ref. 3, fig. B14.15. Courtesy of the Chartered Institute of Building Services.)*

The distance of D_1 between the two limits of the chart indicates the proportion of flash gas at that point. The condenser receives the high-pressure superheated gas, cools it down to saturation temperature, condenses it to liquid, and finally sub-cools it slightly. The energy removed in the condenser is seen to be the refrigerating effect plus the heat of compression.

2-2 HEAT EXCHANGER SIZE

The transfer of heat through the walls of the evaporator and condenser requires a temperature difference, and the larger the surface area of the heat exchanger the lower will be this temperature difference and the closer the evaporator and condenser temperatures will approach those of the load and cooling medium. The closer this approach, the nearer the cycle will be to the ideal reversed Carnot cycle.

	Evaporator		Condenser		Compression
	Temperature	Pressure	Temperature	Pressure	ratio
Ideal reversed Carnot	−5°C	4.24	35°C	13.68	3.23
Modified reversed Carnot, $\Delta T = 5$ K	−10°C	3.54	40°C	15.34	4.33
Modified reversed Carnot, $\Delta T = 10$ K	−15°C	2.96	45°C	17.3	5.85

(Pressures are shown for an R.22 system.)

In the practical cycle, there are three effects resulting from heat exchanger size. First, the larger the evaporator, with consequent higher suction pressure, the denser is the gas entering the compressor and the greater the mass of gas handled by a given swept volume. Second, the larger the condenser, the lower will be the condensing temperature and the colder the liquid entering the expansion valve, giving more refrigerating effect. Third, the larger the evaporator, the lower the compression ratio of the compressor, using less power.

2-3 VOLUMETRIC EFFICIENCY

In a reciprocating compressor, there will be a small amount of clearance space at the top of the stroke, arising from gas ports, manufacturing tolerances, and an allowance for thermal expansion and contraction of the

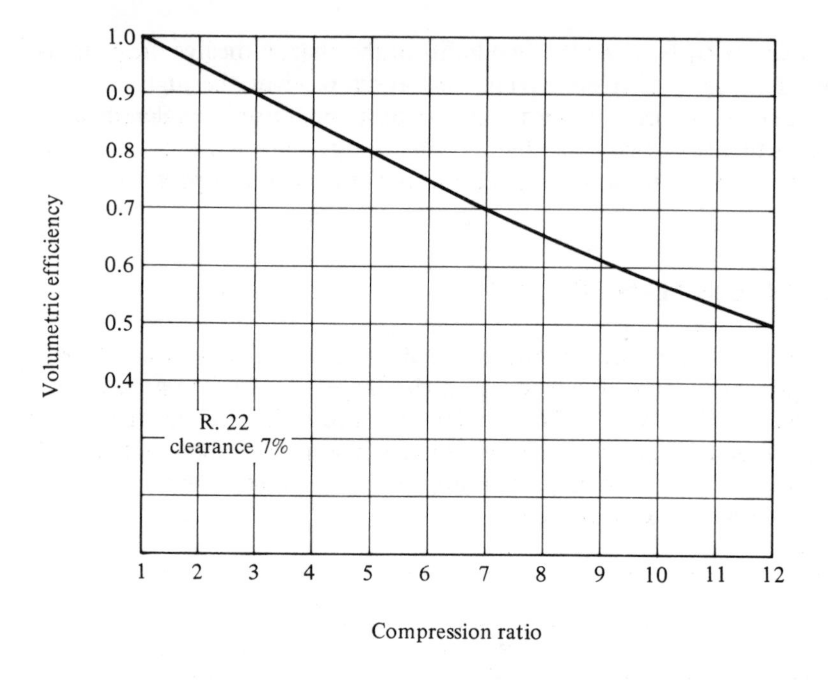

Figure 2-6 Volumetric efficiency.

components in operation. This clearance space is usually of the order of 4 to 7 per cent of the swept volume. High-pressure gas left in this space at the end of the discharge stroke must re-expand to the suction inlet pressure before a fresh charge of gas can be drawn in. This loss of useful working stroke will vary according to the compression ratio, and the compressor pumping efficiency will fall off. This effect is termed the *volumetric efficiency.*[11] Typical figures are shown in Fig. 2-6.

2-4 SUCTION-TO-LIQUID HEAT EXCHANGERS

Cold gas returning from the evaporator to the compressor can be used to pre-cool the warm liquid passing from the condenser to the expansion valve, using a suction-to-liquid heat exchanger (Fig. 2-7). In cooling the liquid and reducing its enthalpy, a greater refrigerating effect will be obtained.

Example 2-1 In an R.22 system evaporating at −15°C and condensing at 35°C, the use of a suction-to-liquid heat exchanger will cool the liquid from 33 to 28.5°C. What is the gain of refrigerating effect?

	Without heat exchanger	With heat exchanger
h_g at −15°C	219.4 kJ/kg	219.4 kJ/kg
h_f at 33°C	76.5	
h_f at 28.5°C		71.3
Refrigerating effect	142.9	148.1

$$\text{Gain of refrigerating effect} = \frac{148.1 - 142.9}{142.9} = 3.6\%$$

This gain is offset to a greater or lesser extent by the superheating of the suction gas and the resultant reduction of mass flow into the compressor. The liquid is cooled by 5.2 kJ/kg and this heat is given to the suction gas, raising its temperature (by reference to refrigerant tables) to 0°C. The resulting loss of mass flow can be determined from the general gas equation

$$pV = mRT$$

in which p, V, and R remain constant with or without the heat exchanger, giving

$$\frac{m_2}{m_1} = \frac{T_1}{T_2}$$

$$= \frac{273.15 + (-15)}{273.15 + 0}$$

$$= 0.945$$

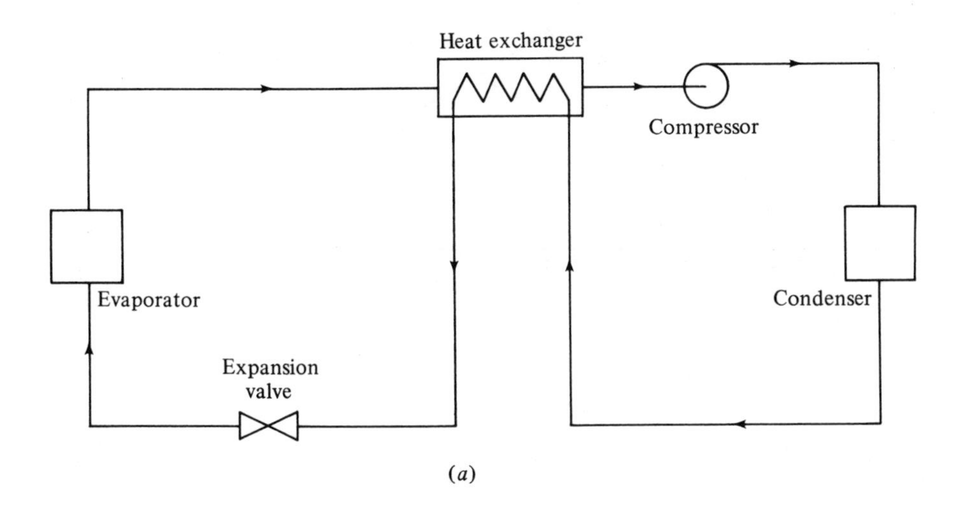

Figure 2-7 Suction-to-liquid heat exchanger: (*a*) circuit.

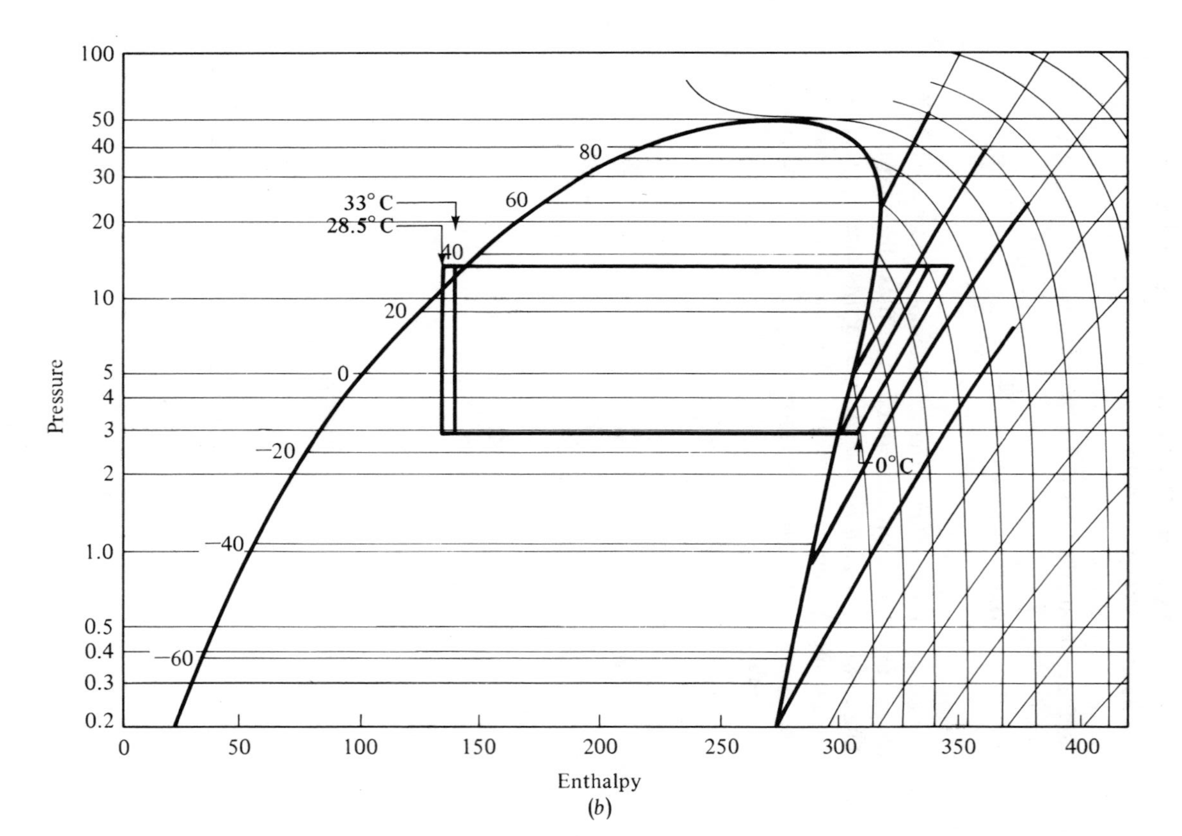

Figure 2-7 (*cont.*) (*b*) thermal effects.

So the loss of mass flow is 5.5 per cent and the overall loss of cooling capacity is 1.9 per cent.

The overall effect of fitting a suction-to-liquid heat exchanger in terms of thermodynamic efficiency will vary with the refrigerant and the operating conditions. Generally speaking, there is a slight gain with R.12 and a slight loss with all other refrigerants. There are other reasons for fitting such heat exchangers (see Chapters 7 and 9).

2-5 MULTISTAGE CYCLES

Where the compression ratio is high enough to cause a serious drop in volumetric efficiency or an unacceptably high discharge temperature, multistage systems are required.

The *cascade* cycle has two separate refrigeration systems, one acting as a condenser to the other (see Fig. 2-8). Different refrigerants may be used in the two systems and high-pressure refrigerants such as R.13 are common in the lower stage of low-temperature plants.

Compound systems use the same refrigerant throughout a common circuit, compressing in two or more stages (see Fig. 2-9). Discharge gas from the first compression stage will be too hot to pass directly to the high-stage compressor, so is cooled in an intercooler, using some of the available refrigerant from the condenser. The opportunity is also taken to sub-cool liquid passing to the evaporator. Small compound systems may cool the interstage gas by direct injection of liquid refrigerant into the pipe.

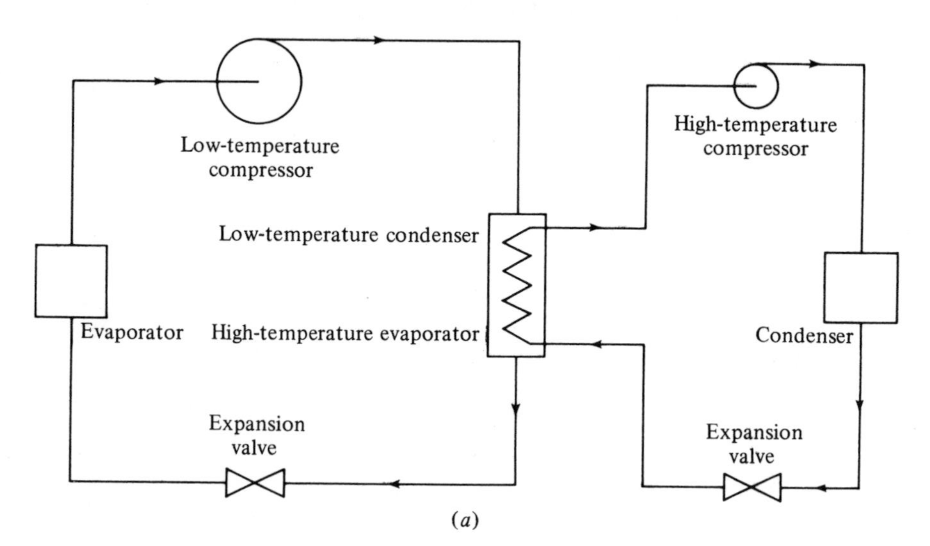

Figure 2-8 Cascade cycle: (*a*) circuits.

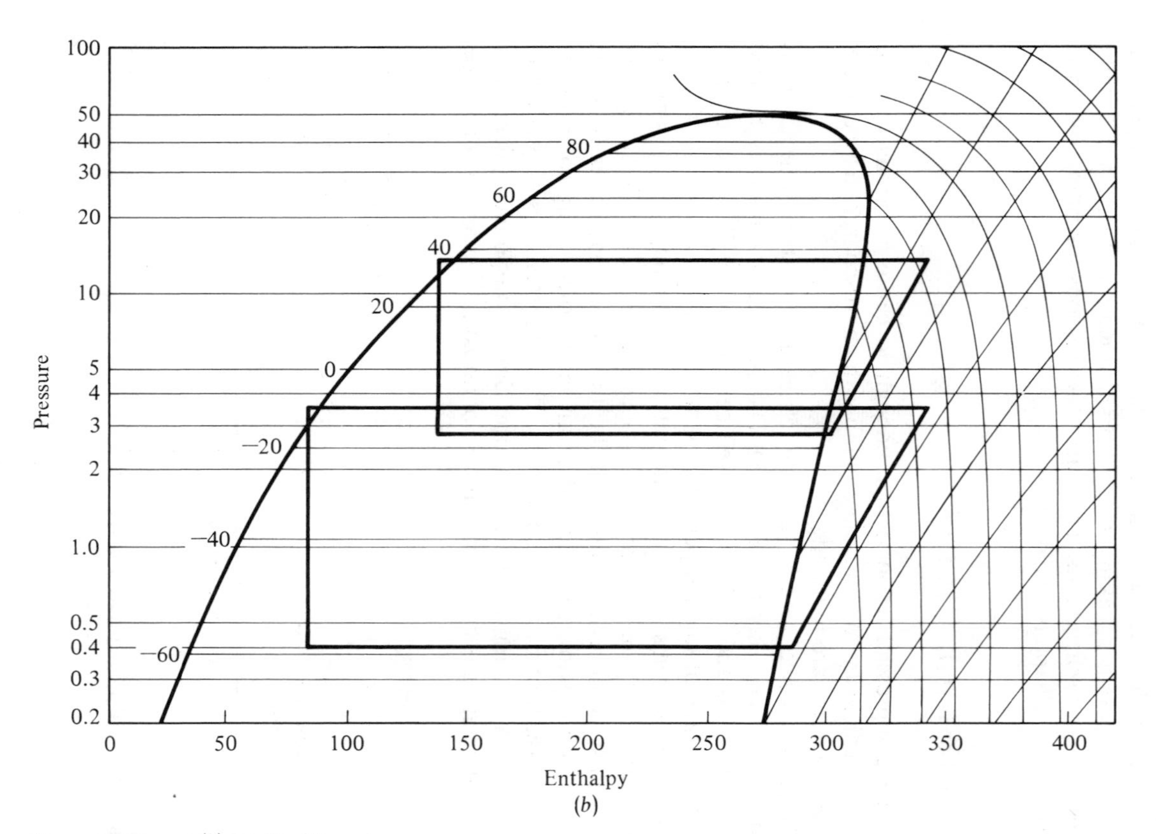

Figure 2-8 (*cont.*) (*b*) Mollier diagrams.

The Mollier diagrams for cascade and compound systems (Figs 2-8*b* and 2-9*b*) indicate the enthalpy change per kilogram of circulated refrigerant, but it should be borne in mind that the mass flows are different for the low and high stages.

2-6 REFRIGERANTS FOR VAPOUR COMPRESSION CYCLES

The requirements for the working fluid are as follows:

1. A high latent heat of vapourization
2. High density of suction gas
3. Non-corrosive, non-toxic, and non-flammable
4. Critical temperature and triple point outside the working range
5. Compatibility with materials of construction, with lubricating oils, and with other materials present in the system
6. Convenient working pressures, i.e., not too high and preferably not below atmospheric pressure
7. High dielectric strength (for compressors having integral electric motors)
8. Low cost
9. Ease of leak detection

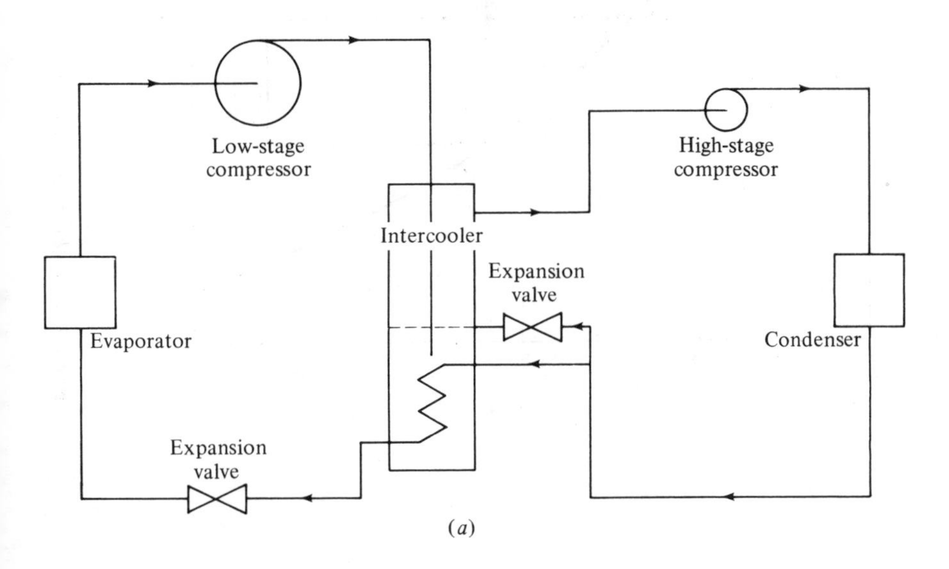

Figure 2-9 Compound cycle: (*a*) circuit.

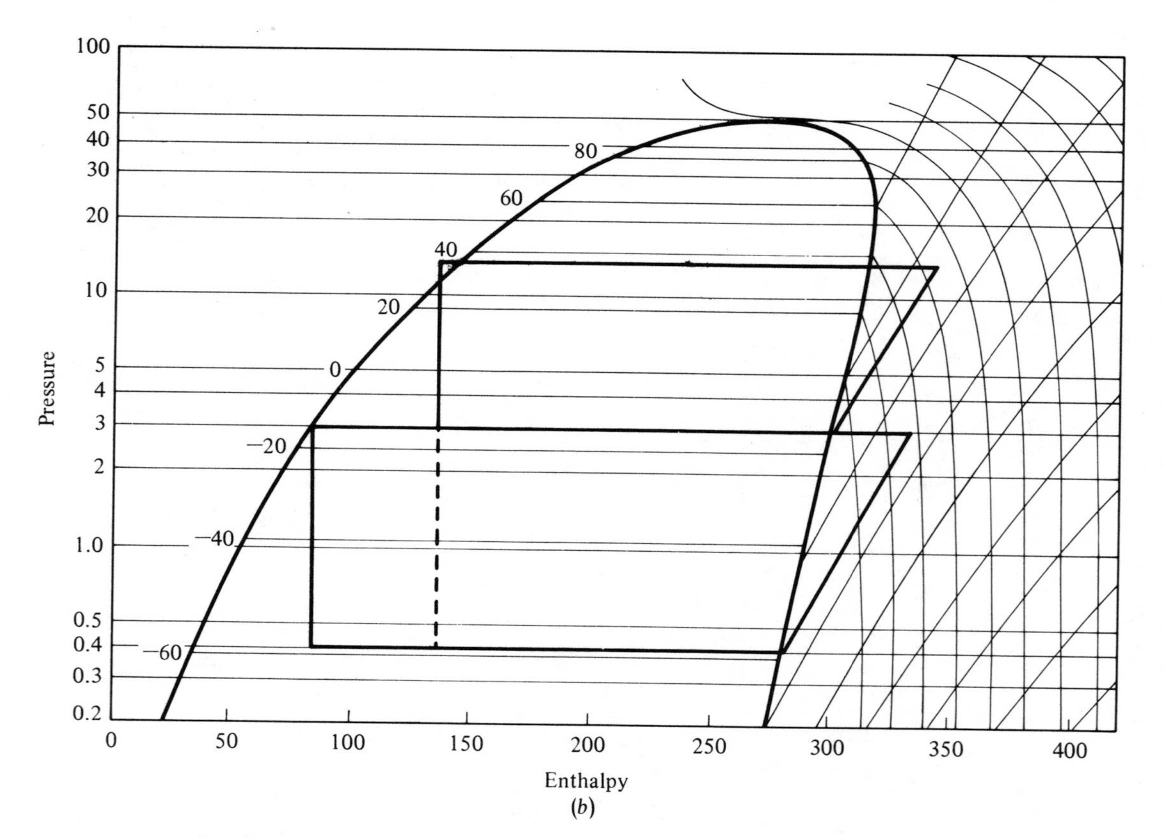

Figure 2-9 (*cont.*) (*b*) Mollier diagram.

No single working fluid has all of these properties, and a large variety of different chemicals has been used since the first compression cycle experiments in 1834. Apart from the special requirements of some manufacturing processes, all systems now use ammonia (R.717) or one of the halocarbons. The latter are known by various proprietary names (Freon, Arcton, Isceon, etc.) and are identified by a number, for example, R.11, R.12, R.22, R.502, etc.

Ammonia is toxic at low concentrations and can form flammable and explosive mixtures with air. These hazards preclude its use in locations accessible to the general public, so it is found mainly in industrial use. A comprehensive guide to these restrictions is given in BS.4434:1980. In the presence of traces of moisture, ammonia attacks copper, and this metal and its alloys cannot be used with the refrigerant.

The *halocarbons* are all non-toxic, non-flammable, and non-corrosive in ordinary use, so may be used without locational constraint. If burnt in air, slight traces of chlorine, fluorine, hydrogen fluoride, and phosgene can be formed. Workshops and plantrooms where the halocarbons may be released should be well ventilated and tobacco smoking should be discouraged.

The release of a large quantity of any chemical into the atmosphere may lead to environmental contamination, and the current release of large quantities of R.12 from aerosols, where it is used as a propellant, is leading to restrictions in this use. Manufacturers are watchful of the possible dangers of their products and will provide information on the subject if requested.

Refrigerants in general use are shown in Fig. 2-10 and Table 2-1. Of these, the four most encountered with positive displacement compressors are as follows:

Refrigerant	Use
R.717 (ammonia)	Commercial and industrial systems within the constraints of its toxicity
R.12	General use, especially for high ambient temperatures
R.22	General purpose
R.502	Especially for low-temperature systems on single-stage compression

R.11, R.12, and R.113 are used in dynamic (centrifugal) compressors on account of the low compression ratios. Other refrigerants will be found in special purpose cooling requirements in chemical manufacturing and processes.

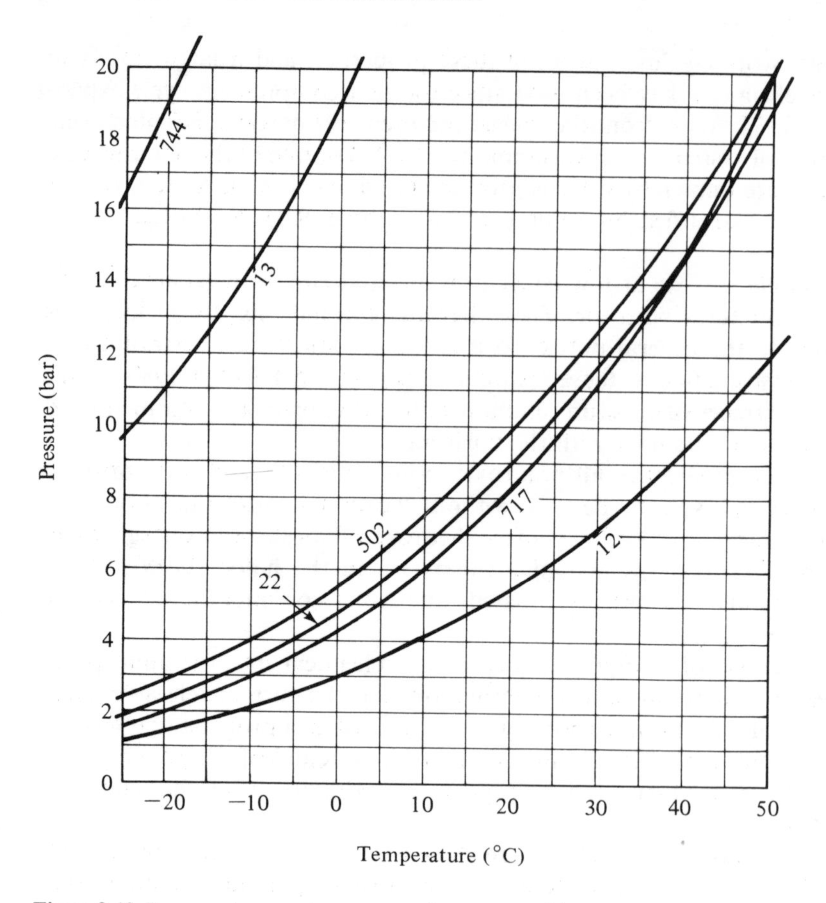

Figure 2-10 Pressure–temperature curves of common refrigerants.

2-7 ABSORPTION CYCLE

Vapour can be withdrawn from an evaporator by absorption (Fig. 2-11) into a liquid. Two combinations are in use, the absorption of ammonia gas into water and the absorption of water vapour into lithium bromide. The latter is non-toxic and so may be used for air-conditioning. The use of water as the refrigerant in this combination restricts it to systems above its freezing point. Refrigerant vapour from the evaporator is drawn into the absorber by the liquid absorbant, which is sprayed into the chamber. The resulting solution (or liquor) is then pumped up to condenser pressure and the vapour is driven off in the generator by direct heating. The high-pressure refrigerant gas given off can then be condensed in the usual way and passed back through the expansion valve into the evaporator. Weak liquor from the generator is passed through another pressure-reducing

Table 2-1 Refrigerants in general use

Refrigerant	Chemical formula	Boiling point at 1.013 bar, °C	Saturation pressure (bar) at		Latent heat (kJ/kg) through		Specific density (kg/m³) of vapour at	
			−15°C	40°C	−15/40°C	+5/40°C	−15°C	5°C
R.11	CCl_3F	23.8	0.20	1.73	146	136	1.31	3.01
R.12	CCl_2F_2	−29.8	1.83	9.61	106	115	10.99	21.1
R.13	$CClF_3$	−81.4	13.04	—	—	—	84	156
R.22	$CHClF_2$	−40.8	2.95	15.36	151	159	12.8	24.7
R.113	$CCl_2F\text{-}CClF_2$	47.6	0.07	0.78	115	128	0.59	1.53
R.502	$CHClF_2$ (48.8%) $CClF_2CF_3$ (51.2%)	−45.6	3.51	16.8	93	102	20.2	37.7
R.717	NH_3	−33.3	2.36	15.5	1055	1078	1.96	4.11

See also Refs. 1, 3, 12, 13, and 14.

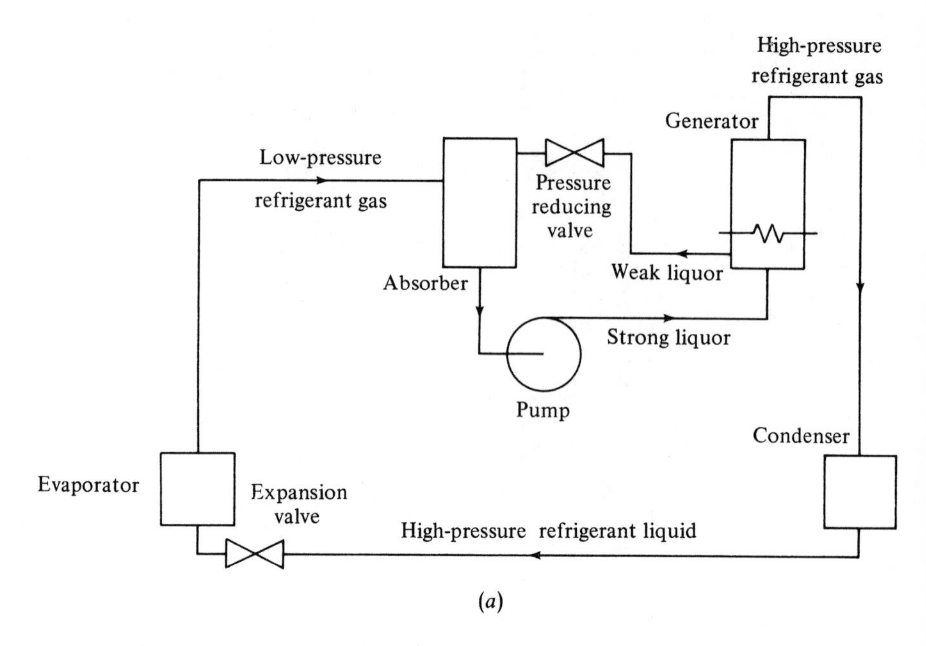

(*a*)

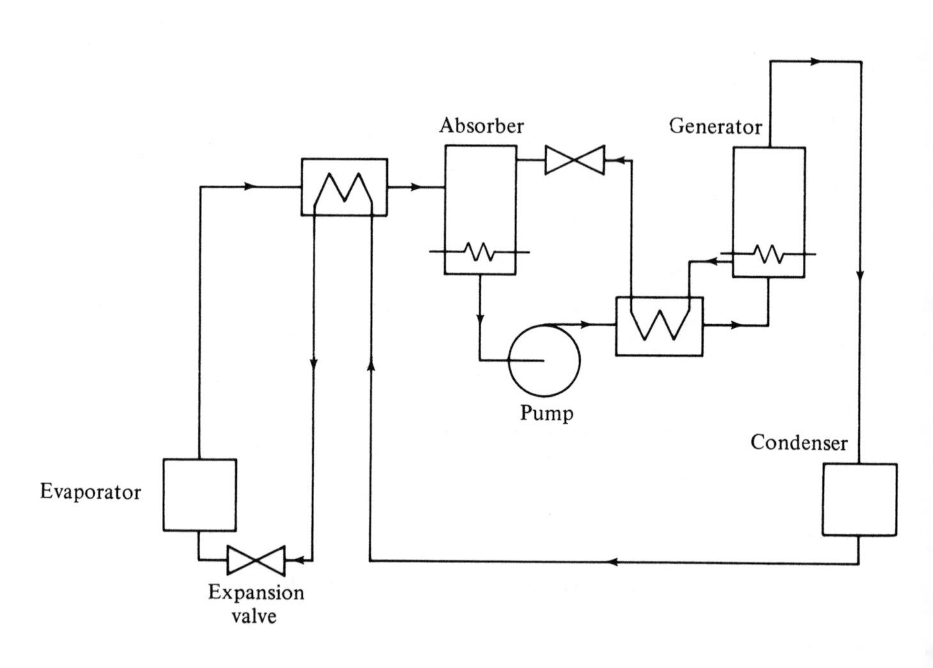

(*b*)

Figure 2-11 Absorption cycle: (*a*) basic circuit, (*b*) circuit with heat interchange.

valve to the absorber. Overall thermal efficiency is improved by a heat exchanger between the two liquor paths and a suction-to-liquid heat exchanger for the refrigerant. Power to the liquor pump will usually be electric, but the heat energy to the generator may be any form of low-grade energy such as oil, gas, or waste steam. (Solar power can also be used.) The overall energy used is greater than with the compression cycle, so the COP (coefficient of performance) is lower. Typical figures are as follows:

Energy per 100 kW cooling capacity at 3°C evaporation, 42°C condensation:

	Absorption	Vapour compression
Load	100.0	100.0
Pump/compressor (electricity)	0.1	30.0
Low-grade heat	165	—
Heat rejected	265.1	130.0

2-8 STEAM EJECTOR SYSTEM

The low pressures (8 to 22 mbar) required to evaporate water as a refrigerant at 4 to 7°C for air-conditioning duty can be obtained with a steam ejector. High-pressure steam at 10 bar is commonly used. The COP of this cycle is somewhat less than with the absorption system, so its use is restricted to applications where large volumes of steam are available when required (large, steam-driven ships) or where water is to be removed as well as cooling, as in freeze-drying and fruit juice concentration.

2-9 AIR CYCLE

Any gas, when compressed, rises in temperature. Conversely, if it is made to do work while expanding, the temperature will drop. Use is made of the sensible heat only (although it is, of course, the basis of the air liquefaction process).

The main application for this cycle is the air-conditioning and pressurization of aircraft. The turbines used for compression and expansion turn at very high speeds to obtain the necessary pressure ratios and, consequently, are very noisy. The COP is lower than with other systems.[15]

2-10 THERMOELECTRIC COOLING

The passage of an electric current through a junction of dissimilar semiconductors can cause a fall in temperature (the Peltier effect). The development of suitable semi-conductors has made possible the construction of small cooling systems for military, aerospace, and laboratory use.

CHAPTER

THREE

COMPRESSORS

3-1 GENERAL

The purpose of the compressor in a vapour compression cycle is to accept the low-pressure dry gas from the evaporator and raise its pressure to that of the condenser.

Compressors may be of the positive displacement or dynamic type. The general form of positive displacement compressor is the piston type, being adaptable in size, number of cylinders, speed, and method of drive. It works on the two-stroke cycle (see Fig. 3-1). As the piston descends on the suction stroke, the internal pressure falls until it is lower than that in the suction inlet pipe, and the suction valve opens to admit gas from the evaporator. At the bottom of the stroke, this valve closes again and the compression stroke begins. When the cylinder pressure is higher than that in the discharge pipe, the discharge valve opens and the compressed gas passes to the condenser. Clearance gas left at the top of the stroke must re-expand before a fresh charge can enter the cylinder (see Fig. 3-2 and also Chapter 2, for theoretical and practical cycles on the Mollier chart and for volumetric efficiency).

The first commercial piston compressors were built in the middle of the last century, and evolved from the steam engines which provided the prime mover. Construction was first double acting, but there was difficulty in maintaining gas-tightness at the piston rod, so the design evolved further into a single-acting machine with the crankcase at suction inlet pressure, leaving only the rotating shaft as a possible source of leakage, and this was sealed with a packed gland.

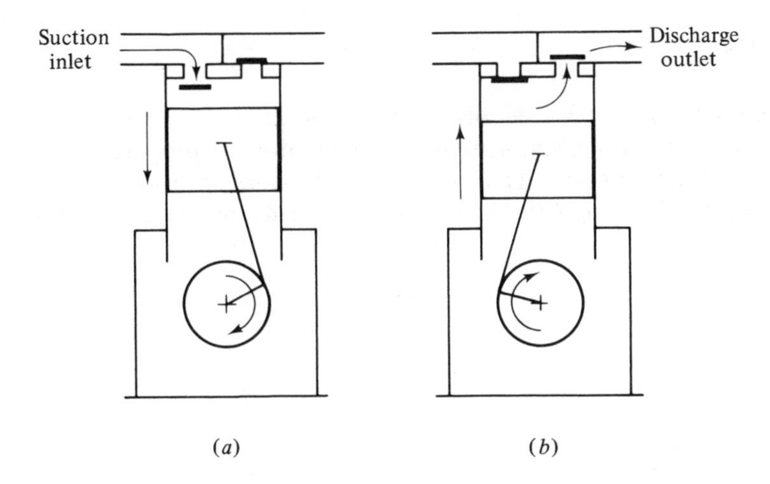

Figure 3-1 Reciprocating compressor: (*a*) suction stroke, (*b*) discharge stroke.

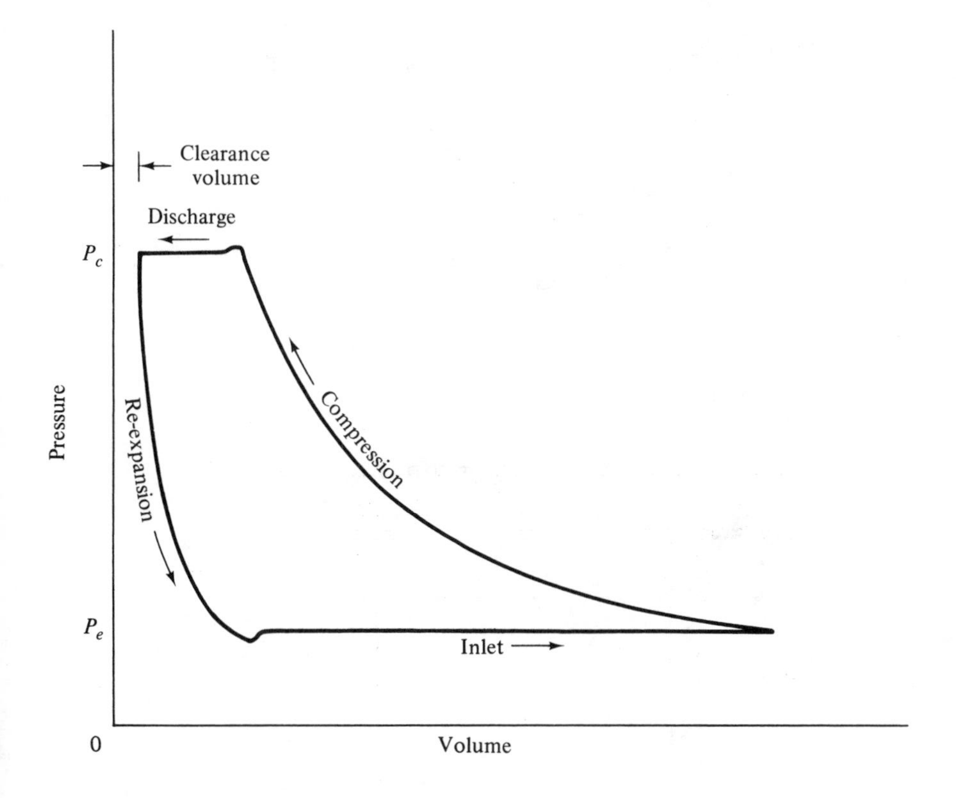

Figure 3-2 Reciprocating compressor, indicator diagram.

3-2 VALVES

Piston compressors may be generally classified by the type of valve, and this depends on size, since a small swept volume requires a proportionally small inlet and outlet gas port. The smallest compressors have spring steel reed valves, both inlet and outlet in the cylinder head and arranged on a

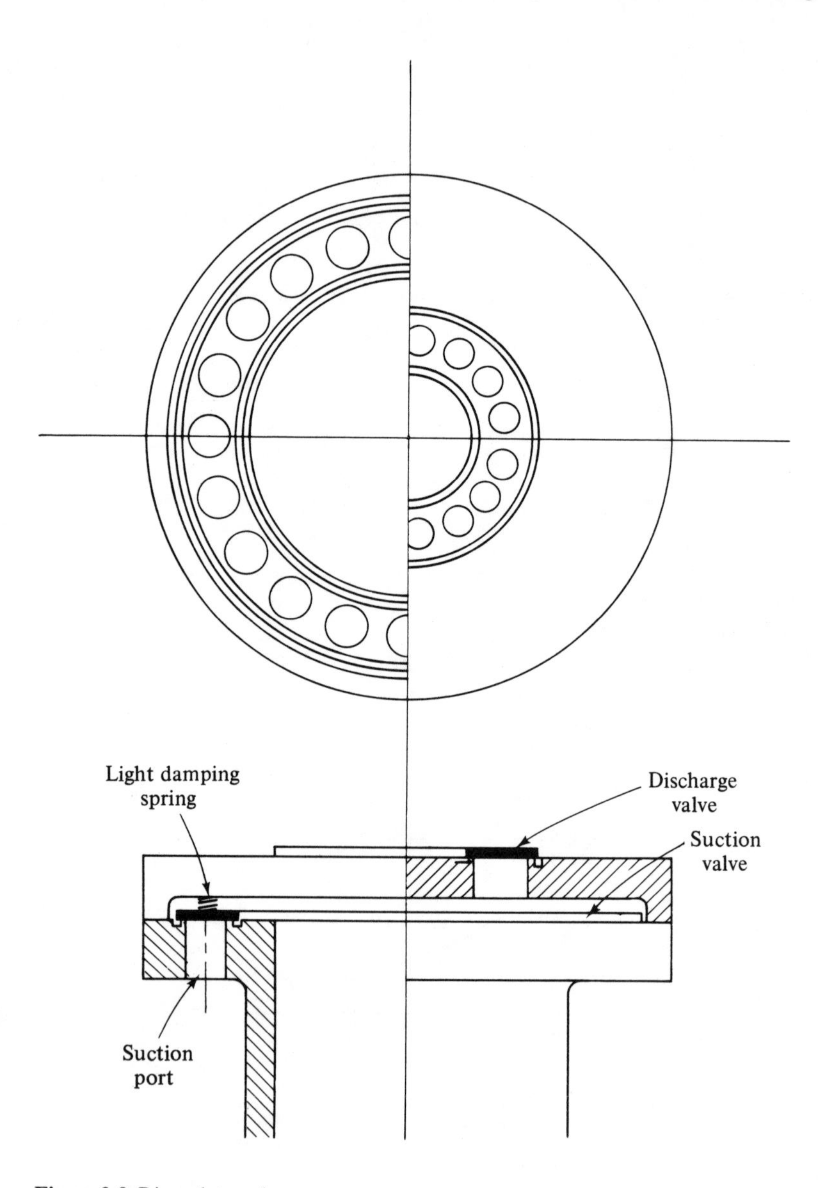

Figure 3-3 Ring plate valves.

valve plate. Above a bore of about 40 mm, the port area available within the head size is insufficient for both inlet and outlet valves, and the inlet is moved to the piston crown or to an annulus surrounding the head. The outlet or discharge valve remains in the central part of the cylinder head. In most makes, both types of valve cover a ring of circular gas ports, so are made in annular form and generally termed *ring plate valves* (Fig. 3-3). Ring plate valves are made of thin spring steel or titanium, limited in lift and damped by light springs to assist even closure and lessen bouncing.

Although intended to handle only dry gas, liquid refrigerant or traces of oil may sometimes enter the cylinder and must pass out through the discharge valves. These are arranged on a spring-loaded head, which will lift and relieve excessive pressures. Some makes also have an internal safety valve to release gas pressure from the discharge back to the suction inlet.

3-3 NUMBER OF CYLINDERS

To take advantage of larger-scale production and provide interchangeability of parts, modern compressors tend to be multicylinder (see Fig. 3-4), and machines of six, eight, and more cylinders are common. A large

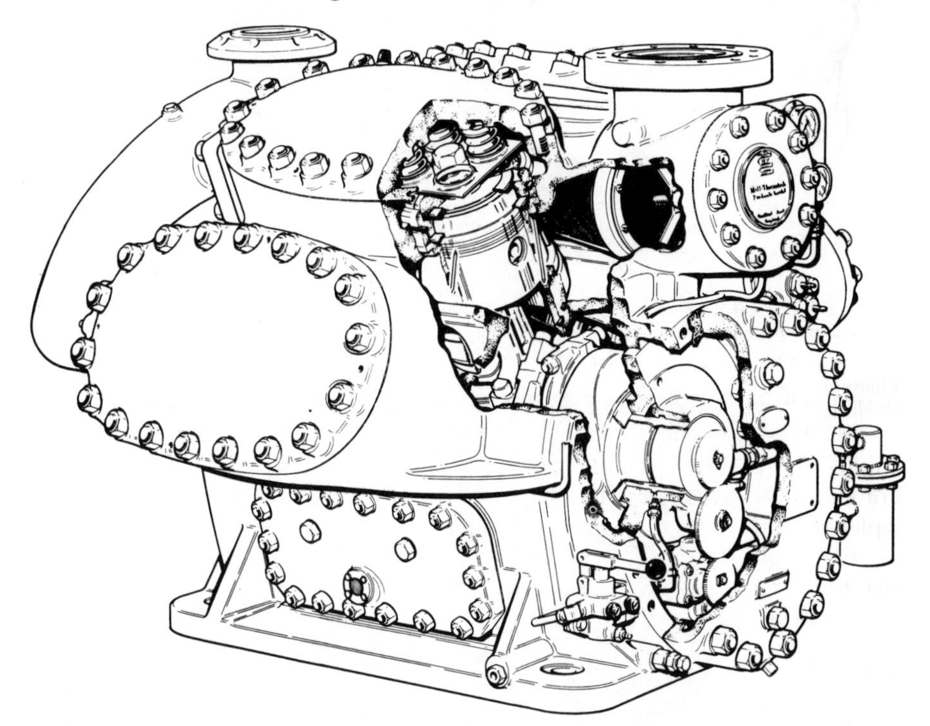

Figure 3-4 Multicylinder compressor. *(Courtesy Hall-Thermotank Products Limited.)*

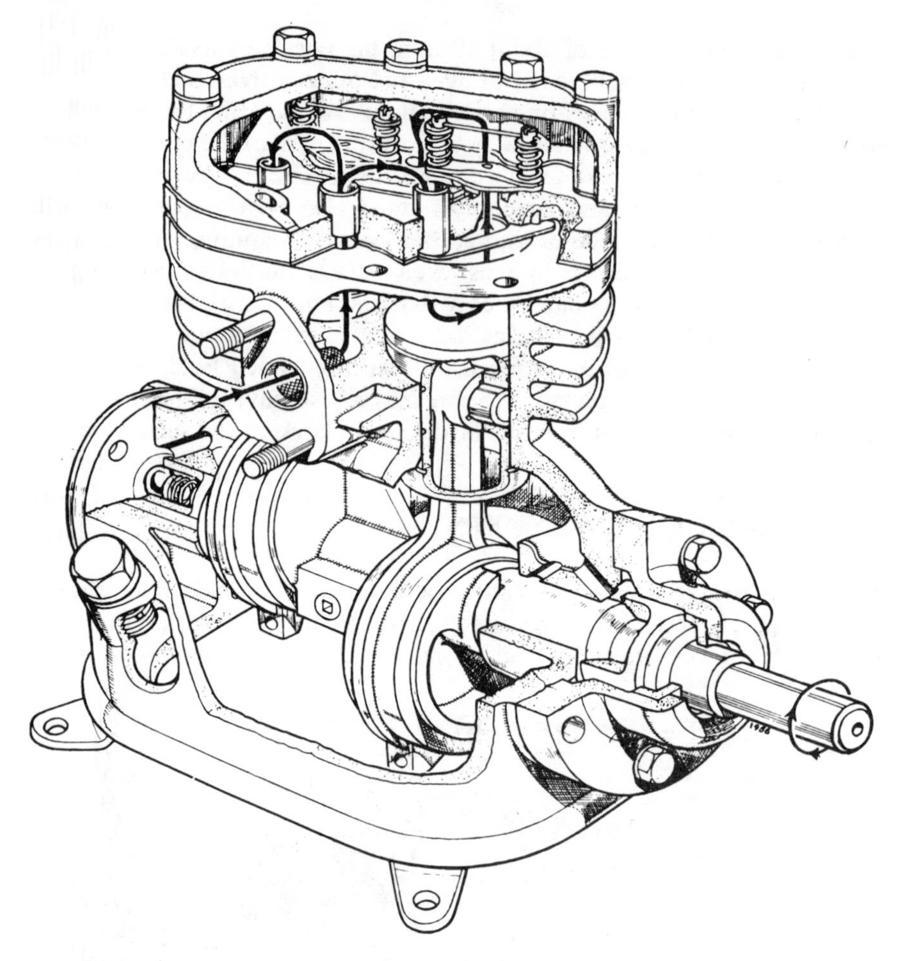

Figure 3-5 A $1\frac{3}{8}$ in bore × 1 in stroke, two-cylinder compressor. *(Courtesy Hall-Thermotank Products Limited.)*

number of parts—pistons, piston rods, loose cylinder liners, and valves—are identical throughout a range of one size of machine and can be quickly replaced when worn or damaged.

Compressors for small systems will be simpler, of two or four cylinders (see Fig. 3-5).

3-4 CAPACITY REDUCTION

A refrigeration system will be designed to have a maximum duty to balance a calculated maximum load, and for much of its life may work at some lower load. Such variations require capacity-reduction devices, originally

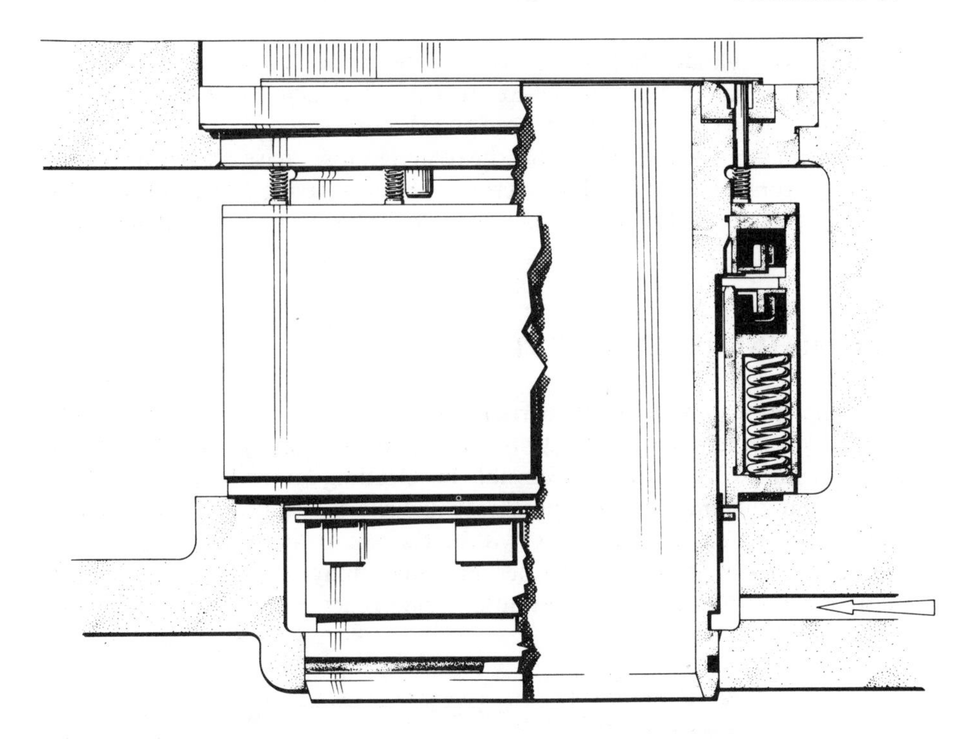

Figure 3-6 Lifting mechanism for ring plate suction valve. *(Courtesy Hall-Thermotank Products Limited.)*

by speed control (when steam driven) or in the form of bypass ports in the cylinder walls.

The construction of multicylinder machines gives the opportunity to change the working swept volume by taking cylinders out of service with valve-lifting mechanisms. The ring plate suction valve which is located at the crown of a loose liner can be lifted by various alternative mechanical systems, actuated by pressure of the lubricating oil and controlled by solenoid valves (see Fig. 3-6). Typically, an annular piston operates push rods under the valves. In this way a multicylinder machine (see Fig. 3-7) can have any number of its cylinders unloaded for capacity reduction and, in addition, will start unloaded until the build-up of oil pump pressure depresses the valve lifters.

3-5 COOLING

Cold suction gas provides cooling for the compressor and is sufficient to keep small machines at an acceptable working temperature. Refrigerants having high discharge temperatures (mainly ammonia) require the use of water-cooled cylinder heads. Oil coolers are needed under some working conditions which will be specified by the manufacturer. These may be water cooled or take refrigerant from the system.

Compressors will tend to overheat under low mass flow conditions resulting from abnormally low suction pressures or lengthy running with capacity reduction. Detectors may need to be fitted to warn against this condition.

3-6 STRAINERS. LUBRICATION

Incoming gas may contain particles of dirt from within the circuit, especially on a new system. Suction strainers or traps are provided to catch such dirt and will be readily accessible for cleaning on the larger machines.

All but the smallest compressors will have a strainer or filter in the lubricating oil circuit. Strainers within the sump are commonly of the self-cleaning slot disc type. Larger machines may also have a filter of the fabric throwaway type, as in automobile practice. Reciprocating com-

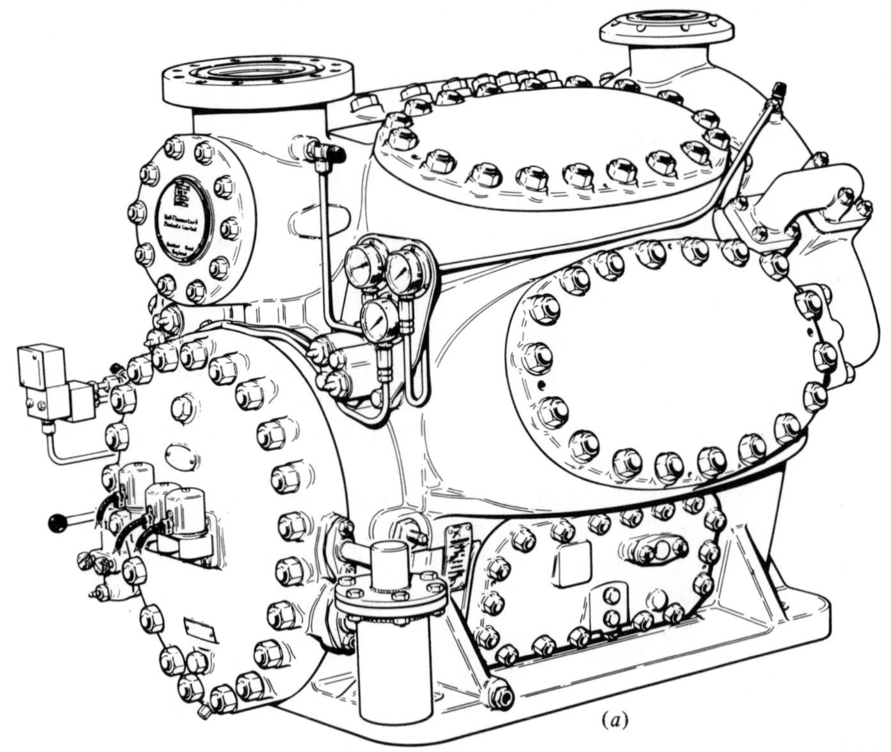

(*a*)

Figure 3-7 Multicylinder compressors, outer views. *(Courtesy Hall-Thermotank Products Limited.)*

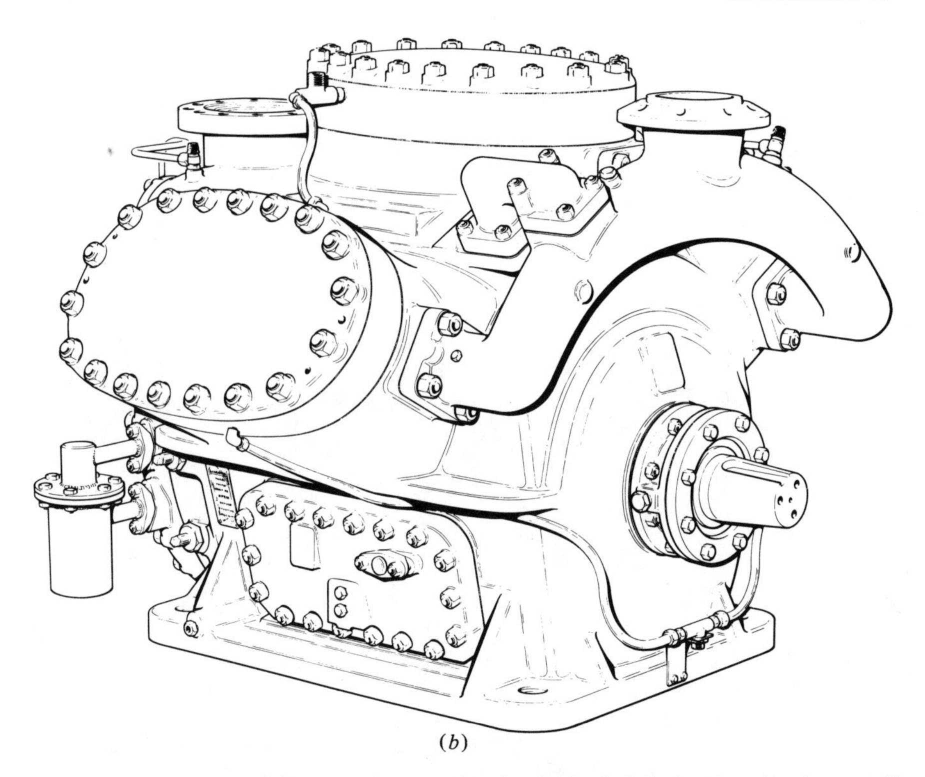

(*b*)

pressors operate with a wet sump, having splash lubrication in the small sizes but forced oil feed with gear or crescent pumps on all others. A sight glass will be fitted at the correct working oil level and a hand pump may be fitted to permit the addition of oil without stopping or opening the plant, the sump being under refrigerant gas pressure.

When the compressor is idle, the lubricating oil may contain a certain amount of dissolved refrigerant, depending on the pressure, temperature, and the refrigerant itself. At the moment of starting, the oil will be diluted by this refrigerant and, as the suction pressure falls, gas will boil out of the oil, causing it to foam.

To reduce this solution of refrigerant in the oil to an acceptable factor, heating devices are commonly fitted to crankcases, and will remain in operation whenever the compressor is idle.

3-7 SHAFT GLANDS. MOTORS

Compressors having external drive require a gland or seal where the shaft passes out of the crankcase, and are termed open compressors. They may be belt driven or directly coupled to the shaft of the electric motor or other prime mover.

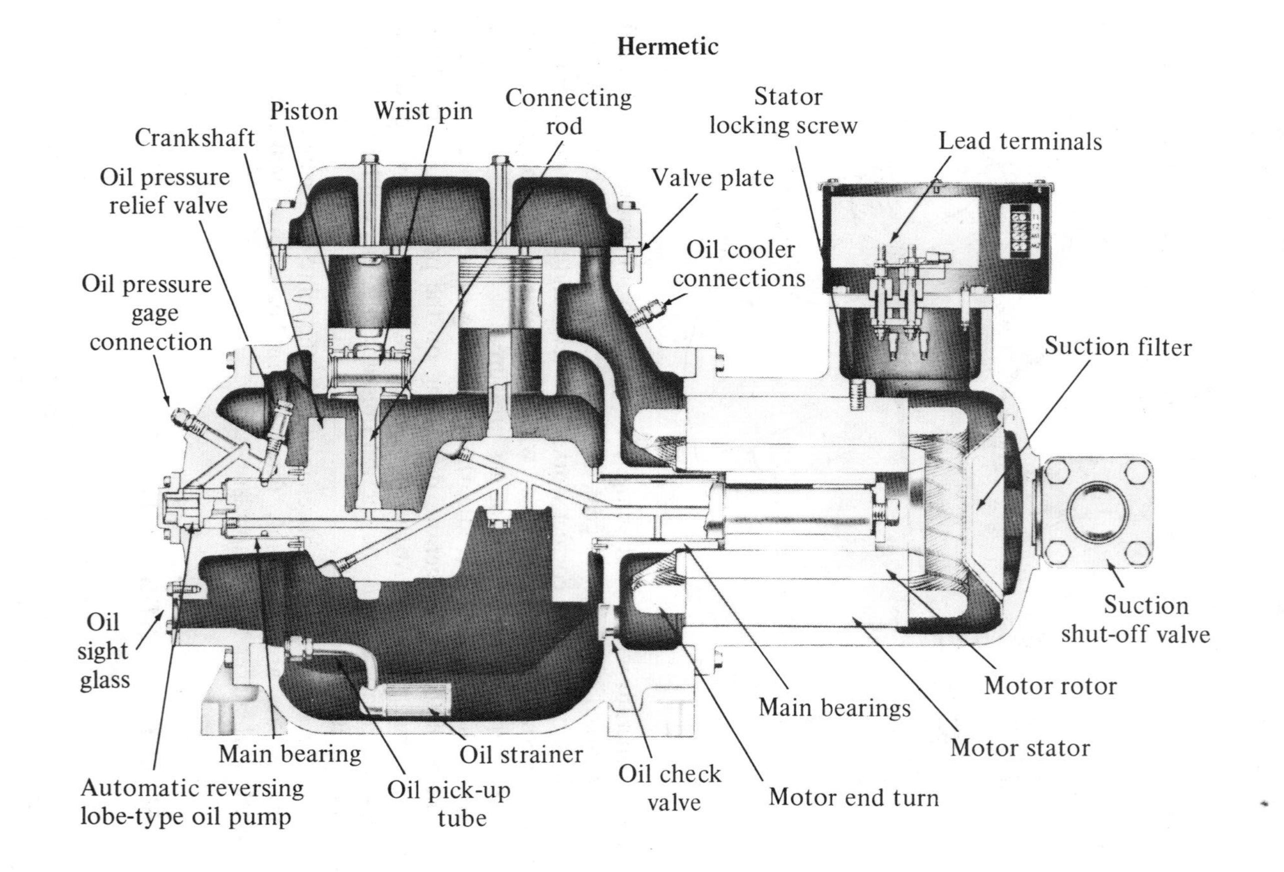
Hermetic
Piston
Wrist pin
Connecting rod
Stator locking screw
Crankshaft
Lead terminals
Oil pressure relief valve
Valve plate
Oil cooler connections
Oil pressure gage connection
Suction filter
Suction shut-off valve
Oil sight glass
Motor rotor
Main bearings
Motor stator
Main bearing
Oil strainer
Oil check valve
Motor end turn
Automatic reversing lobe-type oil pump
Oil pick-up tube

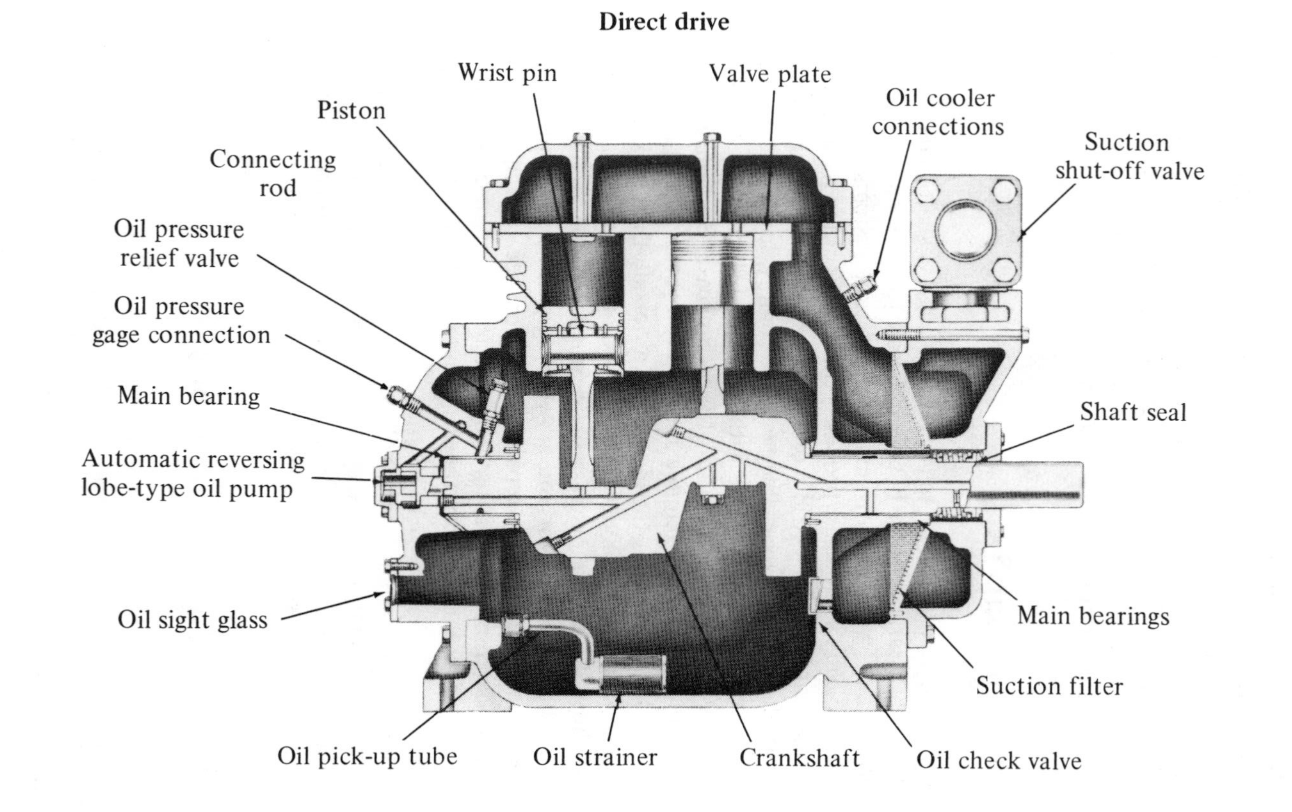

Figure 3-8 Semi-hermetic compressor. *(Courtesy Dunham-Bush, Inc.)*

When first started, a refrigeration system will operate at a higher suction temperature and pressure than normal operating conditions and consequently a higher discharge pressure, taking considerably more power. Drive motors must be sized accordingly to provide this pulldown power, and an allowance of 25 per cent is usual. As a result, the drive motor will run for the greater part of its life at something under 75 per cent rated output, and so at a lower efficiency, low running current, and poor power factor. Electrical protection and safety devices must take this into account and power factor correction should be fitted on large motors. See also Chapter 7 on maximum operating pressure expansion valves.

3-8 HERMETIC DRIVES

The possible slight leakage of refrigerant through a shaft gland may be acceptable with a large system but would lead to early malfunction of a small circuit. The wide use of small refrigeration systems has evolved methods of avoiding shaft seals, provided that the working fluid is compatible with the materials of electric motors and has a high dielectric strength.

The semi-hermetic or accessible-hermetic compressor (Fig. 3-8) has the rotor of its drive motor integral with an extended crankshaft, and the stator is fitted within an extension of the crankcase. Suction gas passes through the motor itself to remove motor waste heat. Induction motors only can be used, with any starting switches outside the crankcase, since any sparking would lead to decomposition of the refrigerant. Electrical leads pass through ceramic or glass seals. Small compressors will be fully hermetic, i.e., having the motor and all working parts sealed within a steel shell, and so not accessible for repair or maintenance. The application of the full hermetic compressor is limited by the amount of cooling by the incoming cold gas, heat loss from the shell, and the possible provision of an oil cooler.

The failure of an inbuilt motor will lead to products of decomposition and serious contamination of the system, which must then be thoroughly cleaned. Internal and external motor protection devices are fitted with the object of switching off the supply before such damage occurs.

3-9 SLIDING VANE COMPRESSORS

The spaces between an eccentric rotary piston and sliding vanes will vary with angular position, to provide a form of positive displacement compressor (Fig. 3-9). Larger models have eight or more blades and do not require inlet or outlet valves. The blades are held in close contact with the outer shell by centrifugal force, and sealing is improved by the injection of lubri-

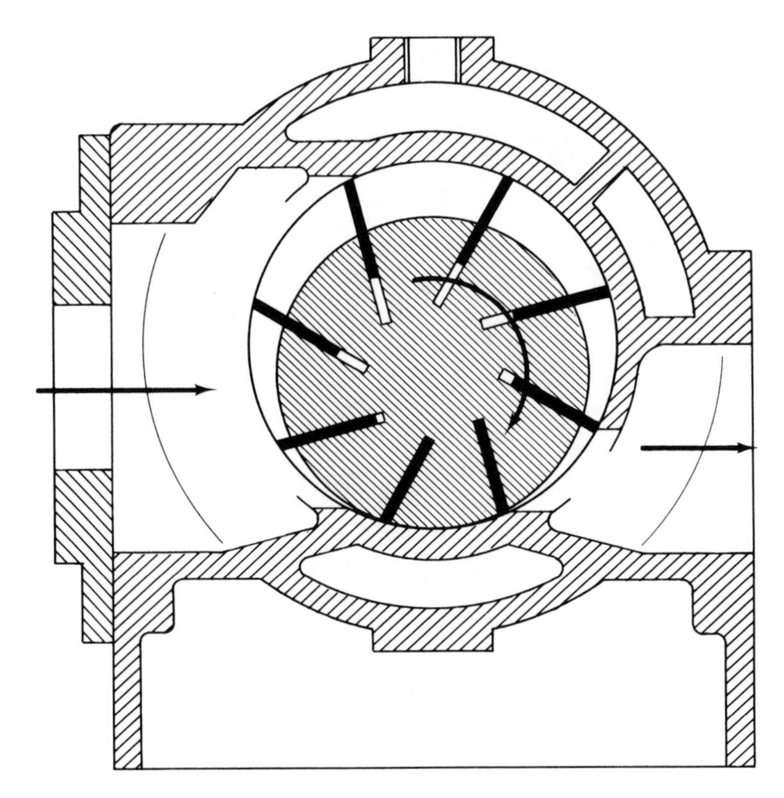

Figure 3-9 Sliding vane compressor. *(Courtesy Hick, Hargreaves & Co. Ltd.)*

cating oil along the length of the blades. This type of compressor is limited in application by the sideways thrust on the sliding blades and by the stiffness of the rotor, so is normally used for compression ratios not exceeding 1 : 4, or for discharge pressures less than 6 bar. This restricts its commercial use to the role of a low-stage or booster compressor on a compound system. Compressors of up to 5 kW capacity for domestic appliances and packaged air-conditioner units may have one or two sliding blades, and require a discharge valve of the reed type.

3-10 SCREW COMPRESSORS

The screw compressor can be visualized as a development of the gear pump. For gas pumping, the rotor shapes are modified to give maximum swept volume, and the pitch of the helix is such that the inlet and outlet ports can be arranged at the ends instead of at the side. The solid portions of the screws slide over the gas ports to separate one stroke from the next, so that no extra inlet or outlet valves are needed.

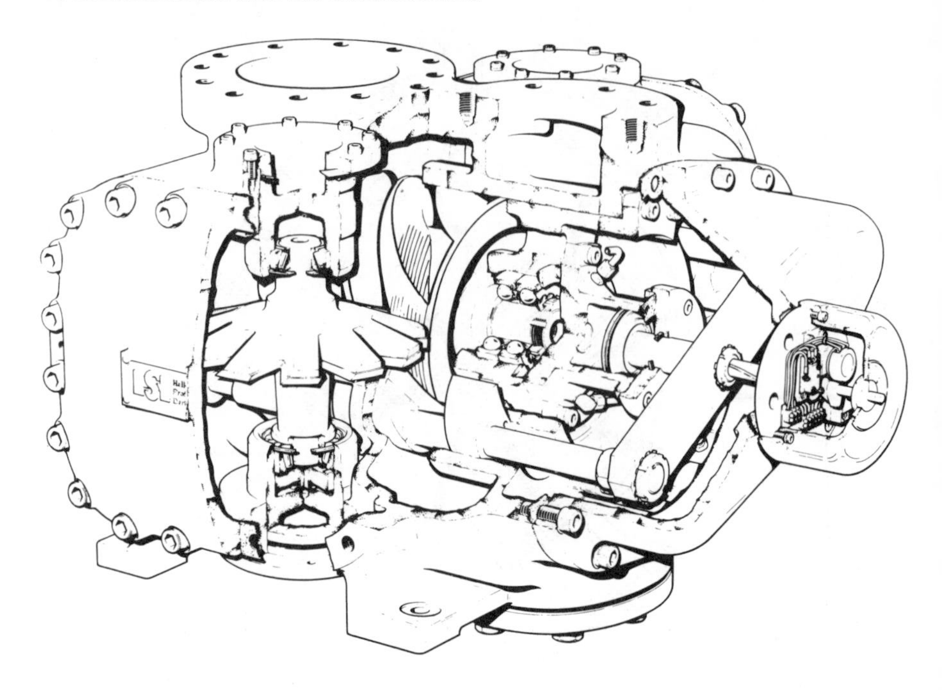

Figure 3-10 Single-screw compressor. *(Courtesy Hall-Thermotank Products Limited.)*

The usual form has twin meshing screws on parallel shafts. Other types have a single screw with rotating seal vanes (Fig. 3-10). Sealing between the working parts is usually assisted by the injection of oil along the length of the barrels. This extra oil must be separated from the discharge gas, and is then cooled and filtered before returning to the lubrication circuit. (See Chapter 4.) Screw compressors have no clearance volume, and may work at high compression ratios without loss of 'volumetric efficiency'. In all screw compressors, the gas volume will have been reduced to a pre-set proportion of the inlet volume by the time the outlet port is uncovered, and this is termed the *built-in compression ratio*. At this point, the gas within the screws is opened to condenser pressure and gas will flow inwards or outwards through the discharge port if the pressures are not equal.

The absorbed power of the screw compressor will be at its optimum only when the working compression ratio is the same as that of the built-in one. This loss of efficiency is acceptable since the machine has no valves and no working parts other than the screws and sealing vanes.

Capacity reduction of the screw compressor is effected by a sliding block covering part of the barrel wall, which permits gas to pass back to the suction, so varying the working stroke. Variation down to 10 per cent of maximum is usual.

3-11 CAPACITY RATINGS

For the convenience of users, the refrigerating effect of compressors is usually tabulated (Table 3-1) or given in graphical form (Fig. 3-11), and is shown as the net cooling capacity based on the evaporating and condensing temperatures or pressures. Such published data will include absorbed power and indicate any limitations of the application.

Ratings of this sort may be standardized to certain conditions at the suction, which may not apply to a particular use and need to be interpreted. (See also Chapter 25.)

3-12 DYNAMIC COMPRESSORS

In the centrifugal compressor, suction vapour enters the centre of the rotor and is impelled outwards by centrifugal force, leaving the tip at a higher speed—of the order of 100 m/s. The gain in dynamic energy is converted to the pressure difference between suction and discharge, and is a function of the tip speed and the gas density. The available pressure is low, so this compressor uses low-pressure refrigerants and works at low compression ratios. The common refrigerant is R.11, but R.12, R.113, and R.114 are also used (see Table 3-2).

The very low pressures used should be noted. The R.113 system works with both the evaporator and condenser below atmospheric pressure. Since the final gas velocity is very high, it is not practical to build small machines, and the centrifugal is limited to sizes of 300 kW and upwards, and usually for air-conditioning duty. In industrial applications it may be used as the low stage for a compound or cascade system. Some makes are two stage, having two impellors on the same shaft.

The performance is affected very sharply by changes in the compression ratio, and is shown in typical capacity curves. In particular, if a

Table 3-1 Model No. PLE08, R.502

Suction		Condensing at							
		30°C		35°C		40°C		45°C	
Temperature	Pressure	Capacity	Power	Capacity	Power	Capacity	Power	Capacity	Power
−50	0.82	5.64	5.70	4.78	5.59	3.87	5.41	3.01	5.18
−45	1.04	8.49	7.00	7.49	7.00	6.48	6.94	5.48	6.81
−40	1.31	11.9	8.29	10.7	8.41	9.56	8.48	8.40	8.47
−35	1.63	15.9	9.65	14.5	9.92	13.1	10.1	11.9	10.3
−30	2.00	20.6	11.0	19.1	11.5	17.6	11.6	15.9	11.9

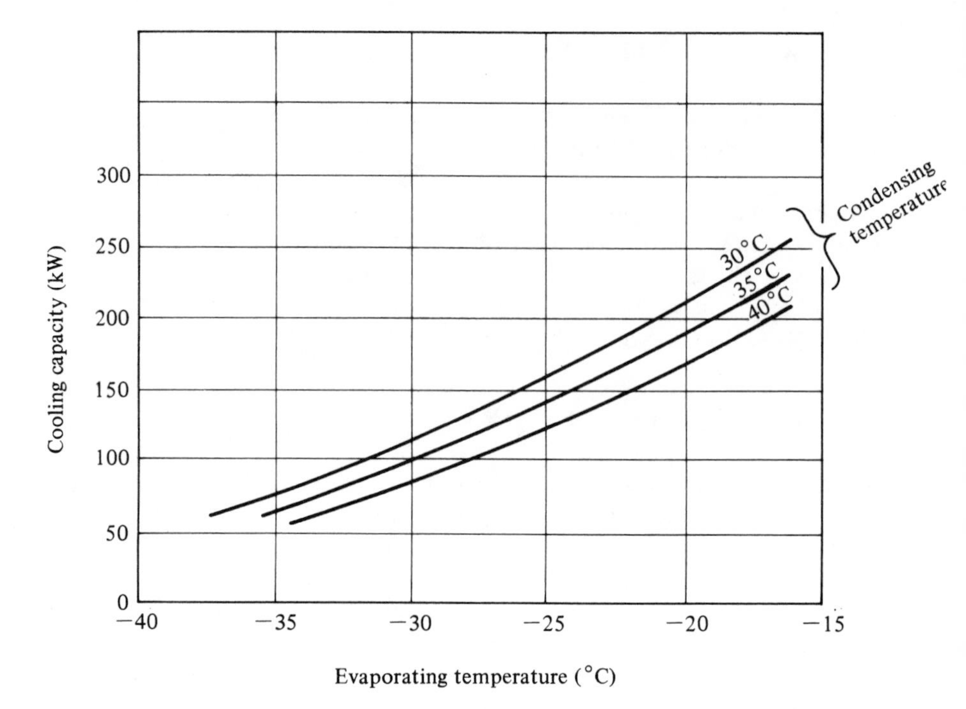

Figure 3-11 Compressor capacity ratings in graph form.

condenser pressure is imposed which is higher than the dynamic pressure of the leaving gas, the mass flow will decrease, reducing the velocity pressure so that the flow pattern collapses. Gas then flows back through the impellor blades, reducing the compression ratio, and the machine starts pumping again. This phenomenon is known as surging and can be compared with the stalling of an aircraft which tries to fly too slowly or climb too fast. It will occur if the condensing temperature rises too high or the evaporator too low, and may damage the machine if allowed to persist. Since the centrifugal compressor's performance is very sensitive to the correct function of other system components, it is usually supplied by the manufacturer with the correct condenser and evaporator in a factory-built

Table 3-2

Refrigerant	*p* at 2.2°C	*p* at 30°C	Compression ratio
R.11	0.44 bar	1.25 bar	2.8
R.12	3.32	7.45	2.2
R.113	0.16	0.54	3.3
R.114	0.95	2.5	2.6

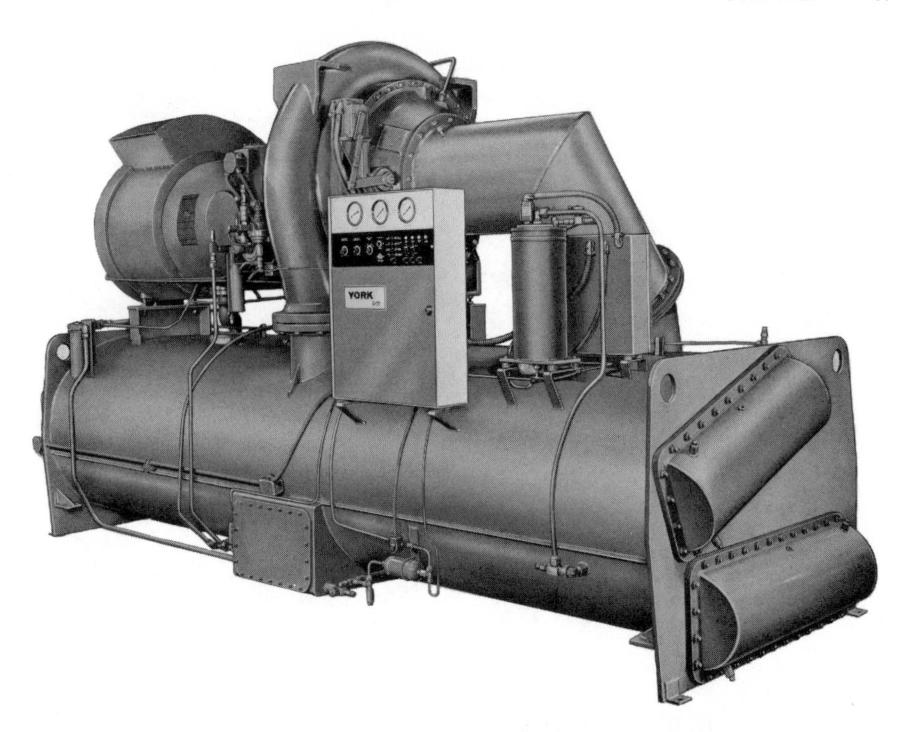

Figure 3-12 Centrifugal compressor unit. *(Courtesy York Division of Borg-Warner Ltd.)*

package (Fig. 3-12). Machines operating below 1 bar will have inbuilt air-purging devices to remove any gas which enters the system from outside.

Capacity reduction of the centrifugal machine is effected by directional blades at the rotor inlet port. When set to assist the gas to enter the runner, maximum pressure gain will result. If the blades are rotated to change this pattern, reduction down to 10 or 15 per cent is possible. The centrifugal compressor has a gland or oil seal, keeping the shaft-bearing lubricant away from the refrigerant, and so is almost oil free.

CHAPTER

FOUR

OIL IN REFRIGERANT CIRCUITS

4-1 OIL SPECIFICATIONS

The behaviour of lubricating oil in a refrigerant circuit and its physical interaction with the refrigerant itself is a dominant factor in the design of circuits in general and evaporators in particular.

Refrigeration compressors are mechanical devices with component parts which slide together, so requiring lubrication to reduce friction, remove frictional heat, and assist with gas sealing.

Lubricants for general commercial systems are based on mineral oils, and the following properties are required of the lubricant selected:

1. It must be compatible with the refrigerant, i.e., not form any compounds or promote chemical activity.
2. The mixture with the refrigerant in the lubrication circuit must provide adequate lubrication of the working parts.
3. It must not solidify or throw out any solids such as waxes, within the working range, or clog strainers or driers.
4. It must be free of water or other contaminants which will affect the system.
5. It must not be prone to foaming.
6. It must be resistant to oxidation (high flash point).
7. It must have a low vapour pressure.
8. For hermetic and semi-hermetic compressors, it must have a high dielectric strength.

A large variety of oils is available, and recommendations for any set of conditions, compressor type, and refrigerant can be obtained from the refiners. They are naphthene or paraffin-based oils. Synthetic lubricants have been developed for ultra-low- and high-temperature systems, especially for process heat pumps.

4-2 OIL SEPARATORS

During the compression stroke of a reciprocating machine, the gas becomes hotter and some of the oil on the cylinder wall will be vapourized and pass out with the discharge gas. To reduce the amount of this oil which will be carried around the circuit, an oil separator is frequently fitted in the discharge line (see Fig. 4-1). The hot entering gas is made to impinge

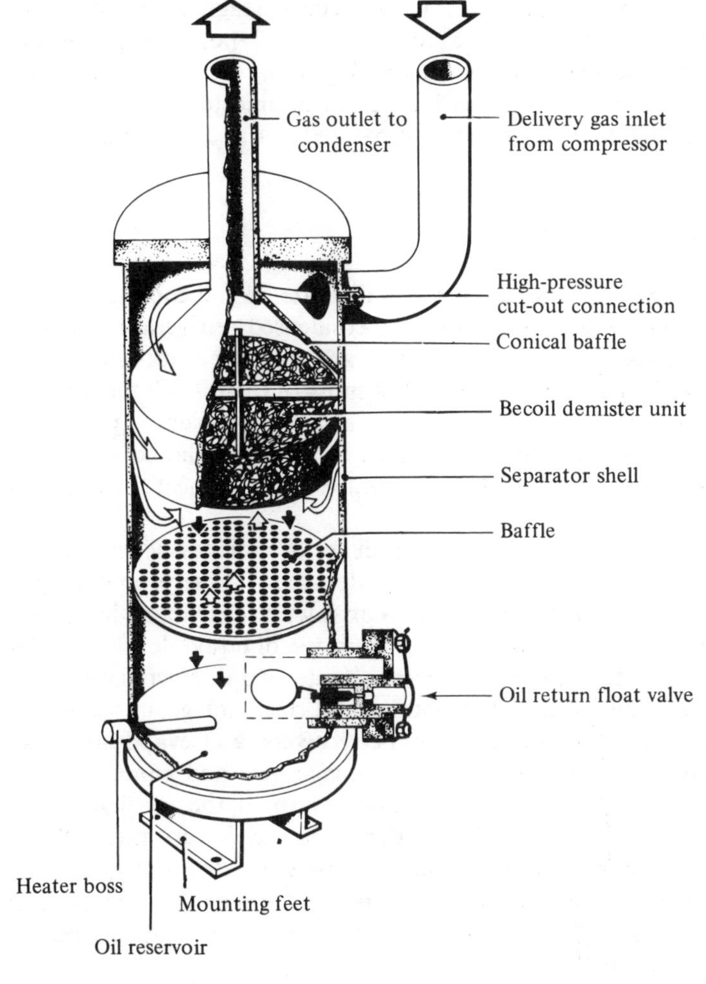

Figure 4-1 Oil separator. *(Courtesy Hall-Thermotank Products Limited.)*

on a plate, or may enter a drum tangentially to lose much of the oil on the surface by centrifugal force. Some 95 to 98 per cent of the entrained oil will be separated from the hot gas and will fall to the bottom of the drum, and can be returned to the crankcase. The oil return line will be controlled by a float valve, or may have a bleed orifice. In either case, this metering device must be backed up by a solenoid valve to give tight shut-off when the compressor stops, since the separator is at discharge pressure and the oil sump at suction.

On shut-down, high-pressure gas in the separator will cool and some will condense into liquid, to dilute the oil left in the bottom. When the compressor re-starts, this diluted oil will pass to the sump. In order to limit this dilution, a heater is commonly fitted into the base of the separator.

For installations which might be very sensitive to accumulations of oil, a two-stage oil separator can be fitted. The second stage cools the gas to just above condensing temperature, and up to 99.7 per cent of the entrained oil can be removed. Even so, a small quantity will be carried over. Sliding vane and screw compressors have extra oil injected into the casing to assist with sealing, and this must be separated out and re-cooled.

4-3 OIL CIRCULATION

Traces of oil which enter the condenser will settle on the cooling surfaces and fall to the bottom as a liquid with the condensed refrigerant. The two liquids will then pass to the expansion valve and into the evaporator. Here, the refrigerant will change to a vapour but most of the oil will remain as a liquid, slight traces of the latter passing out as a low-pressure vapour with the suction gas. It is necessary to limit the build-up of liquid oil in the evaporator, since it would quickly concentrate, reducing heat transfer and causing malfunction.

Methods of limiting oil accumulation in the evaporator depend on the ease with which the liquids mix, and their densities. These properties (see Table 4-1) indicate that different problems exist with refrigerants in general use. The extent of miscibility and the consideration of liquid density divides the problem of oil separation and circulation into two distinct classes.

With *ammonia*, oil sinks to the bottom and does not go into solution with the refrigerant. Ammonia condensers, receivers, and evaporators can be distinguished by the provision of oil drainage pots and connections at the lowest point. Automatic drainage and return of the oil from these would have to depend on the different densities, and is very rarely fitted. The removal of oil from collection pots and low-point drains is a periodic manual function and is carried out as part of the routine maintenance. The *halocarbons* are all sufficiently miscible with oil to preclude the possibility of separate drainage in this way.

Table 4-1 Miscibility of oil with liquid refrigerants

Refrigerant	At 0°C	At 35°C	Specific mass, kg/m³
R.12	Fully miscible	Fully miscible	1295
R.22	Separates into oil-rich mixture at top and refrigerant-rich mixture at bottom	Fully miscible	1177
R.502	Low miscibility	Low miscibility	1195
R.717	Non-miscible	Non-miscible	596
Oil			910

Evaporators containing a large body of R.22 will have a greater concentration of oil in the upper layers. By bleeding off a proportion of the mixture (about 10 per cent of the mass flow) and separating the oil from this by distillation, the concentration can be held to an acceptable working limit (see Fig. 4-2). Since the addition of outside heat for this distillation would be a direct waste of energy, the heat is obtained from the warm liquid passing from the condenser to the expansion valve.

The alternative method of returning oil from the evaporator to the compressor is to keep it moving, by ensuring a minimum continuous fluid velocity in all parts of the circuit. This dynamic circulation method is the decisive factor in the design of nearly all halocarbon evaporators, the exceptions being 'flooded' evaporators (see Chapter 6).

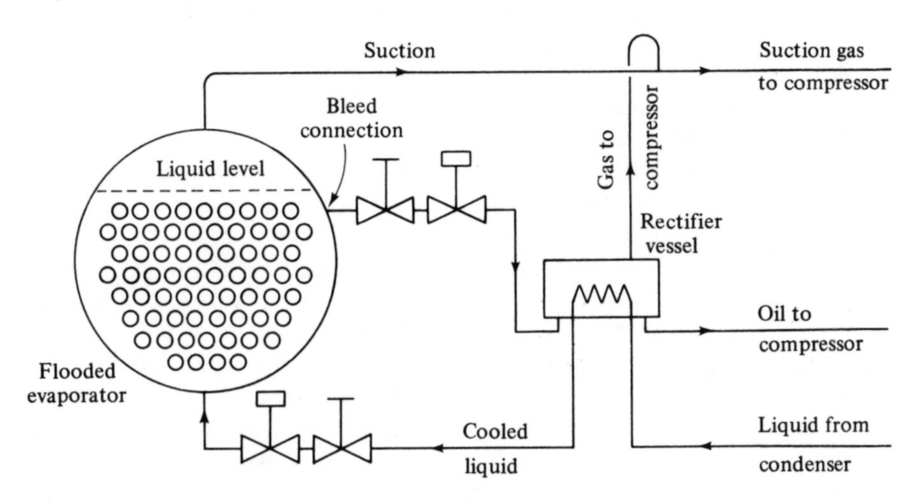

Figure 4-2 Oil bleed and rectifier for R.22 flooded evaporator.

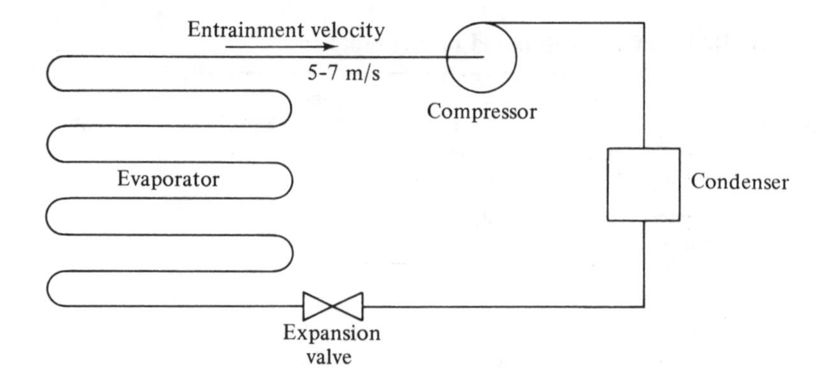

Figure 4-3 Dry expansion circuit.

The critical section of the circuit (Fig. 4-3) is where there is no liquid refrigerant left to help move the oil, i.e., the evaporator outlet and the suction pipe back to the compressor. Entrainment velocities of 5 to 7 m/s are required to ensure that oil droplets will be carried back by the dry refrigerant gas to the compressor. The principle of continuous fluid velocity means that the evaporator will be in a continuous circuit. This does not imply that it has to be one pipe, since many pipes may be arranged in parallel to get the required heat transfer surface, providing the minimum velocity criteria are met.

4-4 CONTAMINANTS IN OIL

The oil in a refrigeration system should remain as clean as it is when it enters the compressor (unlike that of the automobile engine which is quickly contaminated by fuel, water, carbon, and atmospheric dust). The condition of the compressor oil is, therefore, a direct indication of the physical and chemical cleanliness of the system.

Lubricating oil should be kept in tightly sealed containers to exclude atmospheric moisture. Oil drained from oil pots and drains is not used again unless it can be properly filtered and kept dry.

The oil as seen through the crankcase sight glass should remain transparent. If it takes on a white, emulsified appearance it is wet and should be drained and discarded.

Overheating or an electrical fault in the winding of an hermetic or semi-hermetic compressor motor will produce contaminants including the halogen acids, which can be detected by their acrid smell, litmus paper, or other tests.[16] Eye goggles and rubber gloves should be worn when handling such suspect oil. If shown to be acid, the oil must be removed, carefully disposed of, and the system thoroughly cleaned out.[17, 18]

CHAPTER

FIVE

CONDENSERS AND WATER TOWERS

5-1 GENERAL

The purpose of the condenser in a vapour compression cycle is to accept the hot, high-pressure gas from the compressor and cool it to remove first the superheat and then the latent heat, so that the refrigerant will condense back to a liquid. In addition, the liquid is usually slightly sub-cooled. In nearly all cases, the cooling medium will be air or water.

5-2 HEAT TO BE REMOVED

The total heat to be removed in the condenser is shown in the *p–h* diagram (Fig. 5-1) and, apart from comparatively small heat losses and gains through the circuit, will be

Heat taken in by evaporator + heat of compression

This latter, again ignoring small heat gains and losses, will be the net shaft power into the compressor, giving

Evaporator load + compressor power = condenser load

Condenser rating is correctly stated as the rate of heat rejection by it. Some manufacturers give ratings in terms of the evaporator load, together with a 'de-rating' factor, which depends on the evaporating and condensing temperatures.

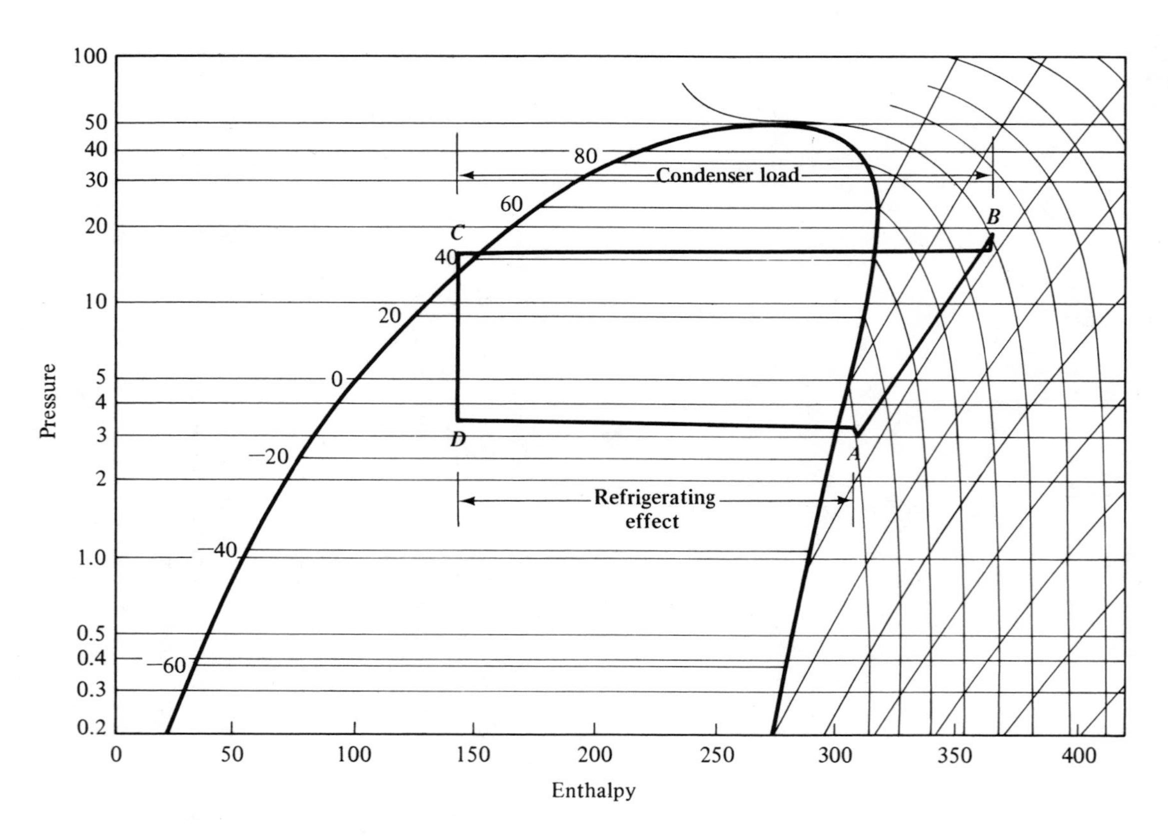

Figure 5-1 Condenser load p–h diagram.

Example 5-1 The following figures from a compressor catalogue give the cooling capacity in British thermal units per hour × 10^{-3} and the shaft horsepower, for a range of condensing temperatures and one evaporator temperature. Calculate the condenser capacities in each case.

	Condensing temperature, °F					
	75	80	85	90	95	100
Btu/h × 10^{-3}	874.6	849.7	824.3	798.3	771.7	744.6
Shaft h.p.	54.3	58.1	61.7	65.1	68.3	71.4

Dividing the British thermal units per hour by 3412 to get kilowatts, and multiplying the shaft horsepower by 0.746, also to get kilowatts, and then adding, we get the condenser capacity:

75°F	80°F	85°F	90°F	95°F	100°F
256	249	242	234	226	218
41	43	46	49	51	53
297	292	288	283	277	271

Example 5-2 A condenser manufacturer gives a 'heat rejection capacity factor' at 26°C wet bulb temperature of 1.22. What is the condenser duty if the cooling capacity is 350 kW?

$$\begin{aligned} \text{Condenser duty} &= \text{cooling capacity} \times \text{factor} \\ &= 350 \times 1.22 \\ &= 427 \text{ kW} \end{aligned}$$

The provision of a separate oil cooler will reduce the condenser load by the amount of heat lost to the oil and removed in the oil cooler. This is of special note with screw compressors, where a high proportion of the compressor energy is taken away in the oil. This proportion varies with the type of screw and the exact method of oil cooling, and figures should be obtained from the compressor manufacturer for a particular application.

5-3 AIR-COOLED CONDENSERS

The simplest air-cooled condenser consists of a plain tube containing the refrigerant, placed in still air and relying on natural air circulation. An example is the condenser of the domestic refrigerator, which may also have some secondary surface in the form of supporting and spacer wires.

Above this size, the flow of air over the condenser surface will be by forced convection, i.e., mechanical fans. The high thermal resistance of the boundary layer on the air side of the heat exchanger leads to the use, in all but the very smallest condensers, of an extended surface. This takes the form of plate fins mechanically bonded onto the refrigerant tubes in most commercial patterns. The ratio of outside to inside surface will be between 5 : 1 and 10 : 1.

Flow of the liquefied refrigerant will be assisted by gravity, so the inlet will be at the top of the condenser and the outlet at the bottom. Rising pipes should be avoided in the design, and care is needed in installation to get the pipes level.

The flow of air may be vertically upwards, or horizontal, and the configuration of the condenser will follow from this (see Fig. 5-2). Small cylindrical matrices are also used, the air flowing radially inwards and out through a fan at the top.

Forced convection of the large volumes of air at low resistance leads to the general use of propeller or single-stage axial flow fans. Where a single fan would be too big, multiple smaller fans give the advantages of lower tip speed and noise, and flexibility of operation in winter (see Sec. 5-12). In residential areas, slower speed fans may be specified to reduce noise levels. The resultant mass flow will de-rate the condenser, and manufacturers will give ratings for 'standard' and 'quiet' assemblies.

It will be recognized that the low specific heat capacity and high

(*a*)

Figure 5-2 (*a*) Air-cooled condenser. *(Courtesy Custom Coils Limited.)*

(b)

Figure 5-2 (*b*) Cylindrical air-cooled condenser. *(Courtesy Friedrich.)*

specific volume of air implies a large volume to remove the condenser heat. If the mass flow is reduced, the temperature rise must increase, raising the condensing temperature and pressure to give lower plant efficiency. In practice, the ΔT between the entering and leaving air should be kept to from 9 to 12 K. The mass flow, assuming a rise of 10.5 K, is then

$$\frac{1}{10.5 \times 1.003} = 0.095 \text{ kg/s kW}$$

As an example of these large air flows required, the condenser for an air-conditioning plant for a small office block, having a cooling capacity of 350 kW and rejecting 430 kW, would need 40.85 kg/s or about 36 m^3/s of air. This cooling air should be as cold as possible, so the condenser needs to be mounted where such a flow of fresh ambient air is available without recirculation.

The large air flows needed, the power to move them, and the resulting noise levels are the factors limiting the use of air-cooled condensers.

Materials of construction are aluminium fins on stainless tube for ammonia, or aluminium or copper fins on aluminium or copper tube for the halocarbons. Aluminium tube is not yet common, but its use is expected to increase.

In view of the high material cost for air-cooled condensers compared with other types, a higher ln MTD is usually accepted, and condensing temperatures may be 5 to 8 K higher for a given cooling medium temperature. Air-cooled condensers must, of course, be used on land and air transport systems. They will also be used in desert areas where the supply of cooling water is unreliable.

5-4 WATER-COOLED CONDENSERS

The higher heat capacity and density of water make it an ideal medium for condenser cooling and, by comparison with the 350 kW plant cited above, the flow is only 9.8 litre/s. Small water-cooled condensers may comprise two concentric pipes ('double pipe'), the refrigerant being in either the inner tube or the annulus. Configurations may be straight, with return bends or headers, or coiled (Fig. 5-3). The double-pipe condenser is circuited in counterflow (media flowing in opposite directions) to get the most sub-cooling, since the coldest water will meet the outgoing liquid refrigerant.

Larger sizes of water-cooled condenser require closer packing of the tubes to minimize the overall size, and the general form is shell-and-tube, having the water in the tubes (Fig. 5-4). This construction is a very adaptable mechanical design and is found in all sizes from 100 mm to 1.5 m

(*a*)

Figure 5-3 (*a*) Water-cooled condensers. *(Courtesy Technocold S.p.A.)*

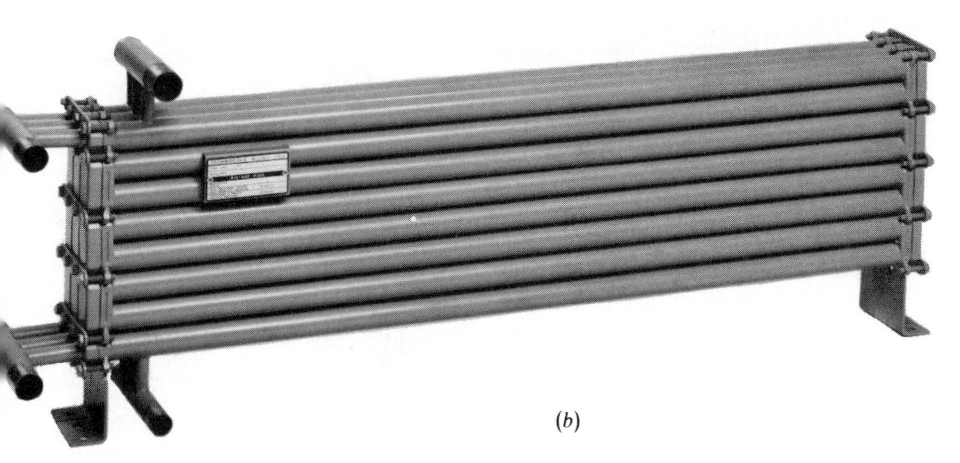

(*b*)

Figure 5-3 (*b*)

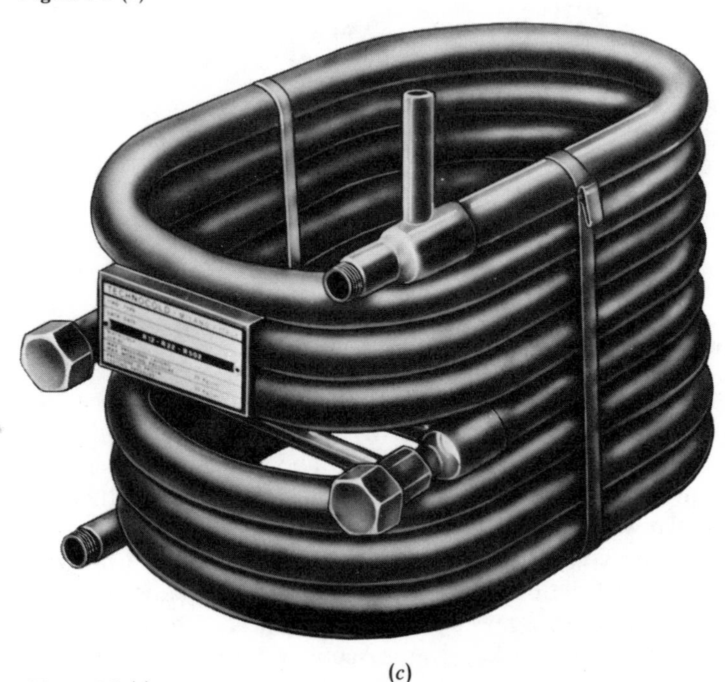

(*c*)

Figure 5-3 (*c*)

diameter and in lengths from 600 mm to 6 m, the latter being the length of commercially available tubing. Materials can be selected for the application and refrigerant, but all mild steel is common for fresh water, with cupronickel or aluminium brass tubes for salt water.

Some economy is size can be effected by an extended surface on the refrigerant side. This can take the form of low integral fins, bounded fins, or porous sleeves. Some gain is also possible on the water side by fitting

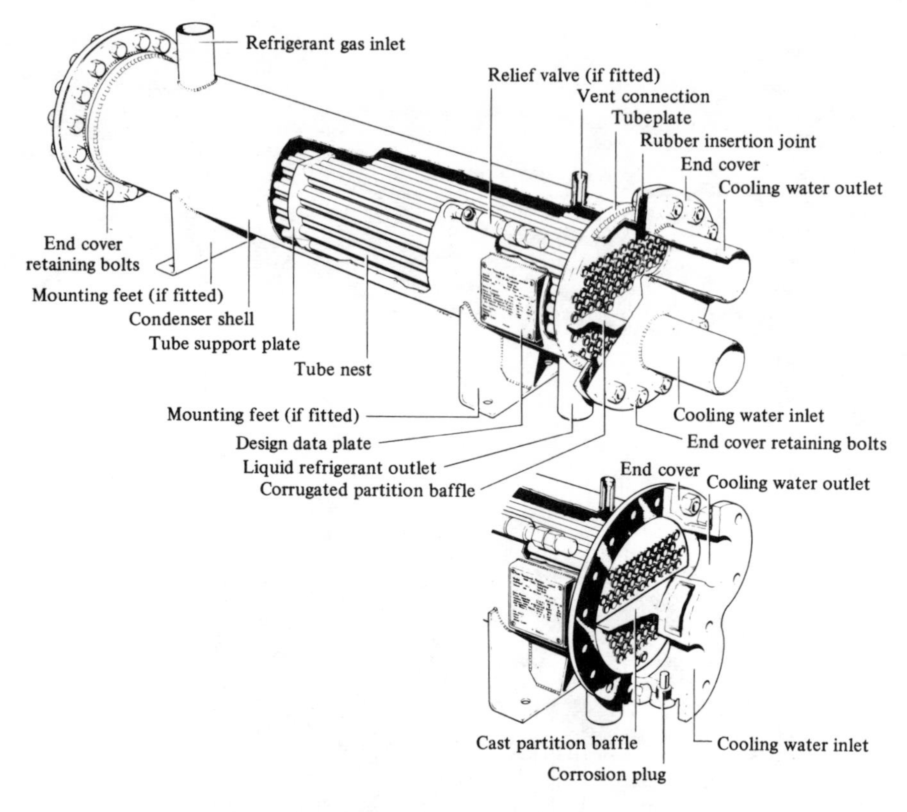

Figure 5-4 Shell-and-tube condenser. *(Courtesy Hall-Thermotank Products Limited.)*

internal swirl strips to promote turbulence, but these interfere with maintenance cleaning and are not much in favour.

Water velocity within the tubes should be about 1 m/s, depending on the bore. To maintain this velocity, baffles are arranged within the end covers to confine the water flow to a number of tubes in each 'pass'. Some condensers have two separate water circuits (double bundle, Fig. 5-5), using the warmed water from one circuit as reclaimed heat in another part of the system. The main bundle rejects the unwanted heat. Where the mass flow of water is unlimited (sea, lake, river, or cooling tower), the temperature rise through the condenser may be kept as low as 5 K, since this will reduce the ln MTD with a lowering of head pressure at the cost only of larger water pumps and pipes.

Example 5-3 A condenser uses water from a river with a temperature rise of 5.2 K. Total duty at the condenser is 930 kW. How much water flow is required?

$$\frac{930}{5.2 \times 4.187} = 43 \text{ kg/s}$$

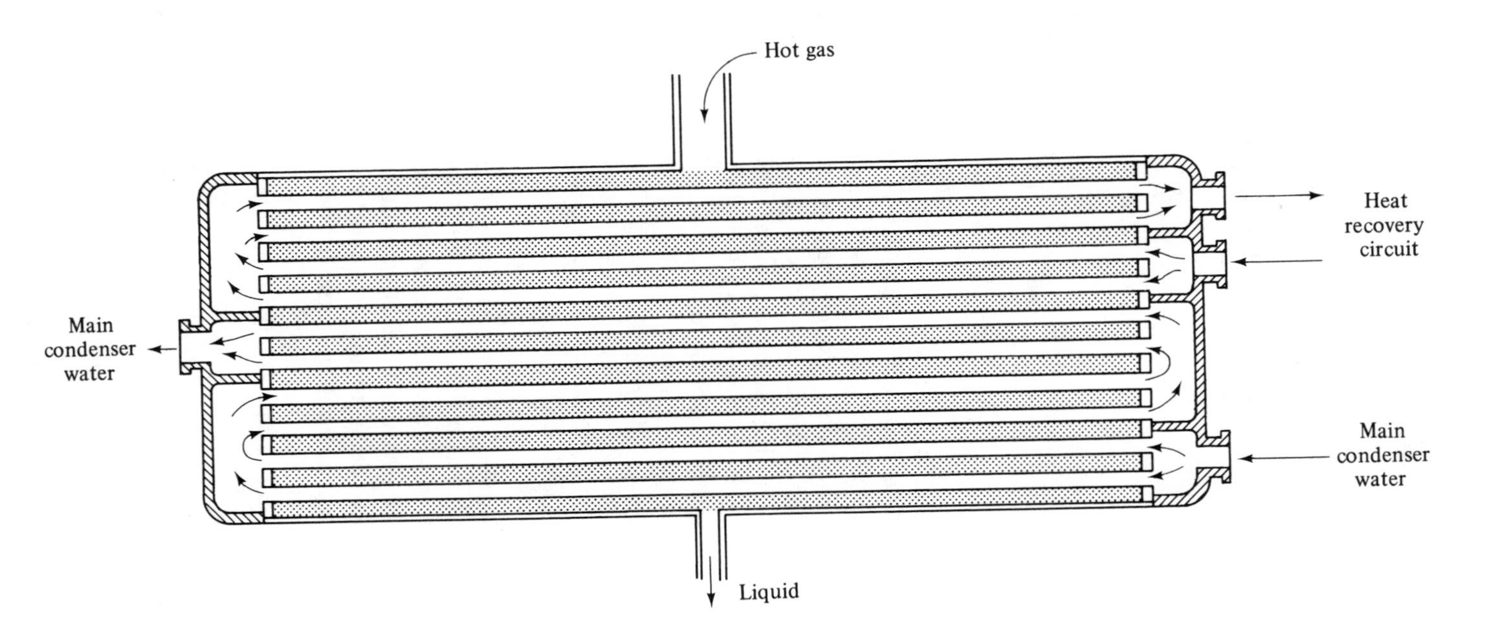

Figure 5-5 Double-bundle shell-and-tube condenser.

If, however, water is used once through only, and is then rejected to a drain, the range will be much higher, possibly 11 to 15 K.

Example 5-4 A small water-cooled condenser uses mains water at 13°C and heats this to 27°C before it goes to waste. The evaporator duty is 4.2 kW and the motor output is 1.7 kW. What is the water mass flow?

$$\text{Condenser load} = 4.2 + 1.7$$

$$= 5.9 \text{ kW}$$

$$\text{Mass flow} = \frac{5.9}{(27 - 13) \times 4.187}$$

$$= 0.1 \text{ kg/s}$$

The supply of water is usually limited and requires the use of a cooling tower. Other possibilities are worth investigation; for example, in the food industries, large quantities of water are used for processing the product, and this could be passed first through the condensers if precautions are taken to avoid contamination. Also, where ground water is present, it could be taken from a shallow borehole and afterward returned to the ground at some distance from the suction. In both these cases, water would be available at a steady temperature and some 8 to 10 K colder than summer water from a cooling tower.

5-5 COOLING TOWERS

In a cooling tower, cooling of the main mass of water is obtained by the evaporation of a small proportion into the airstream. Cooled water leaving the tower will be 3 to 8 K warmer than the incoming air *wet bulb* temperature. (See also Chapters 16 and 18.) The quantity of water evaporated will take up its latent heat equal to the condenser duty, at the rate of about 2430 kJ/kg evaporated, and will be approximately

$$\frac{1}{2430} = 0.41 \times 10^{-3} \text{ kg/s kW}$$

and, for the plant capacity in Example 5-2, would evaporate 0.18 kg/s.

Cooled water from the drain tank is taken by the pump and passed through the condenser, which may be built up with the compressor as part of a compressor/condenser package (condensing unit). The warmed water then passes back to sprays or distribution troughs at the top of the tower and falls in the upgoing airstream, passing over packings which present a

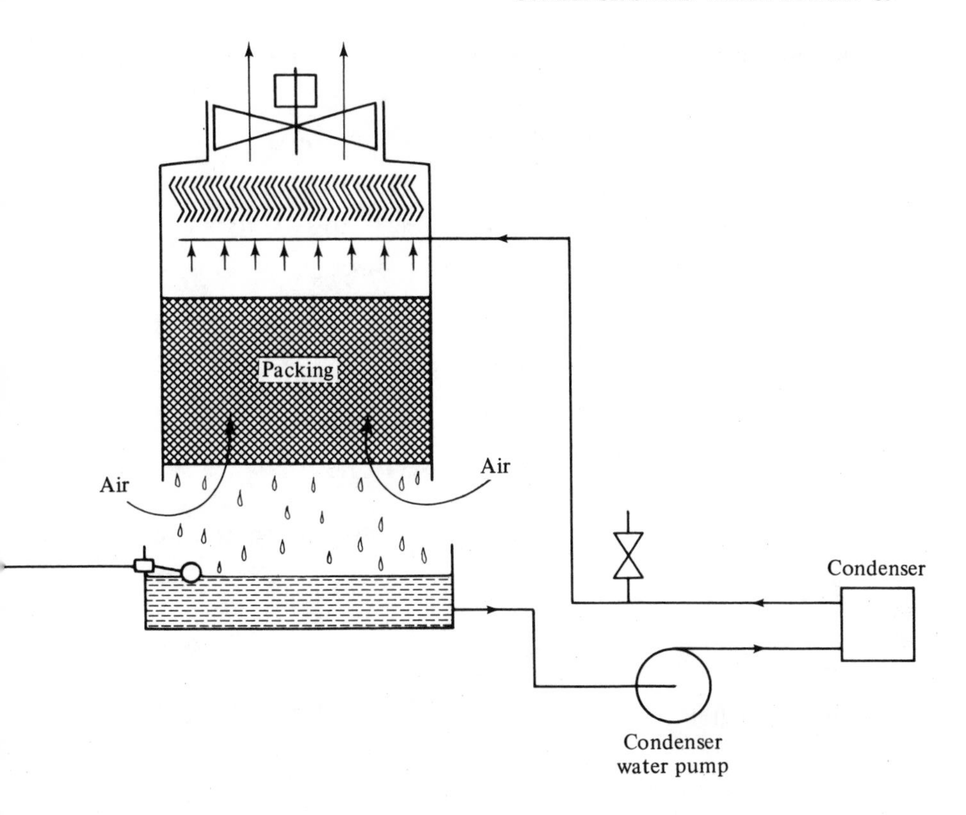

Figure 5-6 Water tower circuit.

large surface to the air. Evaporation takes place, the vapour obtaining its latent heat from the body of the water, which is therefore cooled (see Fig. 5-6).

5-6 EVAPORATIVE CONDENSERS

This cooling effect of the evaporation of water can be applied directly to the condenser refrigerant pipes in the evaporative condenser (Fig. 5-7). The mass flow of water over the condenser tubes must be enough to ensure wetting of the tube surface, and will be of the order of 80 to 160 times the quantity evaporated. The mass flow of air must be sufficient to carry away the water vapour formed, and a compromise must be reached with expected variations in ambient conditions. An average figure is 0.06 kg/s kW.

Example 5-5 A water tower serves a condenser rated at 880 kW and the water circulating pump takes another 15 kW. What will be the evaporation rate, the approximate circulation rate, and the air mass flow?

$$\text{Total water tower duty} = 880 + 15$$
$$= 895 \text{ kW}$$
$$\text{Evaporation rate} = 895 \times 0.41 \times 10^{-3}$$
$$= 0.37 \text{ kg/s}$$
$$\text{Circulation rate, 80 times} = 30 \text{ kg/s } (\Delta T = 7.1 \text{ K})$$
$$\text{160 times} = 60 \text{ kg/s } (\Delta T = 3.6 \text{ K})$$
$$\text{Air flow} = 895 \times 0.06$$
$$= 54 \text{ kg/s}$$

It will be seen that the water and air mass flow rates over a cooling tower are roughly equal.

(*a*)

Figure 5-7 Evaporative condensers. (*a*) *Courtesy Baltimore Air Coil Limited.*

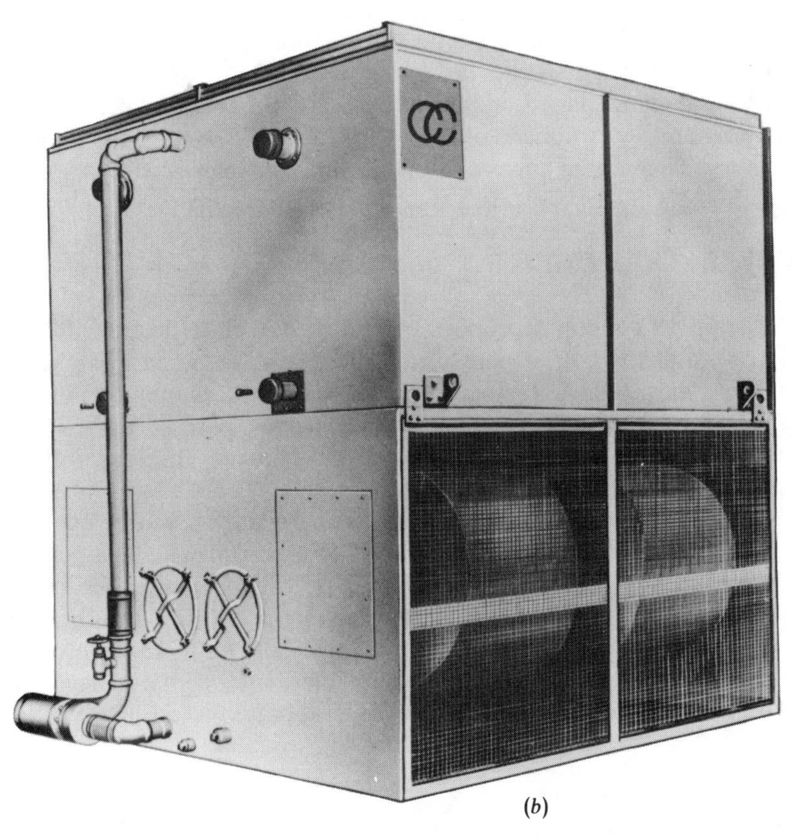

(b)

Fig. 5-7 (b) *Courtesy Custom Coils Ltd.*

Evaporative condensers have a higher resistance to air flow than cooling towers and centrifugal fans are generally used, ganged together to obtain the required mass flow without undue size. This arrangement is also quieter in operation than axial flow fans. Most types use forced draught fans (Fig. 5-7).

Cooling towers and evaporative condensers may freeze in winter if left operating on a light load. A common arrangement is to switch off the fan(s) with a thermostat, to prevent the formation of ice. The water-collection tank will have an immersion heater to reduce the risk of freezing when the equipment is not in use or the tank can be located inside the building under the tower structure, if such space is conveniently available.

Materials of construction must be corrosion resistant. Steel should be hot galvanized, although some resin coatings may suffice. G.r.p. casings are used by some manufacturers. The water-dispersal packing of a cooling tower is made of treated timber or corrugated plastic sheet.

The *atmospheric condenser* is a simplified form of evaporative condenser, having plain tubes over a collecting tank and relying only on natural air draught. This will be located on an open roof or large open space to ensure a good flow of air. The space required is of the order of 0.2 m^2/kW, and such condensers are not much used because of this large space requirement. Atmospheric condensers can still be seen on the roofs of old breweries. They are in current use where space is plentiful.

5-7 WATER TREATMENT

All water supplies contain a proportion of dissolved salts. These will tend to be deposited at the hottest part of the system, e.g., the furring of a kettle or of hot water pipes. Also, these impurities cannot evaporate into an airstream, so where water is being evaporated as part of the cooling process, the salts will remain in the circuit and increase in concentration, thus hastening the furring process.

It is possible to remove all solids from the make-up water, but it is much cheaper to check the concentration by other means. Two general methods are employed. The first relies on physical or chemical effects to delay deposition of scale on the hot surfaces; the second restricts the concentration to a level at which precipitation will not occur. In both cases, the accumulation of solids is removed by bleeding off water from the circuit to drain, in addition to that which is evaporated (see Fig. 5-8).

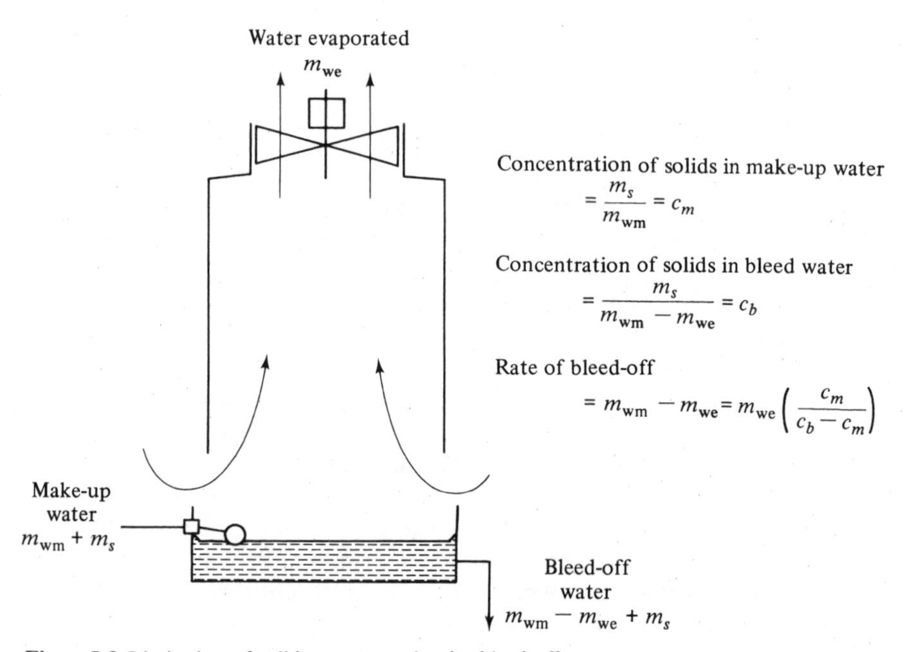

Figure 5-8 Limitation of solids concentration by bleed-off.

It will be seen that, if the concentration of solids in the tower tank is permitted to increase, it will reach a level where the amount carried out in the bleed water will compensate for that not lost in the water which is evaporated. The concentration in the tank must build up to

$$m_s/(m_{wm} - m_{we})$$

where m_{wm} = mass of make-up water
m_s = mass of solids
m_{we} = mass of water evaporated

The concentration of mains make-up water, m_s/m_{wm}, is obtained from the water supply authority. The permissible concentration, $m_s/(m_{wm} - m_{we})$, will be decided by the method of water treatment or the assumed concentration of untreated water which will prevent precipitation.

Example 5-6 The hardness of water in Coventry is given as a maximum of 560 p.p.m. (parts per million) and the water treatment can permit a concentration of solids to 1200 p.p.m. The cooling capacity is 700 kW and the compressor power 170 kW. How much water should be bled to waste and what is the total make-up required?

$$\text{Cooling tower capacity} = 700 + 170 = 870 \text{ kW}$$

$$\text{Latent heat of water vapour} = 2420 \text{ kJ/kg}$$

$$\text{Rate of evaporation} = \frac{870}{2420} = 0.36 \text{ kg/s}$$

$$\text{Rate of bleed-off} = 0.36\left(\frac{560}{1200 - 560}\right) = 0.32 \text{ kg/s}$$

$$\text{Total make-up required} = 0.36 + 0.32 = 0.68 \text{ kg/s}$$

In all cases where water is used for cooling, but more especially where it is being evaporated, the hardness figure should be obtained from the local water supply authority. Enquiries should also be made as to possible variations in the supply, since many cities draw their water from two or more catchment areas, and the type and quantity of hardness may change.

Many suppliers now offer water treatment for use in refrigeration condenser circuits, and the merits of different methods need to be assessed before making a choice. The reader is referred to specialist works on the subject.[3, 17, 19]

There are several methods of providing a percentage 'bleed-off' from the water circuit:

1. The make-up ball valve can be set a little high so that some water always goes down the overflow pipe. This is rather difficult to set initially, but is reliable and cannot easily be tampered with. It will work at all times, so will waste water if the plant is not running.
2. A small bleed-off pipe is taken from the pump discharge, with an adjusting valve, and led to waste. This can be more easily adjusted, works only when the condenser is running, but is subject to interference by unauthorized persons.
3. A tundish, having an area some one-eightieth of the cross-sectional area of the tower, is located just above the water level and is led to the drain, possibly forming part of the overflow fitting. This will bleed off one-eightieth part of the water falling through the tower.

All these methods provide the maximum required rate of bleed-off at all times of the year, so will waste water at light load conditions. The user should be aware of the essential nature of bleed-off, since cases often occur in dry weather of misguided persons closing off the bleed to 'save water'.

In some locations, it is necessary to drain the tank frequently to clear other contaminants. With careful control, this can be used as the necessary bleed-off.

5-8 RATING AND SIZING OF CONDENSERS

Catalogue ratings show heat rejected at a stated condensing temperature and related to the following:

Ambient dry bulb temperature for air-cooled condensers

Available water temperature for water-cooled condensers

Ambient wet bulb temperature for evaporative types

Choice of equipment based on first cost only will almost certainly result in an undersized condenser and a high head pressure.

Example 5-7 In Example 5-1, the required plant capacity is 218 kW and the running time is 2000 hours per year at an electricity cost of 3.2 p/kW h and a motor efficiency of 75 per cent. In order to achieve the condensing temperature of 80°F (26.7°C) the condenser would cost

£8250, while the smaller condenser for a temperature of 100°F (37.8°C) costs £5600. Estimate the break-even time.

Condensing temperature	26.7°C	37.8°C
Rated capacity of plant, kW	249	218
Running time for 218 kW × 2000 h	1751	2000
Power needed, kW	43	53
Electricity cost per year, £	3213	4523
Electricity saving per year, £	1310	

$$\text{Break-even time} = \frac{8250 - 5600}{1310} = 2.02 \text{ years}$$

This is a rough calculation, based on direct capital cost and not on interest rates, and needs to be analysed in terms of the general plant economics. It should also be borne in mind that this is based on present-day electricity costs, and a greater saving will be made as fuel costs rise.

Tendering contractors and prospective users should make themselves aware of alternatives of this sort.

5-9 CONDENSER MAINTENANCE

As with all mechanical equipment, condensers should never be located where they are difficult of access, since there will then be less chance of routine maintenance being carried out. Periodic maintenance of a condenser is limited to attention to the moving parts—fans, motors, belts, pumps—and cleaning of water filters, if fitted.

The overall performance will be monitored from the plant running log (see Chapter 23) and the heat exchange surfaces must be kept clean for maximum efficiency—meaning the lowest head pressure and lowest power.

Air-cooled surfaces may be cleaned by brushing off the accumulation of dust and fluff where the air enters the coil, by the combination of a high-pressure air hose and a vacuum cleaner, or, with the obvious precautions, by a water hose.

Advance warning should be had from the plant running log of any build-up of scale on water-cooled surfaces. Scale within the tubes of a straight double-pipe or shell-and-tube condenser can be mechanically removed with suitable wire brushes or high-pressure water lances, once the end covers have been removed. Tubes which cannot be dealt with in this way must be chemically cleaned (see also Chapter 23).

It will be appreciated that, where air and water are present, as in a water cooling tower or evaporative condenser, the apparatus will act as an air washer, removing much of the dust from the air passing through it. Such dirt may be caught in a fine water filter, but is more commonly allowed to settle into the bottom of the tank and must be flushed out once or twice a year, depending on the severity of local contamination. Where heavy contamination is expected, it is good practice to provide a deeper tank than usual, the pump suction coming out well clear of the bottom, and tanks 3 m deep are in use. Where plant security is vital, the tank is divided into two parts, which may be cleaned alternately.

Algae and other organisms will tend to grow on wet surfaces, in particular those in daylight. Control of these can be effected by various proprietary chemicals.[19]

5-10 CONDENSER FITTINGS

The inlet pipe bringing high-pressure gas from the compressor must enter at the top of the condenser, and adjacent piping should slope in the direction of flow so that oil droplets and any liquid refrigerant which may form will continue in the right direction and not back to the compressor.

The outlet pipe must always be from the lowest point, but may have a short internal upstand so that any dirt such as pipe scale or metal swarf will be trapped and not taken around the circuit.

Condensers for ammonia systems may have an oil trap, usually in the form of a drain pot, and the liquid outlet will be above this.

Water connections to a shell-and-tube condenser must always be arranged so that the end covers can easily be removed for inspection, cleaning, and repair of the tubes. Heavy end covers require the use of lifting tackle, and supports above the lifting points should be provided on installation to facilitate this work.

Condensers having a gross volume of more than 285 litre are required under BS.4434:1980 to have two pressure relief valves or two bursting discs, one always in service. Vessels between 85 and 285 litre must have one such relief device. Those below 85 litre but larger than 76 mm inside diameter may have a fusible plug to relieve pressure in a fire. Manufacturers will be aware of the requirements of this BS and similar Standards, and proprietary products will be correctly equipped.

5-11 OTHER FORMS OF CONDENSER

In a cascade system, the evaporator of the high stage is the condenser for the low stage (see Fig. 2-8). Construction of this heat exchanger will be a combination of the design factors for evaporators and condensers, and no general rules apply apart from these. The intercooler of a two-stage or compound system (see Fig. 2-9) de-superheats the discharge gas from the first stage so that it will not be too hot on entering the high stage. In practice, it will leave the intercooler only slightly superheated above the interstage saturation point. The normal fluctuations in an operating system may lead to actual condensation at times, but is not so intended.

The small condensing surface required by a domestic appliance such as a deep-freeze may allow the use of the outside metal skin of the appliance itself as a surface condenser. In such a construction, the condenser tube is held in close mechanical contact with the skin, so that heat is conducted through to the outside air, where it is lost by natural convection. This system is restricted to a few hundred watts.

5-12 WINTER OPERATION

Condensers are sized so that they can reject the system heat load under maximum conditions of air or water temperature. In colder weather, the condensing temperature will fall with that of the cooling medium and this may cause difficulties in correct operation of the plant. In particular, the pressure across the expansion valve (see also Chapter 7) may be too low to circulate required mass flow of the refrigerant. Under such circumstances, artificial means must be used to keep the head pressure up, always remembering that the condensing pressure should be kept as low as practical for power economy.

Various systems are used:

1. Air-cooled condensers having two or more fans (Fig. 5-9) may have a pressure switch or thermostatic control to stop the fans one by one. This method is simple, cheap, and effective.
2. The fans on such condensers may be fitted with two-speed motors or other speed control. It should be borne in mind that, if one fan of a pair stops, the noise level will fall by 3 dB, but if both fans drop to half speed, the noise drops by 15 dB. This method is also of special use in residential areas where the greater noise level will be tolerated in the daytime when condensing air is warmest, but a lower fan speed can be used at night.
3. Evaporative condensers and water cooling towers with two or more fans on separate drive may be controlled in the same way. If a single motor drives several fans of the same shaft, speed control will be required. In any case, such equipment should be fitted with antifreeze thermostats which will stop all fans before the water reaches freezing point.
4. Cooling air flow can be restricted by blanking flaps, baffles, or winter enclosures providing that, if not automatic, the operating staff are aware

Figure 5-9 Air-cooled condenser. *(Courtesy Custom Coils Ltd.)*

of their presence and will restore the air flow when the weather turns warm again.

5. Water flow may be restricted by throttling valves. One such device is operated directly by head pressure, but electric or pneumatic throttling or flow diversion valves can be applied for the purpose (see Chapter 8).
6. Proprietary devices are available which allow the condenser to partially fill with liquid refrigerant, thus decreasing the heat transfer surface available for condensation. Sufficient refrigerant must be available for this, without starving the rest of the circuit (see Chapter 8).
7. Where a complex system is served by two or more condensers, a complete condenser can be taken off-line by a pressure switch.

Apart from such requirements for head pressure control, winter precautions are needed to prevent freezing of the water while the plant is not rejecting heat to it. These commonly take the form of an electric immersion heater in the water tank, together with lagging and possible trace heating of exposed pipes. In some systems, the evaporative condenser itself may be within the building, with air ducts to the outside. In severe climates, external tanks need to be lagged to conserve the heat provided by the immersion heater.

5-13 RECEIVERS

The total refrigerant charge required in a circuit will vary with different operating loads and ambients, and must be sufficient at all times so that only liquid enters the expansion valve. This implies that, at times, the circuit would have too much charge, which would back up in the condenser and reduce its efficiency. A drain tank is required directly after the condenser which can hold this reserve of liquid, and is termed the *receiver* (see Fig. 5-10).

Receivers also act as pump-down tanks, and should be capable of holding enough of the total refrigerant charge to permit evacuation of any one vessel for maintenance, inspection, or repair. They should never be more than 85 per cent full, to allow for expansion and safety.

Receivers are commonly made of steel tube with welded dished ends, and are located horizontally. The liquid drain pipe from the condenser to the receiver should be amply sized, and any horizontal runs sloped to promote easy drainage. Shut-off valves in this line should not be in a horizontal outlet from the condenser, since their slight frictional resistance will cause liquid back-up in the condenser. Outlet pipes from the receiver may be from the bottom or, by means of an internal standpipe, may leave at the top. A valve is invariably fitted at this point.

Ammonia receivers may have an oil drain pot, and the receiver will slope slightly down towards this.

Receivers are pressure vessels covered by the provisions of BS.4434:1980 and require safety pressure relief devices as outlined above. In cases where there is no shut-off valve between the condenser and receiver, such protection may be fitted to one or the other, providing the total volume is considered.

In practice, receivers will operate about one-sixth full during normal running. Some means are usually provided to indicate the liquid level inside. These are as follows:

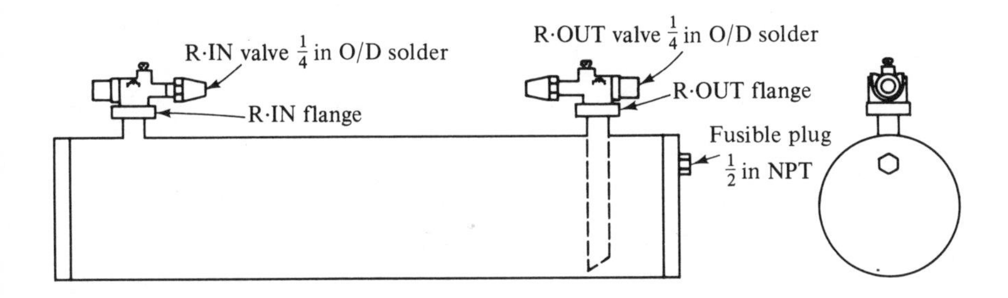

Figure 5-10 Liquid receiver. *(Courtesy Dunham-Bush Limited.)*

1. An external, vertical sight glass, of suitable pattern having self-closing shut-off valves.
2. A number of bull's-eye glasses arranged at different heights in the shell.
3. A pair of bull's-eye glasses, arranged on the same cross section and some 45° up from the horizontal diameter. A light is shone through one and the observer looks through the other.

Example 5-8 The evaporator and condenser of a system hold a total of 115 kg of R.717. Determine the receiver size and dimensions, pressure relief specification, and the total refrigerant charge for the plant.

$$\text{Required working refrigerant mass} = 115 \text{ kg}$$

[This must be accommodated in a space 68% (85% less one-sixth) of the proposed receiver shell.]

$$\text{Gross capacity of receiver} = \frac{115}{68\%}$$

$$= 169 \text{ kg of R.717}$$

$$\text{Specific mass of liquid R.717} = 596 \text{ kg/m}^3$$

$$\text{Volume of receiver for 169 kg gross} = 0.283 \text{ m}^3$$

$$= 283 \text{ litre}$$

The nearest catalogue standard receiver is 240 mm diameter by 3.25 m long and has a gross capacity of 314 litre. Being over the limit of 285 litre it must have dual relief valves.

CHAPTER

SIX

EVAPORATORS

6-1 GENERAL

The purpose of the evaporator is to receive low-pressure, low-temperature fluid from the expansion valve and to bring it in close thermal contact with the load. The refrigerant takes up its latent heat from the load and leaves the evaporator as a dry gas. Evaporators are classified according to their refrigerant flow pattern and their function.

6-2 FLOW PATTERN AND FUNCTION

The refrigerant flow pattern is dependent on the method of ensuring oil removal from the evaporator and, possibly, its return to the crackcase.

Flooded evaporators (Fig. 6-1) have a body of fluid boiling in a random manner, the vapour leaving at the top. In the case of ammonia, any oil present will fall to the bottom and be drawn off from the drain pot or oil drain connection. With the halocarbons, a proportion of the fluid is bled off and rectified (see Fig. 4-3).

Evaporators which keep the oil moving back to the compressor suction by means of continuous gas velocity are termed *dry expansion*, and the refrigerant is totally evaporated in a single pass.

The function of an evaporator will be to cool gas, liquid, or other product load. In most cases air or a liquid is first cooled, and is then used as a secondary refrigerant to cool the load. For example, in a cold room,

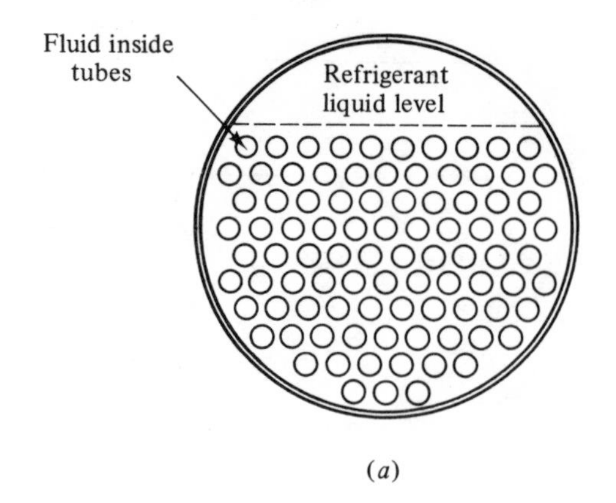

(a)

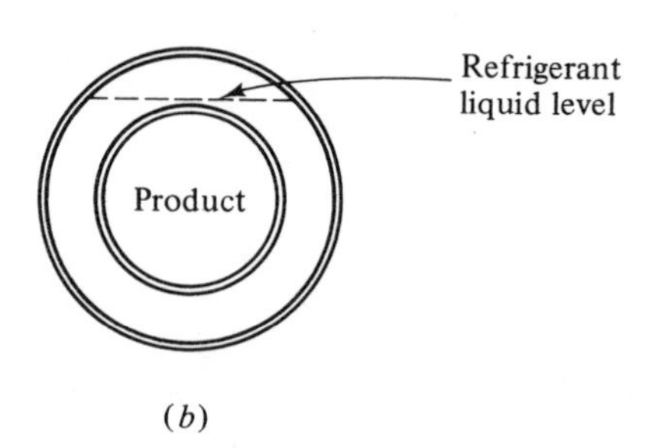

(b)

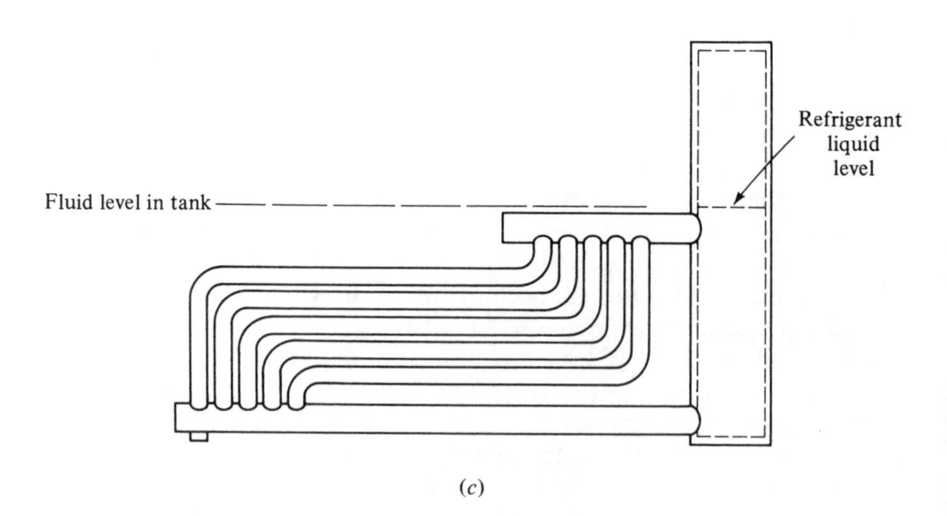

(c)

Figure 6-1 Flooded evaporators: (a) shell-and-tube, (b) jacketted, (c) raceway.

air is cooled and this air cools the stored produce and carries away heat leaking through the structure; in a water chiller system, the water is circulated to cool the load, etc.

6-3 AIR COOLING EVAPORATORS

Air cooling evaporators for cold rooms, blast freezers, air-conditioning, etc., will have finned pipe coils (see Fig. 6-2). In all but very small coolers, there will be fans to blow the air over the coil.

Construction materials will be the same as for air-cooled condensers. Aluminum fins on copper tube is the most common for the halocarbons, with stainless steel tube for ammonia. Frost or condensed water will form on the fin surface and must be drained away. To permit this, fins will be vertical and the air flow horizontal, with a drain tray provided under.

The size of the tube will be such that the velocity of the boiling fluid within it will cause turbulence to promote heat transfer. Tube diameters will vary from 9 to 32 mm according to the size of coil.

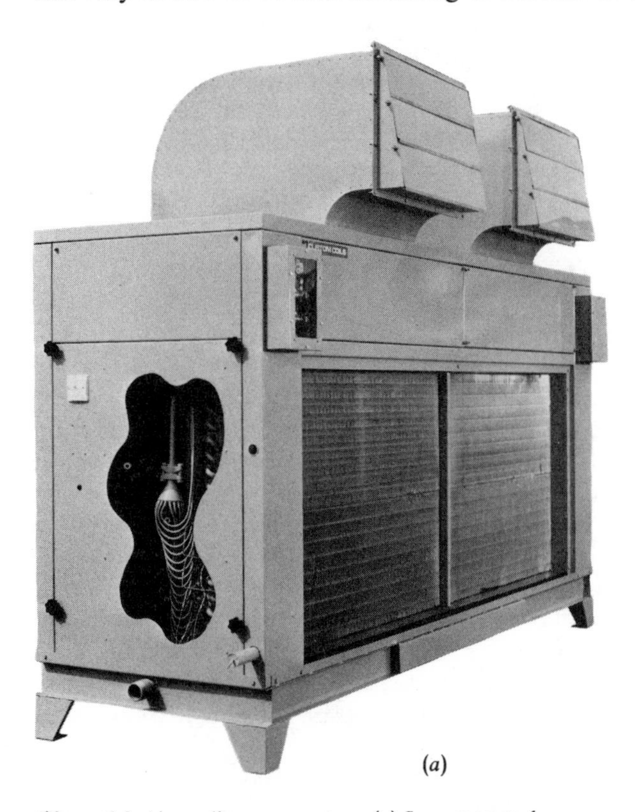

(*a*)

Figure 6-2 Air cooling evaporators: (*a*) floor mounted.

(b)

Figure 6-2 (*cont.*) (*b*) ceiling mounted. (*Courtesy Custom Coils Ltd.*)

Fin spacing will be a compromise between compactness (and cost) and the tendency for the interfin spaces to block with condensed moisture or frost. Spacings will vary from 2 mm on a compact air-conditioner to 12 mm on a low-temperature cold-room coil.[9]

6-4 LIQUID COOLING EVAPORATORS

Liquid cooling is mostly in shell-and-tube or shell-and-coil evaporators.

In the shell-and-tube type, the liquid is usually in the pipes and the shell is some three-quarters full of the liquid, boiling refrigerant. A number of tubes is omitted at the top of the shell to give space for the suction gas to escape clear of the surface without entraining liquid. Further features such as multiple outlet headers, suction trap domes, and baffles will help to avoid liquid droplets entering the main suction pipe. Surface velocities should not exceed 3 m/s and lower figures are used by some designers.

Operated in this manner, the shell-and-tube type is a flooded evaporator (see Fig. 6-3) and has oil drainage pots if using ammonia, or a mixture bleed system if the refrigerant is one of the halocarbons.

The liquid speed should be about 1 m/s or more, to promote internal turbulence, and end cover baffles will constrain the flow to a number of passes, as with the shell-and-tube consenser (q.v.).

Evaporators of this general type with dry expansion circuits will have the refrigerant within the tubes, in order to maintain a suitable continuous

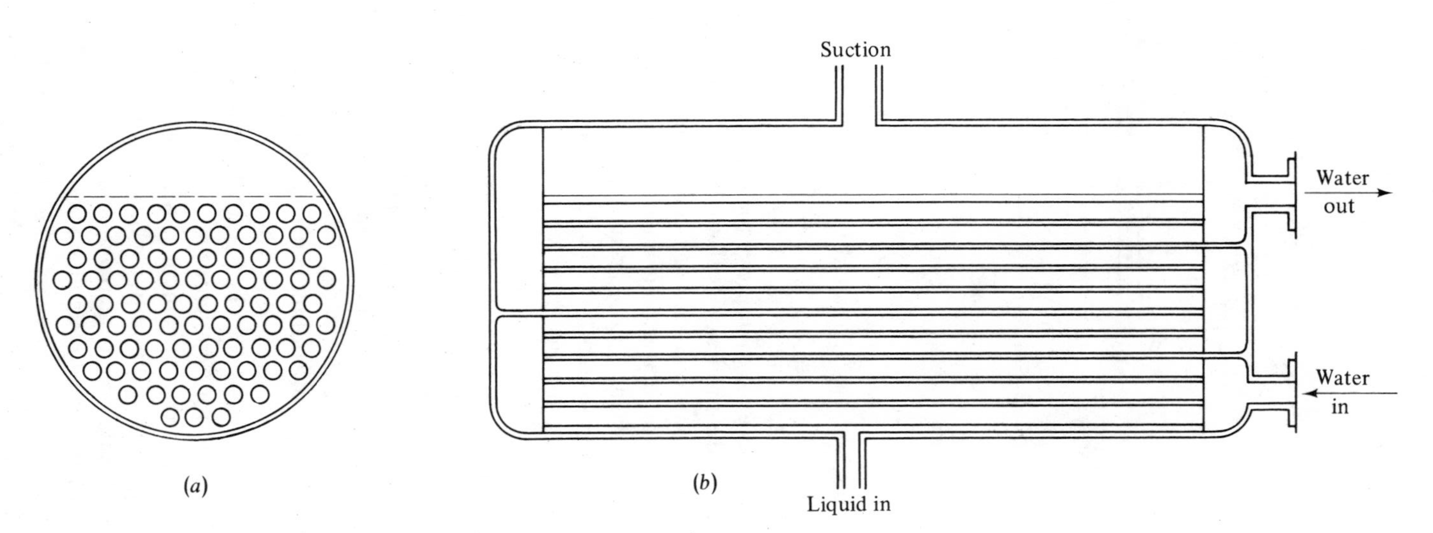

Figure 6-3 Shell-and-tube evaporator, flooded: (*a*) cross section, (*b*) side elevation.

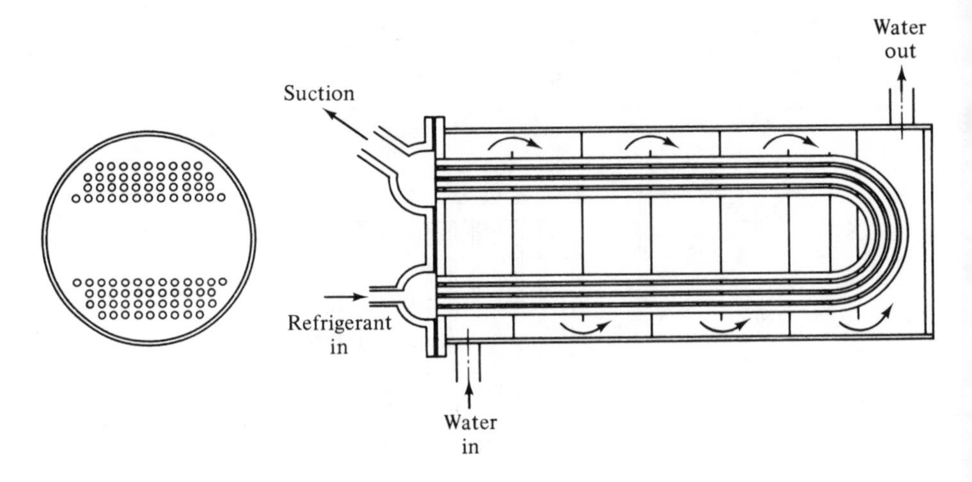

Figure 6-4 Shell-and-coil evaporator.

velocity for oil transport, and the liquid in the shell. These can be made as shell-and-tube, with the refrigerant constrained to a number of passes, or may be shell-and-coil (see Fig. 6-4). In both these configurations, baffles are needed on the water side to improve the turbulence, and the tubes may be finned on the outside, inside, or both.

Liquid cooling evaporators may comprise a pipe coil in an open tank, and can have flooded or dry expansion circuitry. Flooded coils will be connected to a combined liquid accumulator and suction separator (usually termed the *surge drum*), in the form of a horizontal or vertical drum (see Fig. 6-5). The expansion valve maintains a liquid level in this drum and a natural circulation is set up by the bubbles escaping from the liquid refrigerant at the heat exchanger surface. Dry expansion coils for immersion in an open tank will be in a continuous circuit or a number of parallel circuits (see Fig. 6-6). Liquid velocity over such coils can be increased by tank baffles and there may be special purpose agitators, as in an ice-making tank (see Fig. 13-3). Coils within an open tank can be allowed to collect a layer of ice during off-load periods, thus providing thermal storage and giving a reserve of cooling capacity at peak load times (see also Chapter 11).

Where water is to be cooled close to its freezing point without risk of damage to the evaporator, the latter is commonly arranged above the water collection tank and a thin film of water runs over the tubes. Heat transfer is very high with a thin moving film of liquid and, if any ice forms, it will be on the outside, free to expand, and it will not damage the tube. Such an evaporator is termed a Baudelot cooler (Fig. 6-7). It may be open, enclosed in dust-tight shields to avoid contamination of the product (as in surface milk and cream coolers), or may be enclosed in a pressure vessel as

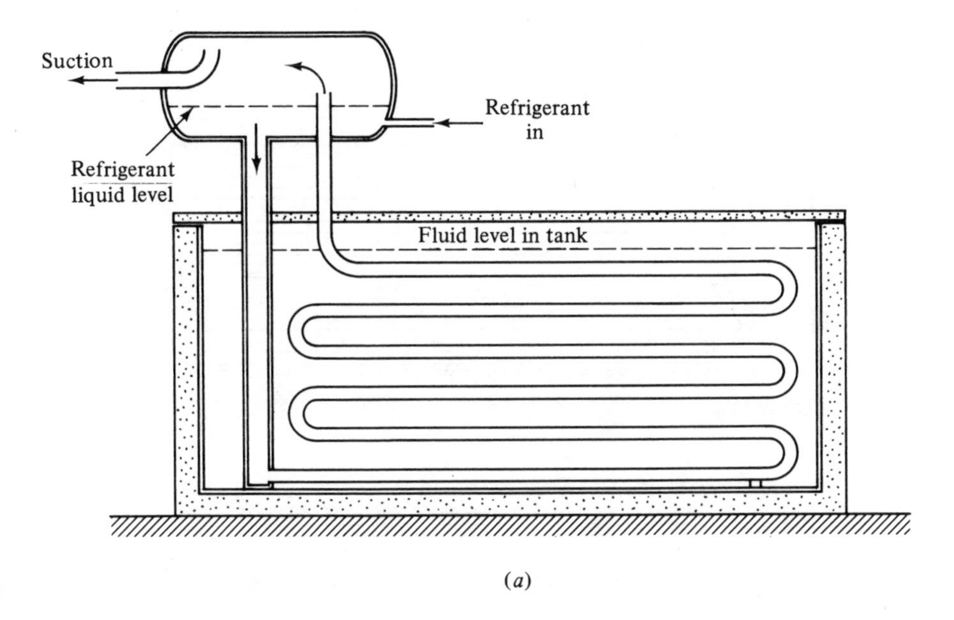

(*a*)

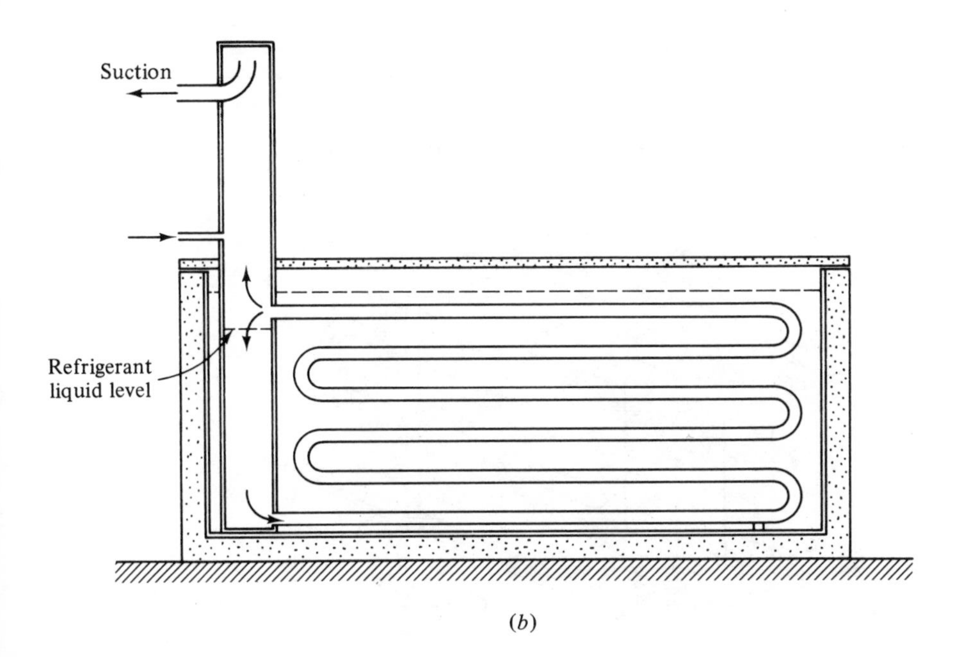

(*b*)

Figure 6-5 Flooded tank evaporators: (*a*) horizontal separator, (*b*) vertical separator.

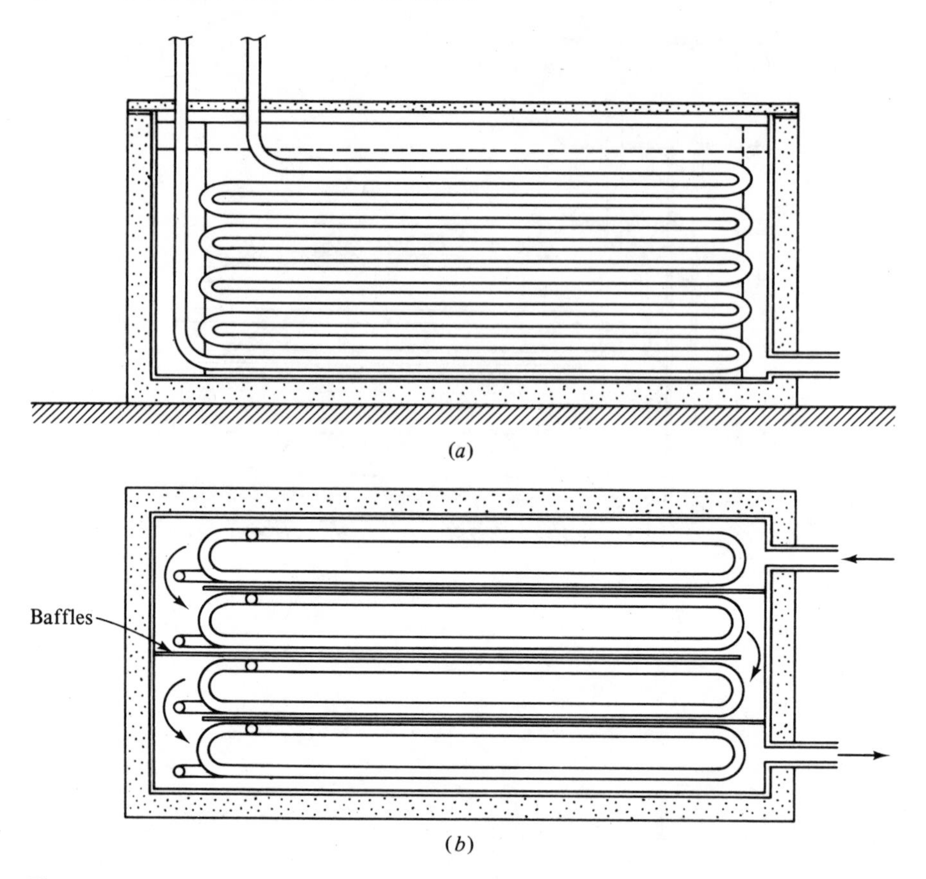

Figure 6-6 Dry expansion tank evaporator: (*a*) section, (*b*) elevation.

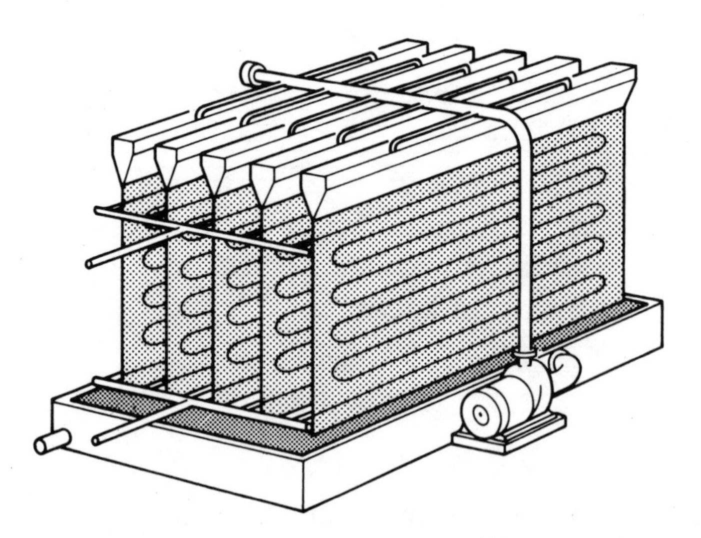

Figure 6-7 Baudelot cooler. *(Courtesy Elliott Turbomachinery Limited.)*

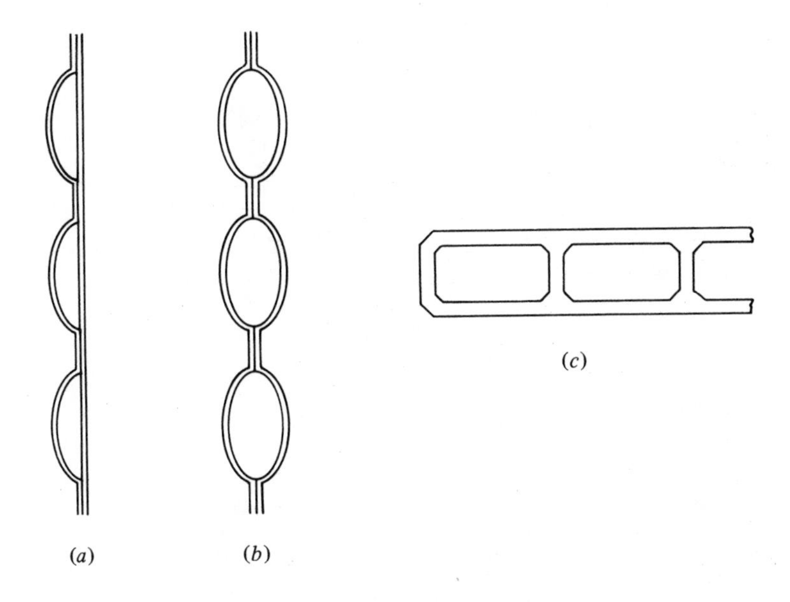

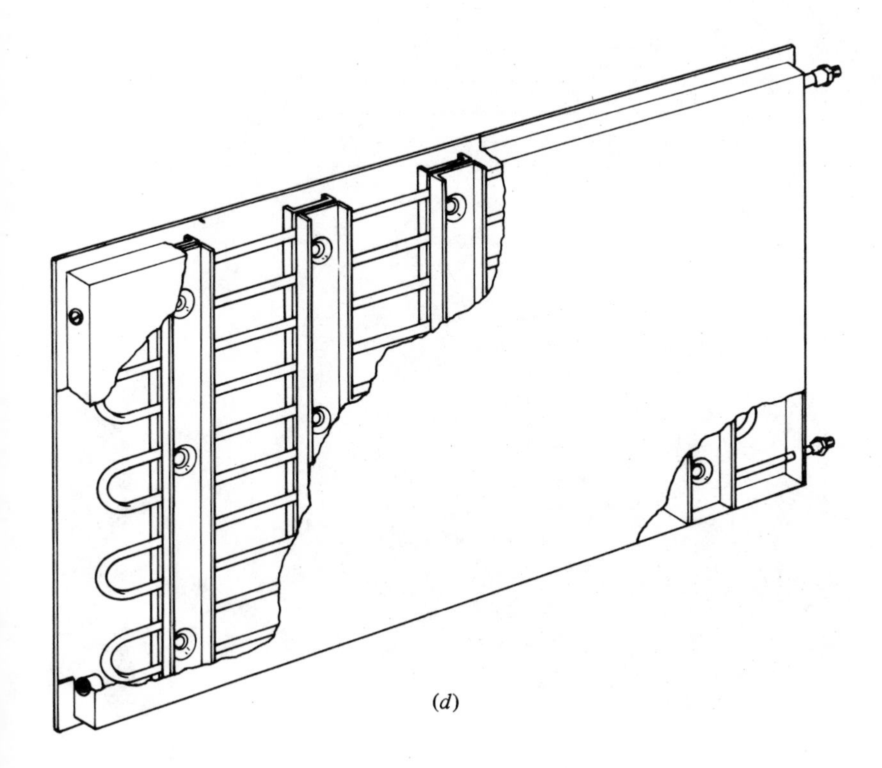

Figure 6-8 Plate evaporators: (*a*) single embossed, (*b*) double embossed, (*c*) extruded, (*d*) holdover (brine filled). *(Courtesy Elliott Turbomachinery Limited.)*

in the Mojonnier cooler for soft drinks, which pressurizes with carbon dioxide at the same time.

Some liquids, such as vegetable fats and ice-cream mixes, increase considerably in viscosity as they are cooled, sticking to the heat exchanger surface. Evaporators for this duty are arranged in the form of a hollow drum, surrounded by the refrigerant and having internal rotating blades which scrape the product off as it thickens, presenting a clean surface to the flow of product and impelling the cold paste towards the outlet.

(*a*)

Figure 6-9 Plate freezers *(Courtesy APV Parafreeze Limited.)* (*a*) horizontal.

6-5 PLATE EVAPORATORS

Plate evaporators (Fig. 6-8) are formed by cladding a tubular coil with sheet metal, welding together two embossed plates, or from aluminium extrusions.

The extended flat face may be used for air cooling, or for liquid cooling if immersed in a tank, or as a Baudelot cooler.

The major use for flat plate evaporators is to cool a solid product by conduction, the product being formed in rectangular packages and held close between a pair of adjacent plates.

In the horizontal plate freezer (Fig. 6-9*a*), the plates are arranged in a

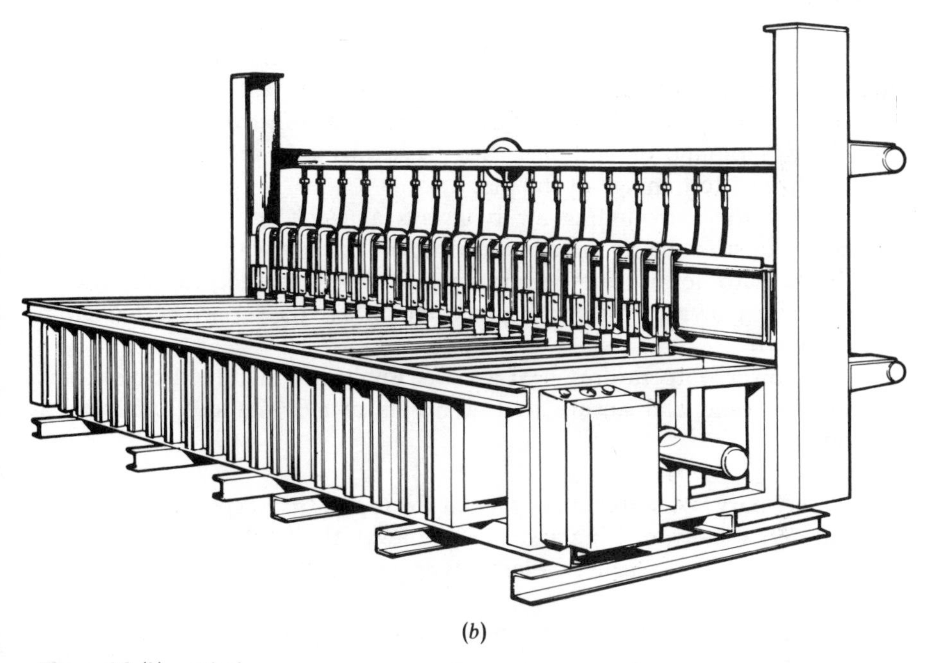

Figure 6-9 (*b*) vertical.

stack on slides, so that the intermediate spaces can be opened and closed. Trays, boxes, or cartons of the product are loaded between the plates and the stack is closed to give good contact on both sides. When the necessary cooling is complete, the plates are opened and the product removed.

The vertical plate freezer (Fig. 6-9*b*) is used to form solid blocks of a wet product, typically fish. When frozen solid, the surfaces are thawed and the blocks pushed up and out of the bank.

To ensure good heat transfer on the inner surface of the plates and achieve a high rate of usage, liquid refrigerant is circulated by a pump at a rate 5 to 12 times the rate of evaporation.

If a plate evaporator is partially filled with brine, this can be frozen down while the plate is on light load, and the reserve of cooling capacity used at other times. The freezing point of the brine can be formulated according to the particular application and the plate can be made as thick as may be required for the thermal storage needed. The major application of this device is the cooling of vehicles. The plates are frozen down at night, or other times when the vehicle is not in use, and the frozen brine keeps the surface of the plate cold while the vehicle is on the road. The refrigeration machinery may be on the vehicle or static.

6-6 DEFROSTING

Air cooling evaporators working below 0°C will accumulate frost which must be removed periodically, since it will obstruct heat transfer.

Evaporators of suitable and robust construction can be defrosted by brushing, scraping, or chipping, but these methods are labour-intensive and may lead to damage of the plant.

Where the surrounding air is always at +2°C or higher, it will be sufficient to stop the refrigerant for a period and allow the frost to melt off (as in the auto-defrost domestic refrigerator). This method can be used for cold rooms, packaged air-conditioners, etc., where the service period can be interrupted.

For lower temperatures, heat must be applied to melt the frost within a reasonable time and ensure that it drains away. Methods used are as follows:

1. Water sparge. A water spray pipe at the top of the evaporator floods the coil periodically, melting and washing away the frost.
2. Electric resistance heaters. Elements are within the coil or directly under it.
3. Hot gas. A branch pipe from the compressor discharge feeds superheated gas to the coil. The compressor must still be working on another evaporator to make hot gas available. Heat storage capsules can be built into the circuit to provide a limited reserve of heat for a small installation.
4. Reverse cycle. The direction of flow of the refrigerant is reversed to make the evaporator act as a condenser. Heat storage or another evaporator are needed as a heat source.

In each of these cases, arrangements must be made to remove cold refrigerant from the coil while defrosting is in progress. Drip trays and drain pipes may require supplementary heating.

CHAPTER

SEVEN

EXPANSION VALVES

7-1 GENERAL

The purpose of the expansion valve is to control the flow of refrigerant from the high-pressure condensing side of the system into the low-pressure evaporator. In all cases, the pressure reduction is achieved through a variable flow orifice, either modulating or two-position. Expansion valves may be classified according to the method of control.

7-2 LOW-PRESSURE FLOAT VALVES

Flooded evaporators require a constant liquid level, so that the tubes remain wetted. A simple float valve suffices, but must be located with the float outside the evaporator shell, since the surface of the boiling liquid is agitated and the constant movement would cause excessive wear in the mechanism. The float is therefore contained within a separate chamber, coupled with balance lines to the shell (see Fig. 7-1).

Such a valve is a metering device and may not provide positive shut-off when the compressor is stopped. Under these circumstances, refrigerant will continue to leak into the evaporator until pressures have equalized, and the liquid level might rise too close to the suction outlet. To provide this shut-off, a solenoid valve is needed in the liquid line.

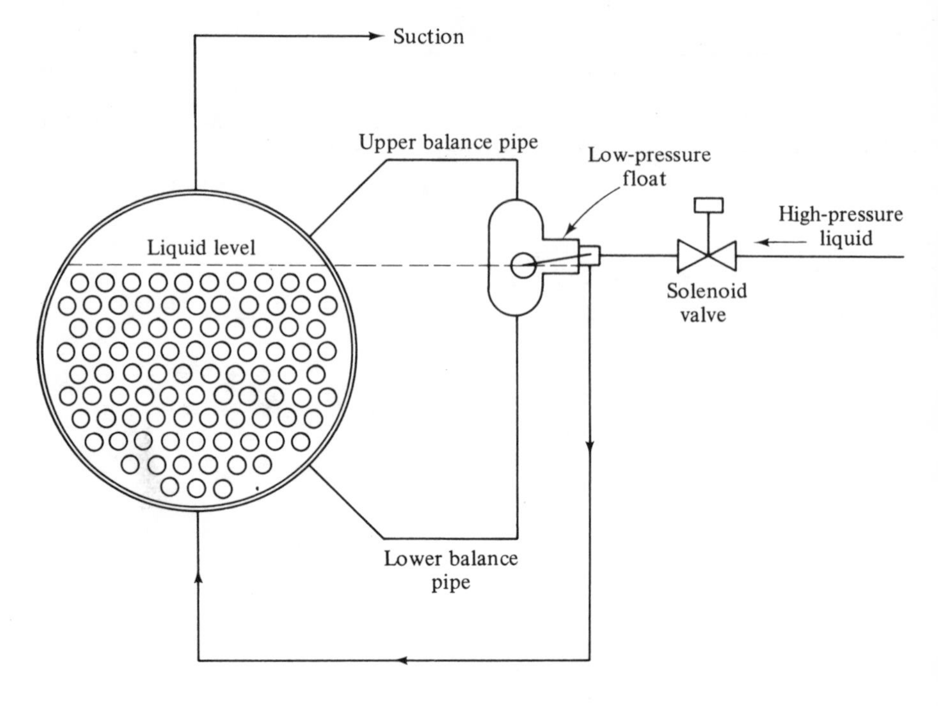

Figure 7-1 Low-pressure float valve on flooded cooler.

7-3 LOW-PRESSURE FLOAT SWITCHES

Since the low-pressure float needs a solenoid valve for tight closure, this can be used as an on–off control in conjunction with a pre-set orifice and controlled by a float switch (Fig. 7-2).

The commonest form of level detector is a metallic float carrying an iron core which rises and falls within a sealing sleeve. An induction coil surrounds the sleeve and is used to detect the position of the core. The resulting signal is amplified to switch the solenoid valve, and can be adjusted for level and sensitivity. A throttle valve is fitted to provide the pressure-reducing device.

Should a float control fail, the level in the shell may rise and liquid pass into the compressor suction. To warn of this, a second float switch is usually fitted at a higher level, to operate an alarm and cut-out.

Where a flooded coil is located in a liquid tank, the refrigerant level will be within the tank, making it difficult to position the level control. In such cases, a gas trap or siphon can be formed in the lower balance pipe to give an indirect level in the float chamber. Siphons or traps can also be arranged to contain a non-volatile fluid such as oil, so that the balance pipes remain free from frost.

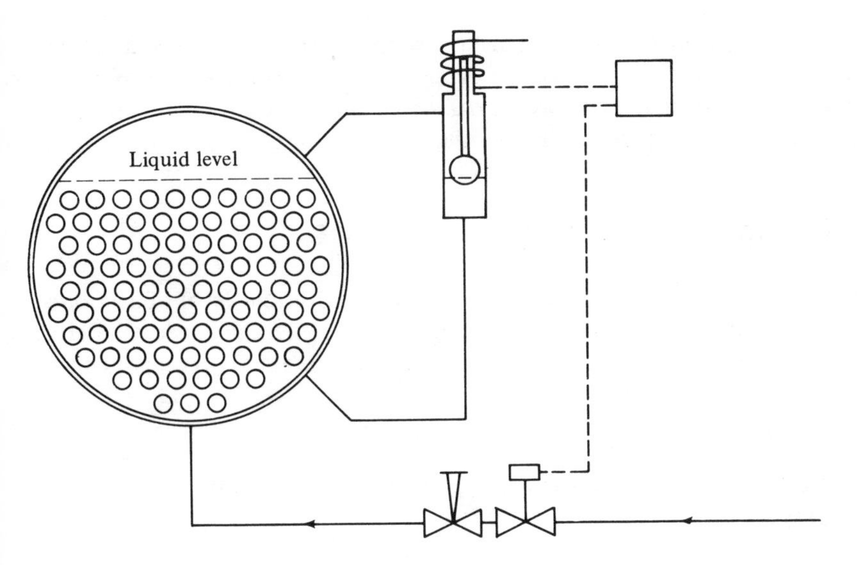

Figure 7-2 Low-pressure float switch.

7-4 HIGH-PRESSURE FLOAT VALVE

On a single-evaporator flooded system, a float valve can be fitted which will pass any drained liquid from the condenser direct to the evaporator. The action is the same as that of a steam trap. The float chamber is at condenser pressure and the control is termed a *high-pressure* float (Fig. 7-3).

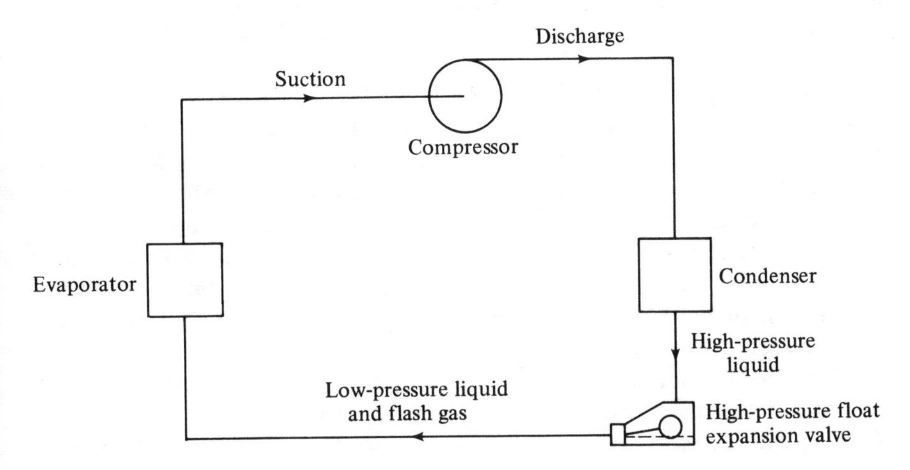

Figure 7-3 High-pressure float valve.

The refrigerant charge of such a system is critical, since it must not exceed the working capacity of the evaporator. It is not possible to have a receiver in circuit and this control cannot feed more than one evaporator, since it cannot detect the needs of either.

7-5 THERMOSTATIC LEVEL CONTROL

If a small heater element is placed at the required liquid level of a flooded evaporator, together with a heat-sensing element, then the latter will detect the greater heat flow if liquid refrigerant is not present. This signal can be used to operate a solenoid valve.

7-6 EXPANSION VALVES FOR DRY EXPANSION CIRCUITS

The dry expansion circuit does not have a liquid level which can be detected, and another type of signal must be used to control the valve. Dry expansion circuits must be designed and installed so that there is no risk of liquid refrigerant returning to the compressor. To ensure this state, extra heat exchange surface is added to that required, in order to heat the dry saturated gas into the superheat region. The amount of superheat is usually of the order of 5 K.

Expansion valves for such circuits embody a mechanism which will detect the superheat of this gas leaving the evaporator (Fig. 7-4). Refrigerant boils in the evaporator at T_e and p_e, until it is all vapour, and then superheats to a condition T_s, p_e, at which it passes to the suction line. A separate container of the same refrigerant at temperature T_s would have a pressure p_s and the difference $p_s - p_e$ is a signal directly related to the amount of superheat.

The basic *thermostatic expansion valve* (Fig. 7-5) has a detector and power element, charged with the same refrigerant as in the circuit. The pressure p_s generated in the phial by the superheated gas passes through the capillary tube to the top of the diaphragm. An adjustable spring provides the balance of $p_s - p_e$ at the diaphragm, and the valve stem is attached at the centre. Should the superheat fall for any reason, there will be a risk of liquid reaching the compressor. The T_s will decrease with a corresponding drop in p_s. The forces on the diaphragm are now out of balance and the spring will start to close the valve.

Conversely, if the load on the evaporator increases, refrigerant will evaporate earlier and there will be more superheat at the phial position. Then p_s will increase and open the valve wider to meet the new demand.

The phial must be larger in capacity than the rest of the power element

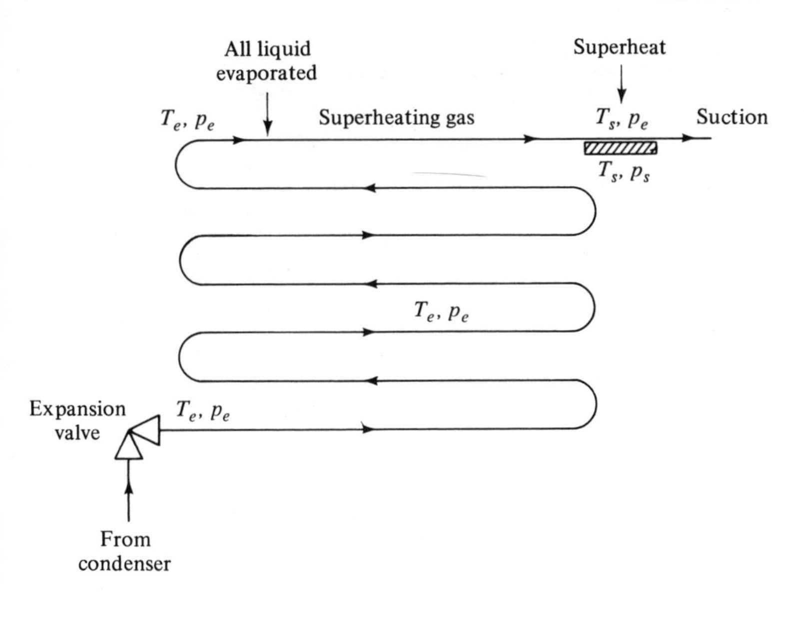

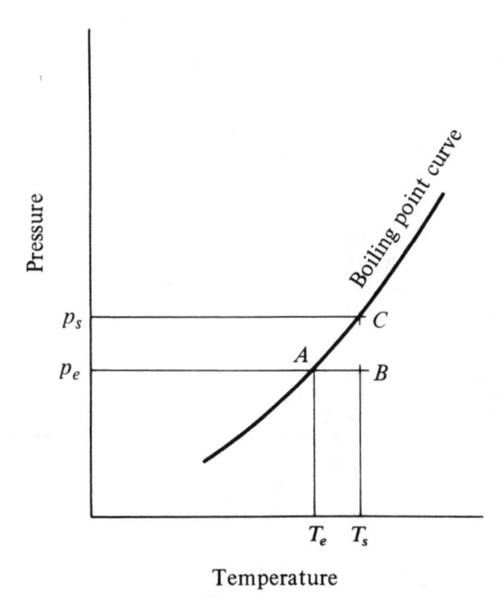

Figure 7-4 Superheat sensor on dry expansion circuit.

or the charge within it may all pass into the valve capsule and tube, if these are colder. If this happened, the phial at T_s would contain only vapour and would not respond to a position T_s, p_s on the T–p curve.

Use can be made of this latter effect. The power element can be 'limit

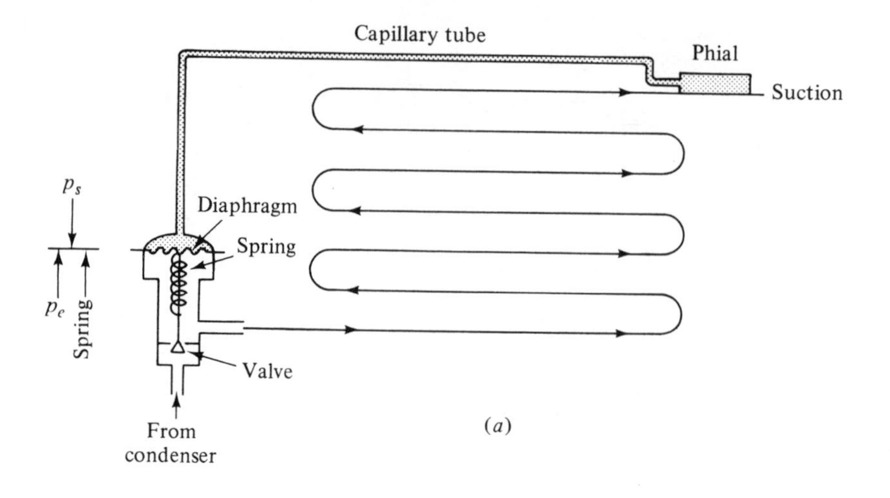

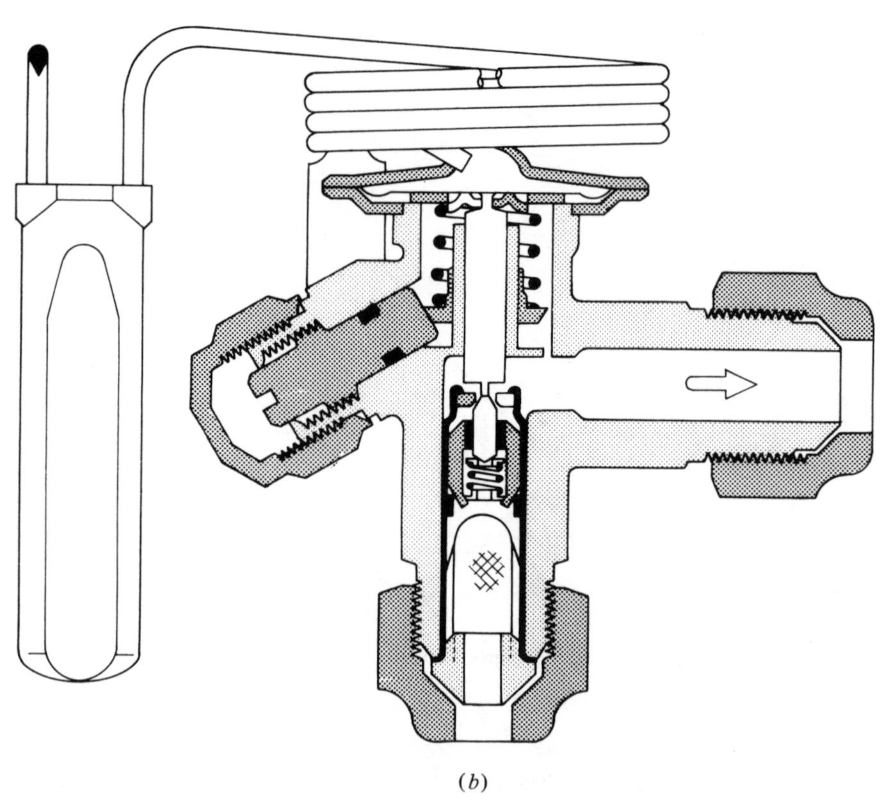

Figure 7-5 Thermostatic expansion valve: (*a*) circuit, (*b*) cross section. *(Courtesy Teddington Refrigeration Controls Ltd.)*

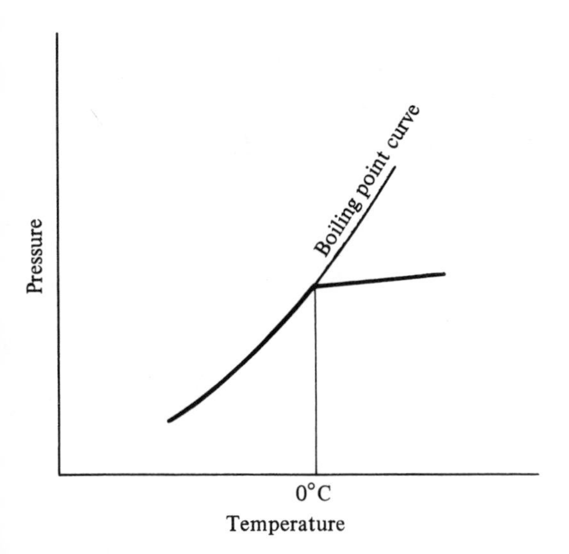

Figure 7-6 Detector pressure for limit charged valve.

charged' so that all the refrigerant within it has vapourized by a predetermined temperature (commonly 0°C). Above this point, the pressure within it will follow the gas laws:

$$\frac{p_1}{p_2} = \frac{T_1}{T_2}$$

and the valve will remain closed. This is done to limit the evaporator pressure when first starting a warm system, which might overload the drive motor. This is termed *limit charging* or *maximum operating pressure*. Such valves must be installed so that the phial is the coldest part (see Fig. 7-6).

The slope of the T–p curve is not constant, so that a fixed spring pressure will result in greater superheat at a higher operating temperature range. To allow for this and provide a valve which can be used through a wide range of applications, the phial may be charged with a mixture of two or more volatile fluids to modify the characteristic curve.

Some manufacturers use the principle of the adsorption of a gas by a porous material such as silica gel or charcoal. Since the adsorbent is a solid and cannot migrate from the phial, these valves cannot suffer reversal of charge.

7-7 EXTERNAL EQUALIZER

The simple *thermostatic expansion valve* relies on the pressure under the diaphragm being approximately the same as that at the coil outlet, and small coil pressure drops can be accommodated by adjustments to the spring setting.

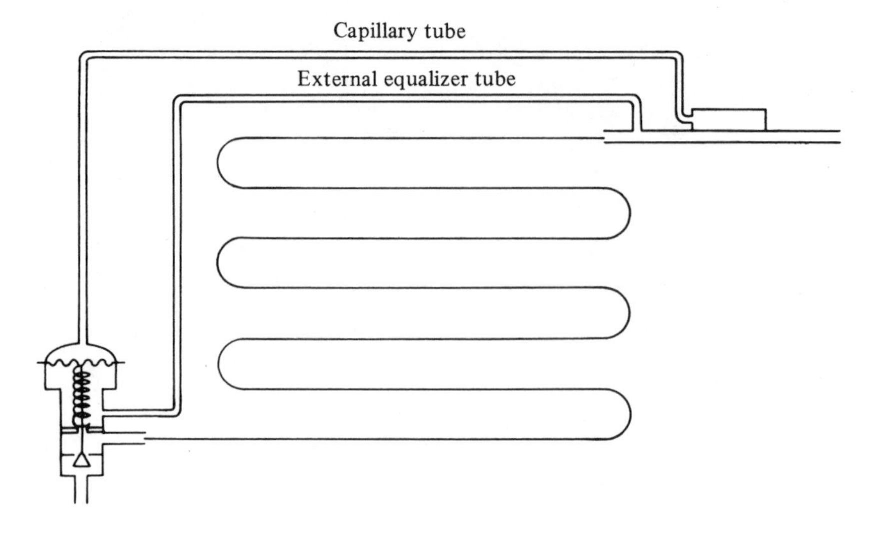

Figure 7-7 Thermostatic expansion valve with external equalizer.

Where an evaporator coil is divided into a number of parallel passes, a distribution device with a small pressure loss is used to ensure equal flow through each pass. Pressure drops of 1 to 2 bar are common. There will now be a much larger finite difference between the pressure under the diaphragm and that at the coil inlet. To correct for this, the body of the valve is modified to accommodate a middle chamber and an *equalizing connection* which is taken to the coil outlet, close to the phial position. Most thermostatic expansion valves will have provision for an external equalizer connection (see Fig. 7-7).

The suction-to-liquid heat exchanger (Fig. 2-7) will supply the suction superheat necessary for safe operation of a dry expansion evaporator, and the coil superheat may be less, giving more efficient use of the evaporator surface. The phial should be located before the heat exchanger, in which case the superheat setting is reduced. It can be located after the heat exchanger, but an external equalizer is then necessary to allow for the gas pressure drop through the exchanger.

7-8 THERMOSTATIC LIQUID LEVEL CONTROL

The thermostatic expansion valve can also be used to maintain a liquid level. The phial and a heater element are both clamped to a bulb at the required liquid level. If liquid is not present, the heater warms the phial to a superheat condition and the valve opens to admit more liquid.

7-9 ELECTRONIC EXPANSION VALVE

Superheat can be sensed by an electronic device and used, through an amplifier, to control refrigerant flow. Such methods are used in conjunction with other plant signals such as load temperature, discharge temperature, and motor current to provide an integrated system control. The method is applicable only to factory-built and tested assemblies.

7-10 CAPILLARY TUBE RESTRICTOR

The variable orifice of the expansion valve can be replaced, in small systems, by a long thin tube. This is a non-modulating device and has certain limitations, but will give reasonably effective control over a wide range of conditions if correctly selected and applied. Mass flow is a function of pressure drop and the degree of sub-cooling on entry.

Tube bores between 0.8 and 2 mm with lengths of 1 to 4 m are common. The capillary tube is only fitted on factory-built and tested equipment, with exact refrigerant charges. It is not applicable to field-installed systems.[15, 20]

CHAPTER

EIGHT

CONTROLS AND OTHER CIRCUIT COMPONENTS

8-1 GENERAL

A refrigeration system can be built with only the four essential components:

Evaporator Compressor Condenser Expansion valve

For ease, economy, and safety of operation, and to assist the maintenance function, other system controls and components will be fitted.

8-2 THERMOSTATS

Since the purpose of a refrigeration or air-conditioning system will be to reduce or maintain temperature, a thermostat will usually be fitted to stop the equipment or reduce its capacity when the required condition is reached. The following types are in use:

1. Movement of a bimetallic element
2. Expansion of a fluid
3. Vapour pressure of a volatile fluid

The above produce a mechanical effect which can be used direct to operate an electric switch or modulate the pressure of an air jet (pneumatic system).

4. Electric resistance
5. Electronic—various types

These last two produce an electric signal which must be detected and amplified to operate the controlled device.

8-3 HUMIDISTATS

Where the equipment is required to maintain a predetermined level of humidity, a humidistat may be used instead of, or in addition to, a thermostat. The function will normally be to operate an electrical switch.

Mechanical humidistats employ materials which change dimension with humidity, such as animal hair, plastics, cellulosics, etc. These can work a switch directly.

Electronic humidistats generally depend on the properties of a hygroscopic salt. The measured signal has to be detected and amplified.

8-4 PRESSURE SWITCHES

The compressor, as a pump, will be limited mechanically to maximum safe operating pressures and must be stopped before such pressures are reached (see Fig. 8-1).

High-pressure cut-outs are fitted to all but the smallest of systems. The compressor outlet pressure is brought to one side of a bellows or diaphragm, and balanced by an adjustable spring. A scale on the control indicates the pressure setting to commercial accuracy and is checked on commissioning the system.

If the spring pressure is overcome, the switch will open and stop the compressor. Normally open contacts on the cut-out can then operate a warning. The cut-out point is commonly some 2 bar higher than the design maximum operating pressure.

Since excess pressure indicates malfunction of part of the system—usually a condenser fault or incorrect closure of a valve—the high-pressure switch should be reset manually, not automatically.

Abnormally low suction pressures will lead to high discharge temperatures, owing to the high compression ratio, and possible malfunction of other components. Air cooling coils may frost excessively.

A *low-pressure cut-out* switch is usually fitted to stop the compressor

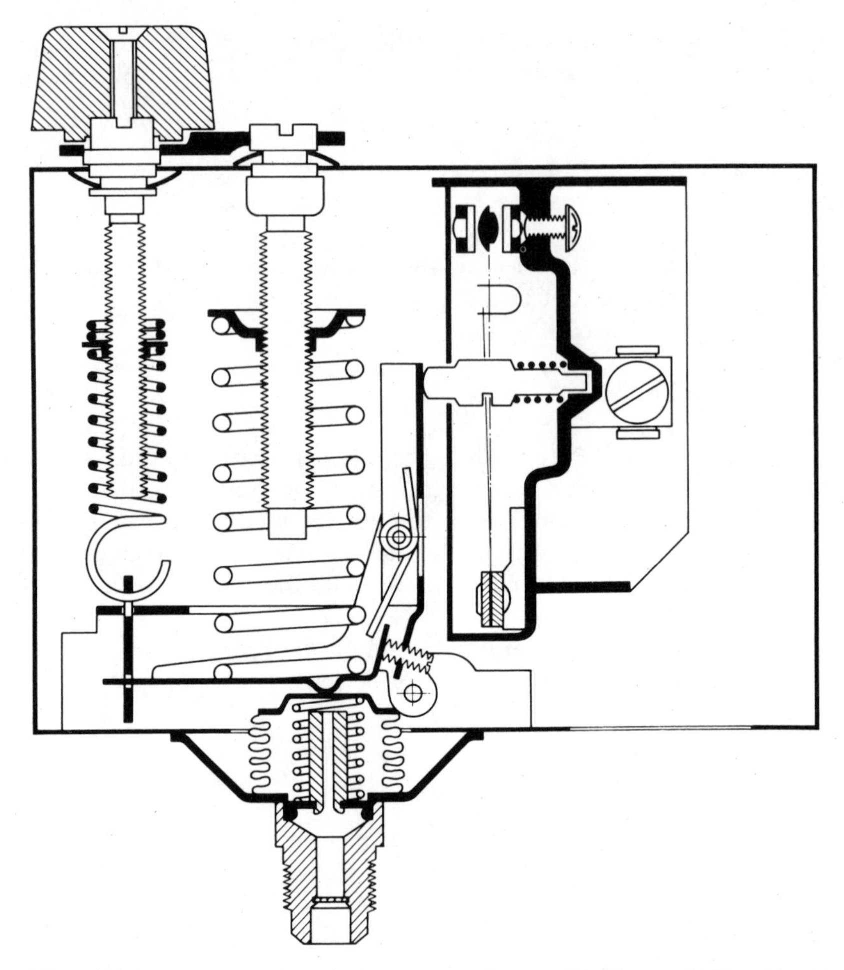

Figure 8-1 Pressure control and thermostat. *(Courtesy Teddington Refrigeration Controls Ltd.)*

under these circumstances. Settings may be 0.6 to 1.0 bar below the design evaporator pressures, but depend very much on the type of system. The cut-out setting should be above atmospheric pressure if possible to avoid the ingress of air through any leaks.

Abnormally low pressure is not usually an unsafe condition and the low-pressure switch may be automatic reset, closing again at a pressure corresponding to a temperature just below that of the load.

If a plant has been shut down long enough for all pressures to equalize and is then restarted, the suction pressure will pull down below normal until the liquid refrigerant has begun to circulate. Under such circumstances the low-pressure switch may operate. This is a normal occurrence, but may require the addition of a delay timer to prevent frequent starting of the compressor motor.

8-5 OIL SAFETY

All compressors except the smallest have mechanical lubrication and will fail if the oil pressure falls through a pump fault or oil shortage. A safety cut-out is required which will stop the compressor. This takes the form of a differential pressure switch with a starting time delay.

Since the oil pump inlet is at sump (suction) pressure, a pressure gauge on the pump discharge will indicate the total pressure at that point above atmospheric, i.e., suction (gauge) plus pump head. Any detection element for true oil pump pressure must sense both suction and pump outlet pressures and transduce the difference. Oil safety cut-outs have pipe connections to both sides of the oil pump and two internal bellows are opposed to measure the difference.

Since there will be no oil pressure at the moment of starting, a time delay must be fitted to allow the oil pressure to build up. This timer may be thermal, mechanical, or electric.

Operation of the oil safety cut-out indicates an unsafe condition and such controls are made with hand reset switches. Normally open contacts on the switch are used to operate an alarm to warn of the malfunction.

8-6 PRESSURE GAUGES

Direct indication of the operating conditions of a compressor is by pressure gauges at suction, discharge, and oil delivery. Such gauges are mounted on or near the compressor.

Since the pressure will also be a direct indication of operating temperatures, suction and discharge gauges are usually marked with the corresponding situation temperature of the working refrigerant, and may have several sets of concentric markings for different refrigerants (see Fig. 8-2).

Gauge mechanisms are mostly of the bourdon tube type, having a flattened tube element, which distorts under pressure change. Gas pulsations from the compressor will be transmitted along the short connecting pipes and may lead to early failure of the needle mechanism. These can be damped by restricting the tube with a valve or orifice, or oil filling the gauge, or both. Gauge needles should not be allowed to flicker noticeably from gas pulsations.

8-7 SOLENOID VALVES

Electrically operated shut-off valves (Fig. 8-3*a*) are required for refrigerant and other circuits. These take the form of a plunger operated by a solenoid and working directly on the valve orifice or through a servo. The usual arrangement is to energize the solenoid to open the valve and de-energize to

Figure 8-2 Pressure gauge with temperature markings for common halocarbon refrigerants. *(Courtesy Refrigerator Components Limited.)*

close. Sizes up to 50 mm bore tube connections are made. Beyond this, the solenoid acts as a pilot to a main servo (see Fig. 8-4).

Solenoid valves are used in refrigeration and air-conditioning systems for refrigerant lines, oil pressure pipes (to control oil return and capacity reducers), and water and compressed air lines. Four-port changeover valves (Fig. 8-3*b*) are used to reverse flow in defrosting and heat pump circuits.

8-8 BACK-PRESSURE REGULATION VALVES

Back-pressure regulation valves (Fig. 8-4) can be used in the suction line, and their function is to prevent the evaporator pressure falling below a predetermined or controlled value, although the compressor suction pressure may be lower.

The application of a back-pressure regulating valve is to:

1. Prevent damage to a liquid chilling evaporator which might result from freezing of the liquid.

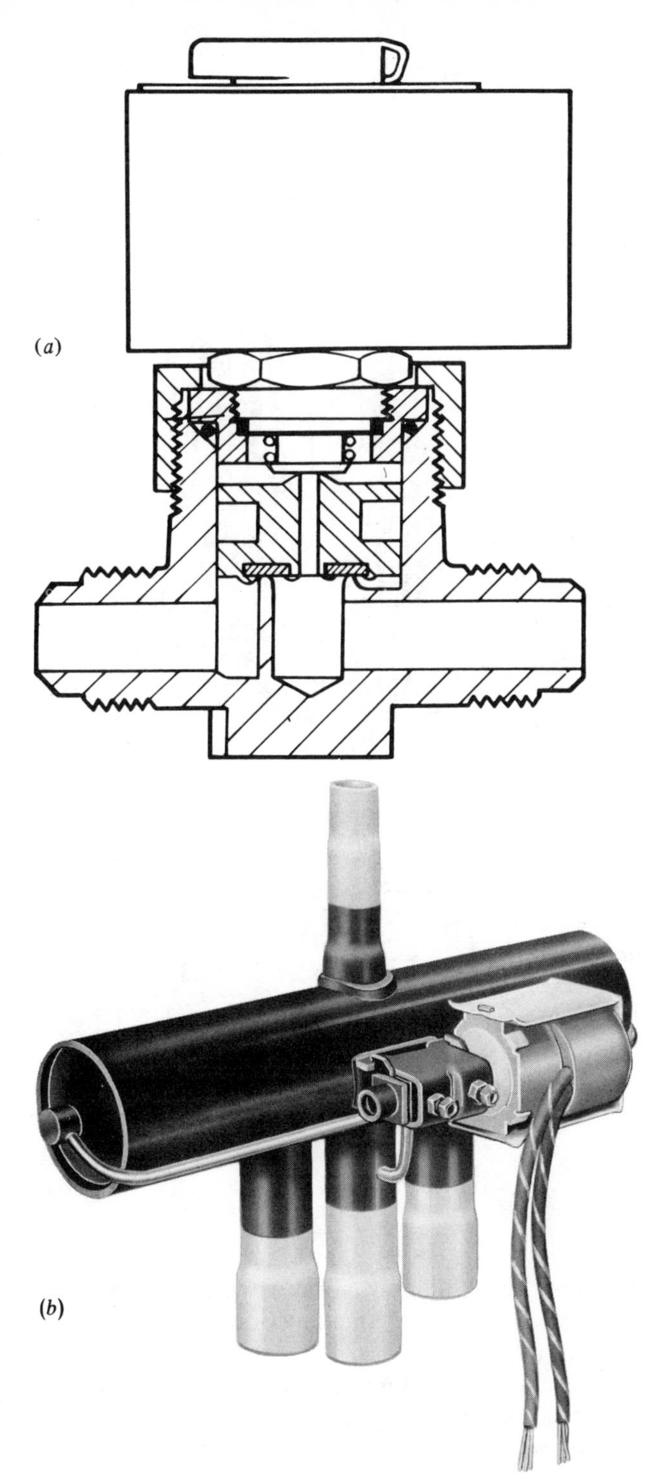

Figure 8-3 Solenoid valves: (*a*) shut-off *(Courtesy Bailey Gill Products Ltd.)*, (*b*) changeover. *(Courtesy Ranco Controls Limited.)*

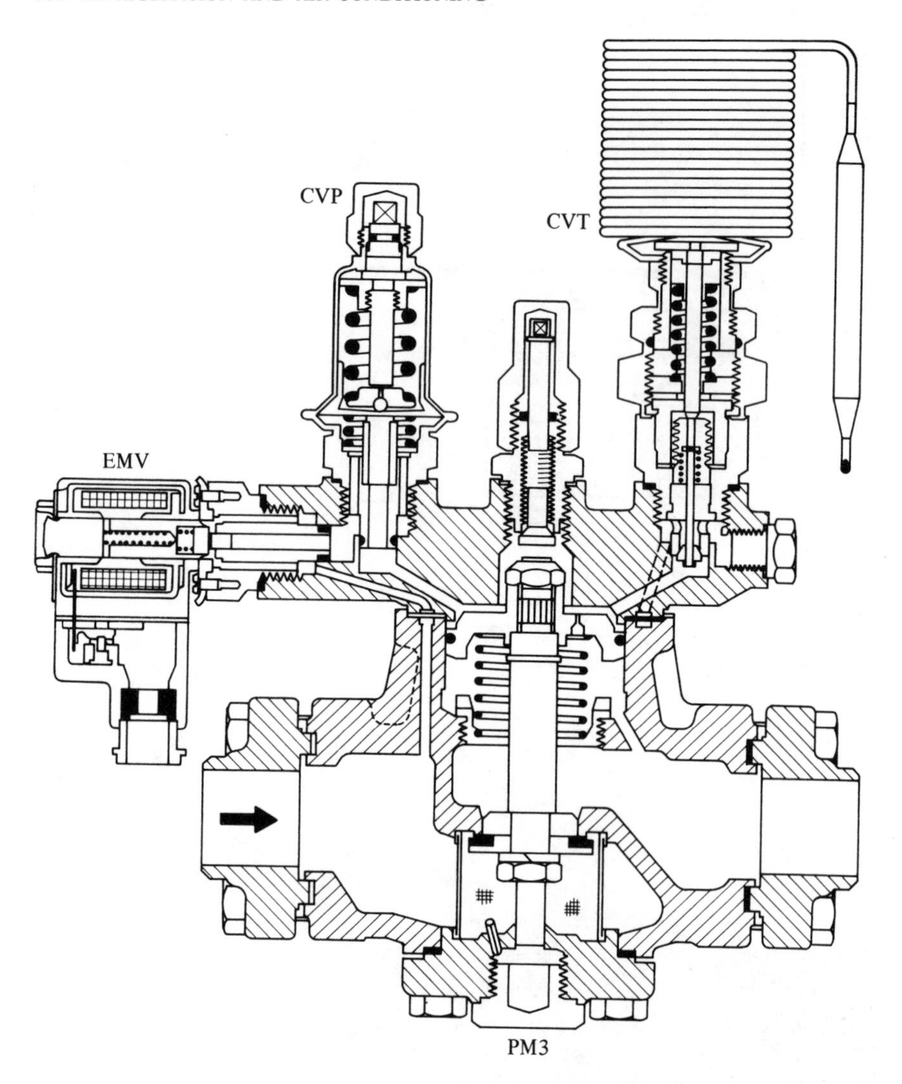

Figure 8-4 Back-pressure regulation valve assembly. *(Courtesy Danfoss.)*

2. Prevent frost forming on an air cooling evaporator, where this is close to freezing point, or where a temporary malfunction cannot be permitted to interrupt operation.
3. Permit two or more evaporators, working at different load temperatures, to work with the same compressor.
4. Modulate the evaporator pressure according to a varying load, controlled by the load temperature.
5. Act as a solenoid valve, controlled by a pilot solenoid valve.

The simplest back-pressure regulating valve is spring loaded, balancing the thrust of the spring, plus atmospheric pressure, on one side of a diaphragm or piston, against the inlet or evaporator pressure. For working pressures below atmospheric, a helper spring is fitted below the diaphragm. Slight variations will result from changes in atmospheric pressure, but these are too small to materially affect a refrigeration control system.

A service gauge is usually fitted adjacent to the valve or as part of the valve assembly, to facilitate setting or readjustment. Above about 40 mm pipe size, the basic back-pressure regulation valve is used as a pilot to operate a main servo valve. Other pilot signals can be used on the same servo. Figure 8-4 shows a main servo controlled by a back-pressure pilot, a thermostatic pilot sensing load temperature, and a solenoid valve.

8-9 CONDENSER PRESSURE REGULATORS

Systems are normally designed to work satisfactorily during maximum ambient conditions, and the condenser will be sized for this. In colder weather, the condensing temperature and pressure will fall and the resulting lower pressure difference across the expansion valve may lead to malfunction, depending on the type of valve. A drop of pressure difference to half the normal figure may reduce mass flow below that required, and it will be necessary to prevent the condenser pressure from falling too low.

With air-cooled condensers and water cooling towers it is possible to reduce the air flow by automatic dampers, fan speed control, or by switching off fans, where two or more drives are fitted. The control should work from pressure but can be made to work from temperature (see Chapter 5).

Water-cooled condensers can be fitted with a directly controlled water regulating valve operated by condenser pressure, or may have a three-way blending valve in the water circuit.

A condenser pressure regulator can be in the form of a pressure-operated bleed valve in a bypass across the condenser, to divert hot gas to the receiver. The valve diaphragm is balanced by a pre-set spring and will open the bypass if the condensing pressure falls. A similar effect can be obtained by a pressure-operated valve between the condenser and the receiver, to restrict the flow and allow liquid to accumulate in the condenser, reducing its efficiency. For operating economy, it is important that such valves are not set at too high a pressure.[21]

Where evaporative condensers and water cooling towers have only one fan (or fan drive motor), coarse control can be effected by on–off switching. The time lag will then depend on the mass of water in the circuit, and the sensing element needs to have a wide differential to prevent frequent motor starts. Towers should have thermostatic control of the fan to prevent water freezing on the packing in winter.

Table 8-1

Condensing temperature, °C	COP	Electricity costs per 40-h week @ 3.2 p/kW h, £
35 (summer maximum)	3.41	164
30	4.00	140
25	4.73	118
20 (probable minimum)	5.76	97

In all forms of condenser pressure control, the minimum maintained pressure should be the lowest which will give satisfactory operation, in the interests of running economy. An indication of the relative electricity costs for a 350 kW air-conditioning plant are as shown in Table 8-1.

8-10 CAPACITY REDUCTION INJECTION VALVES

Where a compressor does not have any capacity reduction device and on–off switching will not give the degree of control required by the process, the cooling capacity can be regulated by injecting discharge gas back into the suction (see Fig. 8-5). It has the effect of keeping the evaporator pres-

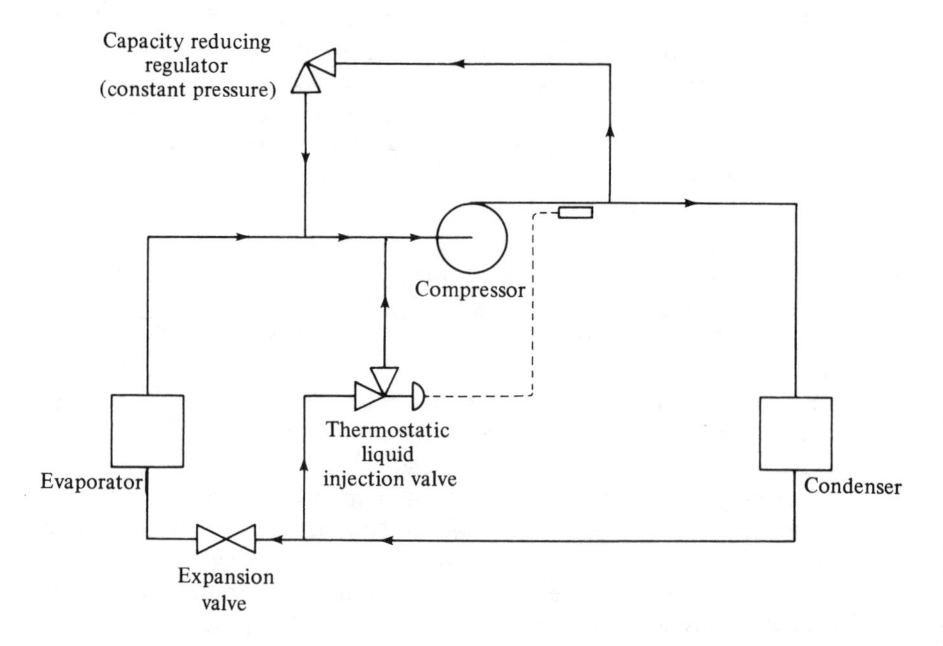

Figure 8-5 Capacity reduction by hot gas injection, with compensating liquid injection.

sure constant, regardless of the load, and can have a wide range of capacity reduction, down to 10 per cent of full load. It is a constant pressure valve, balancing the suction pressure against a pre-set spring.

However, since the suction gas to the compressor would then be hotter than its normal slightly superheated condition, the compressor may overheat and the discharge gas become too hot for correct and safe working. This form of capacity reduction is usually combined with a liquid injection valve, thermostatically operated, which introduces liquid also into the suction to keep it cool. The fitting of dual interdependent controls of this sort, both of which have inherent fail–unsafe possibilities, should be approached with caution.

8-11 RELIEF VALVES

Under several possible conditions of malfunction, high pressures can occur in parts of the system and mechanical relief devices are advised or mandatory. The standard form of relief valve is a spring-loaded plunger valve. No shut-off valve is permitted between the relief valve and the vessel it protects, unless two such valves are fitted, when the shut-off may isolate one at a time.[12]

In all cases, the outlet of the valve must be led to the open air, in a location where the sudden discharge of refrigerant will not cause annoyance or danger. Under certain circumstances, a relief valve from the high-pressure side may enter the low side of the same system. Small vessels may have a plug of a low melting-point metal, which will melt and release the pressure in the event of fire. Plunger-type relief valves, if located outdoors, should be protected from the ingress of rain, which may corrode the seat. Steel valves, when installed, should have a little oil poured in to cover the seat as rust protection.

To prevent overpressure within a compressor, a relief valve or bursting disc is often fitted between the inlet and discharge connections.

8-12 SHUT-OFF VALVES

Manual stop valves are required throughout a circuit to permit isolation during partial operation, service, or maintenance (see Fig. 8-6).

Small valves for refrigerants have a packless gland, either a diaphragm or bellows. Since the valve operating stem can thrust through this flexible seal but not pull, an internal spring is fitted to open the valve. As a further safeguard against leakage, the handwheel is removable and the stem provided with a covering cap which is only removed when the valve is to be operated.

Figure 8-6 Shut-off valve. *(Courtesy Refrigerator Components Limited.)*

Larger valves have a packed gland or '0' rings. Since these may leak in service and require repacking or replacement, such valves are made to back-seat, thus relieving gas pressure from the spindle for servicing. Valves of this type should never be installed with the spindle at the bottom, since any dirt in the pipe will fall into the spindle and damage the gland.

Under low-temperature conditions, ice will form on the spindle and will be forced into the gland if the valve is operated quickly. Under such circumstances, the spindle should be well greased, or the ice melted off first.

Service stop valves on small compressors may also carry a connection for a pressure cut-out or gauge, or for the temporary fitting of gauges or charging lines when servicing. The valve back-seats to close off this port while gauges are being fitted. Valve seats are commonly of soft metal or of a resistant plastic such as p.t.f.e.

8-13 STRAINERS

Piping circuits will usually contain a small quantity of dirt, scale, and swarf, no matter what care is taken to keep these out. A strainer is fitted in the compressor suction to trap such particles before they can enter the machine. Such strainers are of metal mesh and will be located where they can be removed for cleaning. In some configurations two strainers may be fitted.

As an extra safeguard, on new compressors a fabric liner may be fitted inside the mesh strainer to catch fine dirt which will be present. Such liners must be removed at the end of the running-in period, as they create a high resistance to gas flow.

Oil strainers may be of metal mesh and within the sump, in which case the sump must be opened for cleaning. Self-cleaning disc strainers are also used, the dirt falling into a drain pot or into the sump itself. There is an increasing tendency to provide replaceable fabric oil filters external to the compressor body, following automobile practice.

8-14 STRAINER-DRIERS

With the halocarbons, it is essential to reduce the water content of the refrigerant circuit to a minimum by careful drying of components and the fitting of drying agents in the system. The common form of drier is a capsule charged with a solid dessicant such as silica gel, activated alumina, or zeolite (molecular sieve), and located in the liquid line ahead of the expansion valve. These capsules must have strainers to prevent loss of the drying agent into the circuit, and so form an effective strainer-drier to also protect the valve orifice from damage by fine debris.

Large driers are made so that they can be opened, emptied, and re-charged with new drying agent. Small sizes are throwaway.

8-15 SIGHT GLASSES

Pipeline sight glasses can be used to indicate whether gas is present in a pipe which should be carrying only liquid. The main application in refrigeration is in the liquid line from the receiver to the expansion valve. If the equipment is running correctly, only liquid will be present and any gas bubbles seen will indicate a refrigerant shortage (see also Chapters 9 and 23).

Sight glasses for the halocarbons are commonly made of brass, and may have solder or flare connections. For ammonia, they are made of steel or cast iron.

Since the interior of the system can be seen at this point, advantage is taken in most types to insert a moisture-sensitive chemical which will indicate an excess of water by a change of colour. When such an indication is seen, the drier needs changing or re-changing, and the colour should then revert to the 'dry' shade.

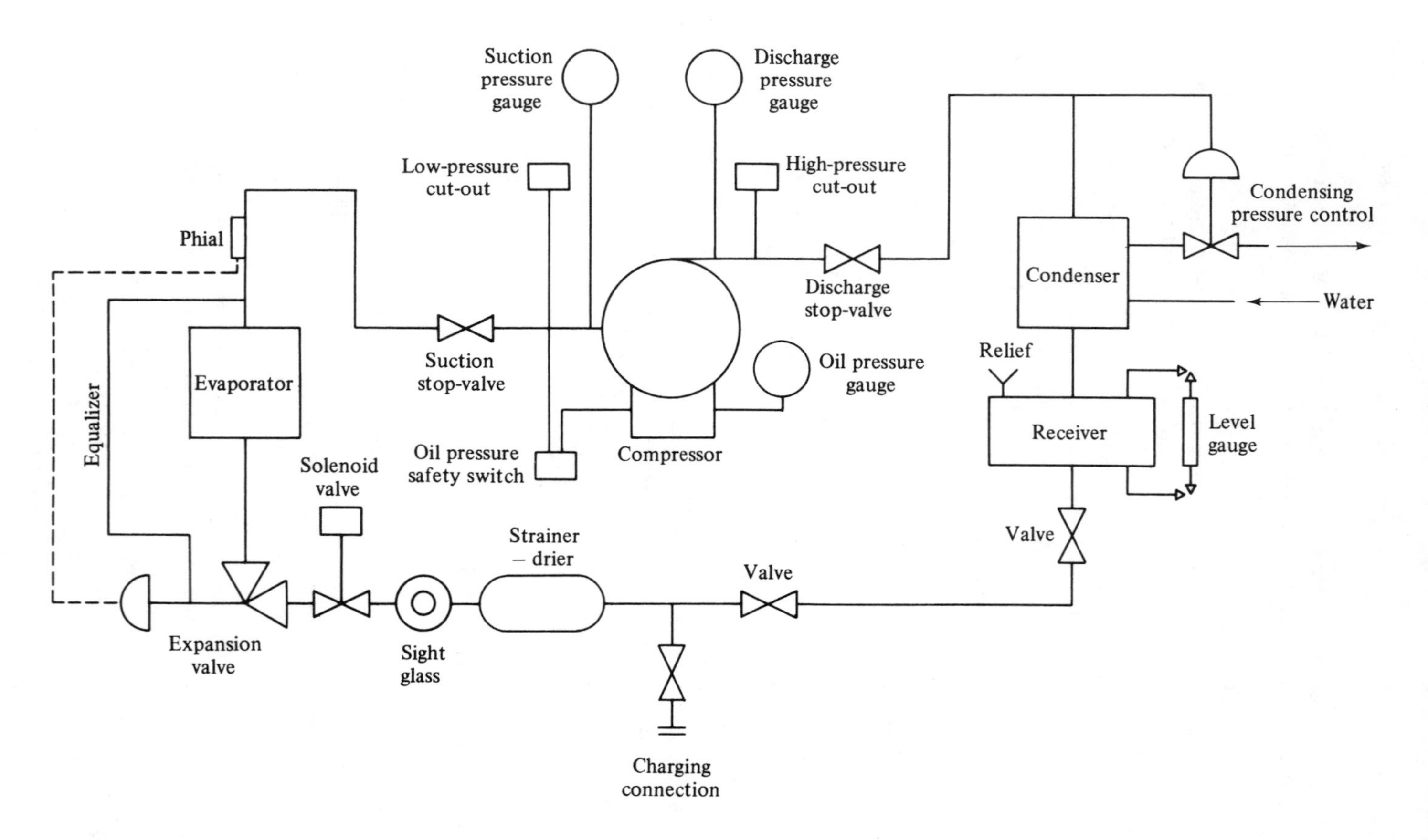

Figure 8-7 Dry expansion circuit showing components.

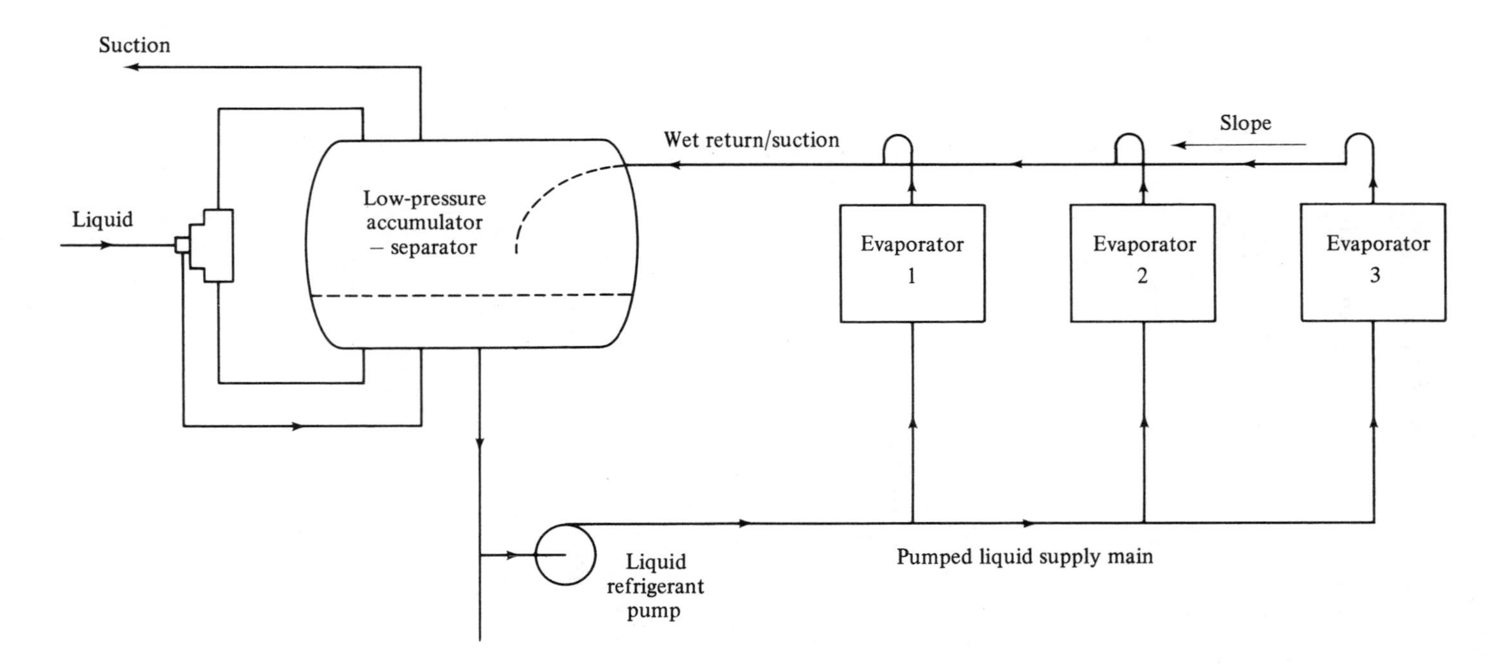

Figure 8-8 Pumped liquid circuit.

8-16 CHARGING CONNECTION

In order to admit the initial refrigerant charge into the circuit, or add further if required, a *charging connection* is usually provided. The safest place to introduce refrigerant will be ahead of the expansion valve, which can then control the flow and prevent liquid reaching the compressor. The usual position is in a branch off the liquid line, and it is fitted with a shut-off valve and a suitable connector with a sealing cap or flange. A valve is needed in the main liquid line, just upstream from the branch and within reach. For the method of use, see Chapter 9.

The relative positions of all these components are shown in the complete circuit in Fig. 8-7.

8-17 AUXILIARY COMPONENTS

More complex refrigeration systems may have components for specific purposes which are not encountered in simple circuits. Non-return or *check*[22] *valves* will be found in the following positions:

1. On heat pump circuits, to prevent flow through expansion valves which are not in service on one cycle
2. On hot gas circuits, to prevent the gas entering another evaporator
3. Where several compressors discharge into one condenser, to prevent liquid condensing back to an idle compressor
4. Where two or more evaporators work at different pressures, to prevent suction gas flowing back to the colder ones

Liquid refrigerant pumps (Fig. 8-8) are used where several flooded evaporators are to be fed from a single accumulator-separator. By circulating an excess of liquid, commonly 5 to 12 times the system mass flow, good turbulence and wetting of the heat exchanger surface can be maintained, improving heat transfer. These pumps are found mainly on low-temperature coldrooms, blast freezers, and process applications.[23]

CHAPTER

NINE

MATERIALS. CONSTRUCTION. SITE ERECTION

9-1 MATERIALS

Materials used in the construction of refrigeration and air-conditioning systems are standard engineering materials, with a few special points of interest.

1. Compressors are generally of gray cast iron, but some makes are fabricated from mild steel.
2. Compressor pistons are of cast iron or aluminium, the latter following automobile practice.
3. Piping for the smaller halocarbon installations is usually of copper, because of the cleanliness and the ease of fabrication and jointing.
4. Some stainless steel pipe is used, mainly because of its cleanliness, although it is difficult to join.
5. Most other piping will be of mild steel. For working temperatures below −45°C, only low carbon steels of high notch strength are used (mainly to BS.3603).
6. Copper and its alloys are not used with ammonia.
7. Sheet steel for ductwork, general air-conditioning components, and outdoor equipment is galvanized.

Specific guidance on materials and their application may be had from various works of reference.[4, 15, 24, 25]

9-2 PRESSURE TESTS FOR SAFETY

Factory-built equipment will be constructed to the relevant Standards and will be pressure tested for safety and leaks at the works. In cases of doubt, a test certificate should be requested for all such items. Design and test pressures will depend on the refrigerant or other fluids used.

Site-assembled plant will be pressure tested for safety and leakage after erection (see Sec. 9-12).

9-3 ERECTION PROGRAMME

Successful site erection of plant demands coordination of the following:

1. Site access or availability
2. Supply on time, and safe storage, of materials
3. Availability of layout drawings, flow diagrams, pipework details, control and wiring circuits, material lists, and similar details
4. Availability at the correct time of specialist trades and services—builders, lifting equipment, labourers, fitters, welders, electricians, commissioning engineers, etc.

Site work is now mostly carried out by a number of subcontractors representing specialist trades. It is essential that authority and executive action are in the hands of a main contractor and that this authority is acknowledged by the subcontractors. If this is not so, delays and omissions will occur, with divided responsibility and lack of remedial action.

The controlling authority must, well before the start of site erection, draw up a material delivery and progress chart and see that all subcontractors (and the customer) are in agreement and that they are kept informed of any changes.[26]

9-4 PIPE JOINING METHODS

Steel pipe is now entirely welded, except for joints which need to be taken apart for service, which will be flanged. It is essential that welding is carried out by competent craftsmen and is subject to stringent inspection.[24, 25]

Flanges for ammonia (and preferably, also, for other refrigerants) must be of the tongued-and-grooved type which trap the gasket. Mechanical joints for copper tube up to $\frac{3}{4}$ in outside diameter can be of the *flare* type, in which the tube end is coned out to form its own gasket. This must be carried out with the proper flare tools, and it may be necessary to anneal the tube to ensure that the resulting cone gasket is soft.

Copper tube can be bent to shape in the smaller sizes and the use of bending springs or formers is advised, to retain the full bore. Where fittings are required, these should be of copper or brass to give a correct capillary joint gap of not more than 0.2 mm, and joined with brazing alloy. This, again, is a craft not to be entrusted to the semi-skilled.

The brazing of copper tube will leave a layer of copper oxide inside, which may become detached and travel around the circuit. The best practice is to pass nitrogen into the pipe before heating, to avoid this oxidation. The use of special grades of oxygen-free or moisture-free nitrogen is not necessary.

9-5 PIPE SIZING

Refrigeration system pipes are sized to offer a low resistance to flow, since this reflects directly on compression ratio, commensurate with economy of pipe cost and minimum flow velocities to ensure oil return with the halocarbons.

Pressure losses due to pipe friction can be calculated from the basic formulas established by Reynolds and others. However, as with the calculation of heat transfer factors, this would be a time-consuming process and some of the parameters are not known accurately. Recourse is usually made to simplified estimates or tables published in works of reference.[1, 27]

Example 9-1 A suction pipe for an R.502 system evaporating at −40°C and having a cooling capacity of 70 kW is to be run in copper tube. What size should it be?

Reference 1 (chap. 26, table 23, 1972 ed.) shows that a copper tube of 3.625 in outside diameter and at a pressure corresponding to −40°F (−40°C), will carry a cooling capacity of 20.15 TR or 70.8 kW, with a pressure drop of 0.9 lbf/in^2 per 100 ft run (0.2 kPa/m), which is given as a commercially acceptable pressure drop.

9-6 PIPING FOR OIL RETURN

The sizing and arrangement of suction and discharge piping for the halocarbons is dominated by the need to ensure proper entrainment of oil, to return this to the compressor. Pipes for these gases usually have a higher velocity at the expense of a greater pressure drop than those for ammonia. Pipe sizes may only be increased in runs where the oil will be assisted by gravity to flow in the same direction as the gas.

Horizontal pipes should slope slightly downwards in the direction of flow, where this can be arranged. If a suction or discharge line has to rise, the size may be decreased to make the gas move fast. In the case of a lift of more than 5 m, a trap should be formed at the bottom to collect any oil which falls back when the plant stops.

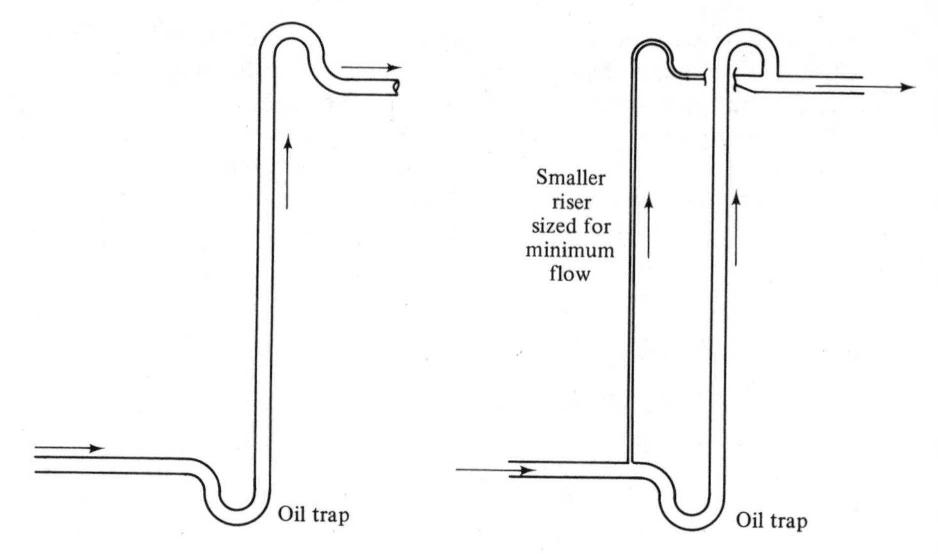

Figure 9-1 Gas risers for oil return.

Suction and discharge risers (Fig. 9-1) will normally be sized for full compressor capacity, and velocities will be too low if capacity reduction is operated. In such installations, double risers are required, the smaller to take the minimum capacity and the two together to carry the full flow. Traps at the bottom and goosenecks at the top complete the arrangement. At part capacity, any oil which is not carried up the main riser will fall back and eventually block the trap at the bottom, leaving the smaller pipe to carry the reduced flow, with its quota of oil. When the system switches back to full capacity, the slug of oil in this trap will be blown clear again.

9-7 PIPE SUPPORTS. VALVE ACCESS

Piping must be properly supported at frequent intervals to limit stress and deflection.[3] Supports must allow for expansion and contraction which will occur in use. In particular, pipework which might form a convenient foothold for persons clambering about the plant should be protected from damage by providing other footholds and guarding insulation.

Stop valves, especially those which might need to be operated in a hurry (and this means most, if not all, of them), should have easy access. Where they are out of reach, reliance should not be placed on movable ladders, which may not be there when needed, but permanent access provided. Chain-operated wheels can be fitted to the larger valves, to permit remote operation.

Emergency stop valves must not be placed in tunnels or ducts, since personnel may be subject to additional danger trying to operate them.

9-8 INSTRUMENTS

Until recently it has been the custom to fit thermometer wells at various points in the pipework, to enable check temperatures to be taken during initial commissioning and also during the life of the plant. The advent of the electronic probe thermometer has simplified commissioning work, and the fitting of thermometer wells is less important. Even so, such facilities are worth considering when the pipe is being erected, and will be necessary with insulated pipes if true temperatures are to be taken without damaging the insulation.

Wells should slope downwards into the pipe, so that they can be part filled with liquid to provide better thermal contact. Where a pipe temperature is a critical factor in the operation of a system, it is usually worth fitting a permanent thermometer.

Pressure gauges should always be fitted on the discharge side of liquid pumps, to check performance and give warning of a possible drop in flow resulting from dirty strainers. Manometer pressure gauges are required across air filters (see Chapter 23).

9-9 RISING LIQUID LINES

If liquid refrigerant has to rise from the condenser or receiver to an expansion valve at a higher level, there will be a loss of static head, and the refrigerant may reach its boiling point and start to flash off. Under such circumstances, bubbles will show in the sight glass and will not be dispersed by adding more refrigerant to the system.

Example 9-2 The R.22 condenses in a system at 34°C and is subcooled to 30°C before it leaves the condenser. How much liquid lift can be tolerated before bubbles appear in the liquid line?

Saturation pressure at 34°C = 16.93 bar

Saturation pressure at 30°C = 15.37 bar

Permissible pressure drop = 1.56 bar (156 000 Pa)

Specific mass of liquid = 1122 kg/m³

$$\text{Possible loss in static head} = \frac{156\,000}{9.81 \times 1122} \text{ (where } g = 9.81 \text{ m/s}^2\text{)}$$

$$= 14 \text{ m approx.}$$

Where a high lift cannot be avoided, the liquid must be sub-cooled enough to keep it liquid at the lower pressure. Sub-cooling can be accomplished by fitting a sub-cooling coil to the condenser, a water-cooled sub-cooling coil, a suction-to-liquid heat exchanger before the lift, or by means of a refrigerated sub-cooler.

The same effect will occur where the liquid line picks up heat on a horizontal run, where it may be in the same duct as hot pipes, or pass through a boilerhouse. If the sight glass flashes even with the addition of refrigerant, the possibility of such extra heating should be investigated. To cure this, insulate the pipe.

To reduce the risk of these troubles, the condenser should always be higher than the evaporator, if this can be arranged.

9-10 VIBRATION

Compressors and pumps will transmit vibration to their connecting pipework.

Water and brine pumps may be isolated with flexible connectors. For small-bore pipes, these can be ordinary reinforced rubber hose, suitably fastened at each end. For larger pipes, corrugated or bellows connectors of various types can be obtained. In all cases, the main pipe must be securely fixed close to the connector, so that the latter absorbs all the vibration. Flexible connectors for the refrigerant usually take the form of corrugated metal hose, wrapped and braided. They should be placed as close to the compressor as possible.

A great deal of vibration can be absorbed by ordinary piping up to 50 or 65 mm nominal bore, providing it is long enough and free to move with the compressor. Three pieces, mutually at right angles and each 20 diameters long, will suffice. At the end of these vibration-absorbing lengths, the pipe must be securely fixed.

In all instances of antivibration mounting of machinery, care must be taken to ensure that other connections—water, electrical, etc.—also have enough flexibility not to transmit vibration.

9-11 CLEANLINESS OF PIPING

All possible dirt should be kept out of pipes and components during erection. Copper pipe will be clean and sealed as received, and should be kept plugged at all times, except when making a joint. Use the plastic caps provided with the tube—they are easily seen and will not be left on the pipe. Plugs of paper and rag tend to be forgotten and left in place. Steel pipe will have an oily coating when received, and it is important that this should be wiped out, since the oil will otherwise finish up in the sump and contaminate the proper lubricating oil. If pipe is not so cleaned, the compressor oil should be changed before the plant is handed over.

Rusty pipe should not be used. The rust and loosened mill scale will travel around the circuit to block the suction strainer and the drier. Other avoidable debris are loose pieces of weld, flux, and the short stubs of welding rod often used as temporary spacers for butt welds. Pipe should only be cut with a gas torch if all the oxidized metal can be cleaned out again before closing the pipe.

It should be borne in mind that all refrigerants have a strong solvent effect and swarf, rust, scale, water, oils, and other contaminants will cause harm to the system, possible malfunction, and shorten the working life.

9-12 SITE PRESSURE SAFETY TESTS

Site-erected pipework, once complete, must be pressure tested for safety and freedom from leaks. Pressures will be 1.3 times the maximum working pressure, and usual figures for the United Kingdom are 28.5 bar for the high side of air-cooled plants, 23.75 bar for water-cooled plants, and 14.75 bar for the evaporator side. These figures are for R.22, R.502, and R.717, and lower pressures apply for R.12 and other refrigerants.[12]

Factory-built components and pressure vessels which have already undergone test should not be retested, unless they form part of the circuit which cannot be isolated, when the test pressure must not exceed the original figure. Site hydraulic testing is considered unnecessary, owing to the extreme difficulty of removing the test fluid afterwards. However, it must always be appreciated that site testing with gases is a potentially dangerous process, and must be governed by considerations of safety. In particular, personnel should be evacuated from the area and test personnel themselves be protected from the blast which would occur if a pressure vessel exploded.[25]

Testing should be carried out with nitrogen or, should this happen to be more readily available, argon. The use of grades of gas having very low levels of water or oxygen is not necessary. Air may be used where no oil is present but cannot be recommended, as it will bring with it a quantity of moisture, which is difficult to remove.

Nitrogen is used from standard cylinders, supplied at about 200 bar, and a proper reducing valve must always be employed to get the test pressure required. A separate gauge is used to check the test pressure, since that on the reducing valve will be affected by the gas flow.

If the high side is being tested, the low side should be vented to the atmosphere, in case there is any leakage between them which could bring excessive pressure onto the low side. It may be necessary to remove relief valves. Other valves within the circuit will have to be open or closed as necessary to obtain the test pressure.

After the test gas has gone in, there may be a slight change in pressure with a change of temperature. In particular, if left overnight, the pressure may drop as much as 1 bar. This is not significant.

The test pressure should be maintained for at least an hour. In this period a thorough test is made of each joint with soapy water. This method is no more tedious than a refrigerant leak test and saves the time and loss of refrigerant. Large leaks will be heard.

The alternative method is to charge the system with refrigerant to 1 to 2 bar and then boost this pressure with nitrogen. Subsequent leak testing is then carried out with refrigerant leak detectors, but requires that personnel come into close contact with piping at high pressure. This method is restricted to the smaller systems.

9-13 EVACUATION

It is now necessary to remove as much as possible of the original air, with its moisture content, and the test gas before introducing the refrigerant.

The test pressure is released and a vacuum pump connected to draw from all parts of the circuit. This may require two connections, to bypass restrictions such as expansion valves, and all valves must be opened within the circuit, requiring electrical supplies to solenoid valves and the operation of jacking screws, where these are fitted.

On small systems, such as factory packages, a final pressure of 50 μm (7 Pa) should be reached, but larger and site installations for air-conditioning temperatures are acceptable at a final vacuum of 170 Pa.[28] Vacuum pumps of this quality can be hired if not immediately available. Evacuation of a large system may take a couple of days. During this time, checks should be made around the pipework for cold patches, indicating water boiling off within, and heat applied to get this away.

If it is quite impossible to obtain a good vacuum pump, a poorer quality pump could be used, providing it can get down to 0.01 bar, together with a double-flushing technique. The circuit is first pulled down to the lowest possible pressure and charged with refrigerant to a slight positive pressure and re-evacuated. At the end of this second rough vacuum, the partial pressure of the original gases will be 0.01×0.01 or 0.0001 bar (10 Pa). This method is expensive in refrigerant for large systems.

Care should be taken that the pump used will tolerate the refrigerant gas.

9-14 CHARGING WITH REFRIGERANT

Refrigerant may be charged as a liquid through the connection shown in Fig. 9-2. The cylinder is connected as shown and the connecting pipe purged through with a little of the gas to expel air from it. For small charges, the bottle may be supported on a weighing machine, or a calibrated charging cylinder may be used.

The compressor must not be started while the system is under vacuum, so refrigerant is admitted first up to cylinder pressure. At this point, the compressor may be started, assuming that all auxillary systems (condenser fan, pump, tower, cooler fan, etc.) are running. The liquid-line valve upstream of the charging connection is partially closed to reduce the line pressure at this point below that of the supply cylinder, and the refrigerant will flow in. While the refrigerant can be safely admitted in this way, the system is not running normally, since the throttle valve is reducing the pressure across the expansion valve. At intervals during charging, the cylinder valve must be closed and the throttle valve opened fully. Only under these conditions can correct running be observed. When fully charged, the sight glass will be clear.

If no receiver is fitted, extra charge may be added, possibly another 5 per cent above that already in the plant, to allow for seasonal and load variations. If a receiver is in circuit, this should be about one-sixth full under normal running conditions.

Systems having high-pressure float expansion valves, and those without sight glasses, must be charged gradually, observing the frost line or using a contact thermometer to measure superheat.

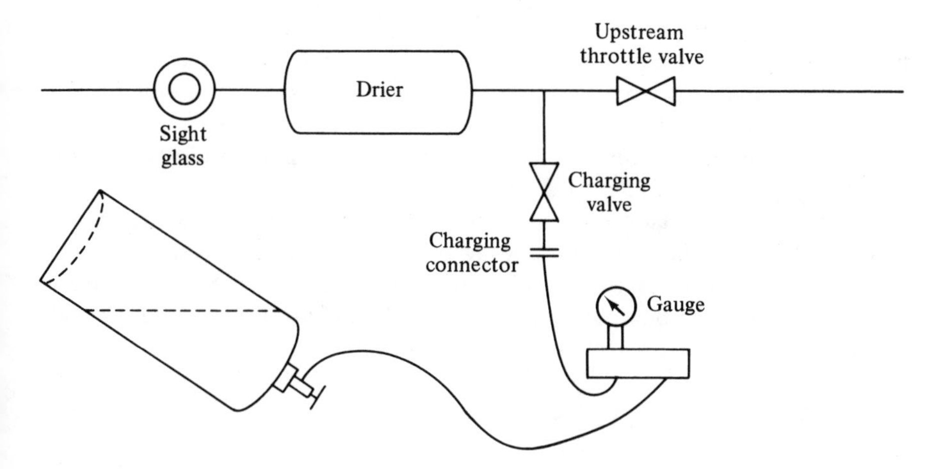

Figure 9-2 Charging connection.

Small packages will have the charge marked on the nameplate and must be carefully charged to this weight, which will be critical.

Systems using the halocarbons will need to have further lubricating oil added, to make up for that which will be carried around with the refrigerant. In the absence of any firm guidance from the supplier, the crankcase must be topped up gradually during normal running, until it is level with the middle of the sight glass under operating conditions.

9-15 INSULATION

Pipework and other components should be insulated after the safety pressure test, but usually before prolonged running of the plant, since it is very difficult to remove water and frost once it has formed. Only the low-pressure piping is insulated, where it does not form part of the evaporator, i.e., after the expansion valve, where this may be outside the cooled space, and from the evaporator back to the compressor. Basic materials are cork and the expanded plastics. These are sufficiently rigid to be moulded to the correct shape, remain firmly in place, and support the external vapour seal which is essential to prevent the ingress of water vapour (see also Chapter 15).

Insulants for pipework and curved pressure vessels can be obtained ready shaped, so that they fit tightly to the surface. All surfaces should be quite dry before the material is applied, even if the adhesive is a water-based emulsion, and the water or other solvent must be given ample time to dry or set before any outer wrapping is applied.

Any air spaces within the insulation should be avoided, since this air will contain moisture, which will condense and freeze, leading to early deterioration of the insulant.

The essential part of the insulation system is the vapour seal, which must be complete and continuous over the outer (warm) surface. Even materials such as coconut fibre, rice husks, sawdust, and wood shavings are successfully used as insulants if the vapour seal is good.

The application of insulating materials is a specialist trade and justifies careful supervision and inspection.

Much use is made of flexible foamed plastic material, which can be obtained in tubular form for piping up to 114 mm diameter and in flat sheets of various thicknesses for tailoring onto other shapes. This material has a vapour-tight outer skin, but must be sealed at all joins and the ends. The manufacturers are helpful in advising users.

9-16 WATER CIRCUITS

Water and other fluid circuits will be pressure tested for safety and leakage, using water at a pressure of 1.5 times the working pressure, or as may be required.

The opportunity is taken while filling for testing to ensure that the circuits can be filled without airlocks. Air vents at high points of the circuit may be automatic or manual. While the pipes are full, pumps should be run if possible to dislodge any dirt before draining down and cleaning the strainers. If a lot of dirt is found, the pipework should be filled again and reflushed. In any case, the pumps should be run at the earliest opportunity and the strainers cleaned out.

Fluids, if other than water, are not put in until this pressure testing and flushing has been carried out.

9-17 NON-CONDENSIBLE GASES

Other gases, mainly ambient air, may enter a refrigeration system as a result of incomplete evacuation before charging, opening of parts for maintenance or repair, or inward leaks on circuits operating below atmospheric pressure. These gases will be circulated with the refrigerant vapour until they are all in the condenser and receiver. They cannot move further around the circuit because of the liquid seal at the outlet to the expansion valve.

Within the confines of the condenser and receiver, the gases will diffuse together and will exist in the same proportions throughout. The non-condensibles may therefore be removed through purge valves on either vessel, but such valves are commonly fitted on or near to the hot gas inlet to the condenser. The presence of non-condensible gas will be shown as an increase of condenser pressure (Law of Partial Pressures) and may be detected during normal operation if the running log is accurate. The effect of this higher condenser pressure is to increase the compression ratio and so reduce the volumetric efficiency and increase the power. There will also be a slight effect with the gas blanketing the condenser surface, reducing heat flow.

Where the presence of such gas is suspected, a cross-check can be made, providing the high-pressure gauge is of known accuracy. The method is to switch off the compressor after a short running period, and so stop the flow of thermal energy into the condenser, but continue to run the condenser until it has reached ambient conditions. The refrigerant vapour pressure can then be determined from the coolant temperature and any increase indicates non-refrigerant gas in the system.

The bleeding of gas from the purge valve will release a mixture which can be estimated from the total pressure.

Example 9-3 A system containing R.22 is cooled to an ambient temperature of 20°C and the condenser gauge then indicates 10.5 bar. What is the partial pressure of the non-condensible gas, and how much R.22 must be lost to purge 1 kg of this gas? Assume it is air.

Vapour pressure of R.22 at 20°C = 9.09 bar

Observed pressure = 10.50 bar

Partial pressure of non-condensible gas = 1.41 bar

Gas	Proportion by pressure	Molecular mass	Proportion by weight	Ratio
Air	1.41	28.97	40.85	1
R.22	9.09	86.5	786.3	19.2

So 19.2 kg of R.22 must be wasted to purge 1 kg of non-condensible gas.

Ammonia has a much lower molecular mass and the proportion by weight in this example would only have been 3.8 kg of ammonia lost. Also, ammonia is much cheaper than R.22.

Wastage of refrigerant can be reduced by cooling the mixture of gases and thus reducing the ratio. By means of a refrigerated purge device, which cools the mixture down to the evaporator temperature (for example, −35°C in a blast freezer plant), the ratio would become

Vapour pressure of R.22 at −35°C = 1.318 bar

Partial pressure of non-condensible gas (10.50 − 1.318) = 9.182 bar

The ratio now becomes 0.43 kg of R.22 lost per kilogram of air.

Automatic gas purgers comprise a collection vessel for the gas mixture with an inbuilt cooling coil connected to the main suction. Condensed refrigerant returns to the condenser, and any gas remaining in the vessel will be non-condensible and can be vented by an inverted bucket trap.

Purging of gases must always be to the open air. The release of ammonia–air mixture is usually made through a flexible tube into a container of water. The water will absorb the ammonia and any bubbles seen to rise to the surface will be other gases.

CHAPTER

TEN

PACKAGED UNITS

10-1 GENERAL

A high proportion of the total cost of a refrigeration or air-conditioning system is made up of work which can be carried out quicker, cheaper, and under better managerial control within a factory rather than on the installation site. This work includes the following:

1. Procurement, inspection, and storage of bought-out items
2. Storage of manufacturing materials
3. Manufacture of in-house components
4. Assembly of parts into systems
5. Piping, wiring, charging, testing

A wide range of factory-built packaged units is now made, and covers most requirements except the larger or more specialized installations.

The advantages of packaged units are as follows:

1. Correct selection and balance of components
2. Assembly, leak testing, processing, and charging under factory conditions
3. Inspection and testing of the complete unit before it leaves the factory
4. Delivery to the site complete and in working order, so no site delays for materials
5. Simplified site installation, with a minimum of disruption, inconvenience, and cost

Disadvantages are that the unit may not be exactly the right size for the duty, since a stock unit may be used, and the risk of misapplication.

10-2 CONDENSING UNITS

The basic condensing unit is a single package comprising the compressor with its drive, the condenser (either air or water cooled) and all connecting piping, and the necessary controls to make the set functional (Fig. 10-1).

Such assemblies might have the compressor and drive only, for site connection to a remote air-cooled condenser. As such, they are correctly termed compressor units. Compressor and condensing units will be site connected to evaporators, and these components must be matched in capacity (see Chapter 25).

Cooling capacity data will be based on various condensing conditions, in terms of air or water temperature onto the condenser, and for a range of evaporating conditions for which the set may be suitable.[29]

Figure 10-1 Air-cooled condensing unit. *(Courtesy Prestcold Limited.)*

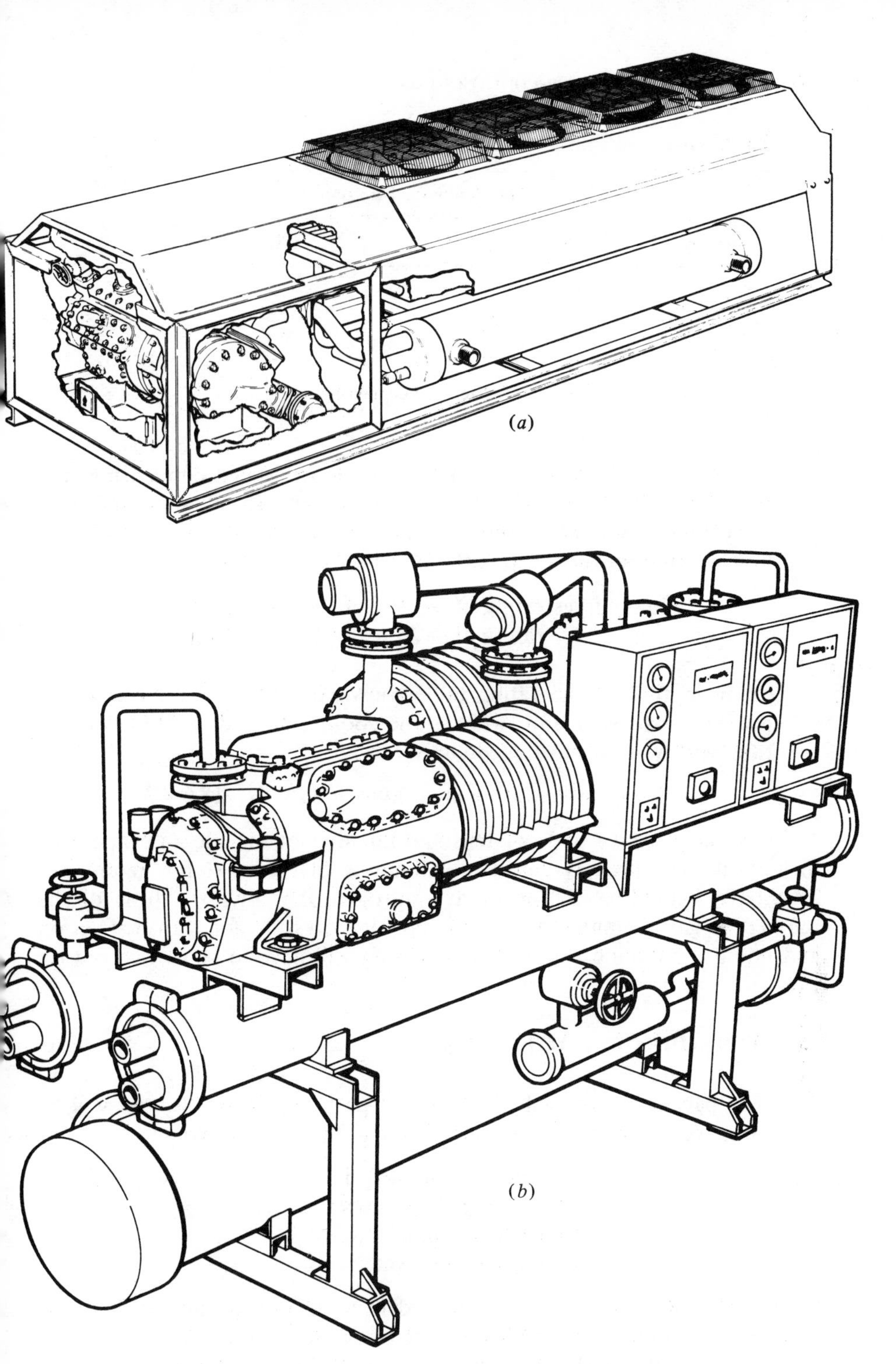

Figure 10-2 Packaged units: (*a*) air cooled, (*b*) water cooled. *(Courtesy Hall-Thermotank Products Limited.)*

Table 10-1

Water onto condenser, °C	Evaporating temperature			
	−30°C	−20°C	−10°C	0°C
25	10.5	18.6	30.6	45.1
30	9.7	17.2	28.1	41.8
35	9.0	15.3	25.1	37.4

Capacity, in kW, of B150HW water-cooled condensing unit on R.12.

Example 10-1 In the rating curves for an air-cooled condensing unit shown in Fig. 10-3, what is the cooling capacity at an evaporating temperature of −25°C and with air onto the condenser at 25°C? By how much does this drop with condenser air at 35°C?

From curves rating at −25 to +25°C is 1 310 W
rating at −25 to +35°C is 1 085 W

Example 10-2 In the rating table shown in Table 10-1 for a water-cooled condensing unit, what is the cooling duty at −20°C evaporation, with water onto the condenser at 25°C?

From table, rating at −20 to +25°C is 18.6 kW

Since compressor and condensing units do not include an evaporator they are not complete systems and will not be charged with refrigerant, but may have a holding charge of dry nitrogen, dry air, or a little of the refrigerant gas to maintain a slight positive pressure for transit. Suction and liquid interconnecting lines and wiring will have to be carried out on site.

10-3 ONE-PIECE PACKAGES

The true packaged unit will have all the parts of the system and will be factory tested in the complete state. There are four basic types:

Air cooling, air cooled
Air cooling, water cooled
Liquid cooling, air cooled
Liquid cooling, water cooled

Ratings for such units will be published in terms of the entering fluids on both the evaporator and condenser side (see also Chapter 25).

The siting of a packaged unit is more critical than separate plant, since all components are together, and a compromise may have to be reached

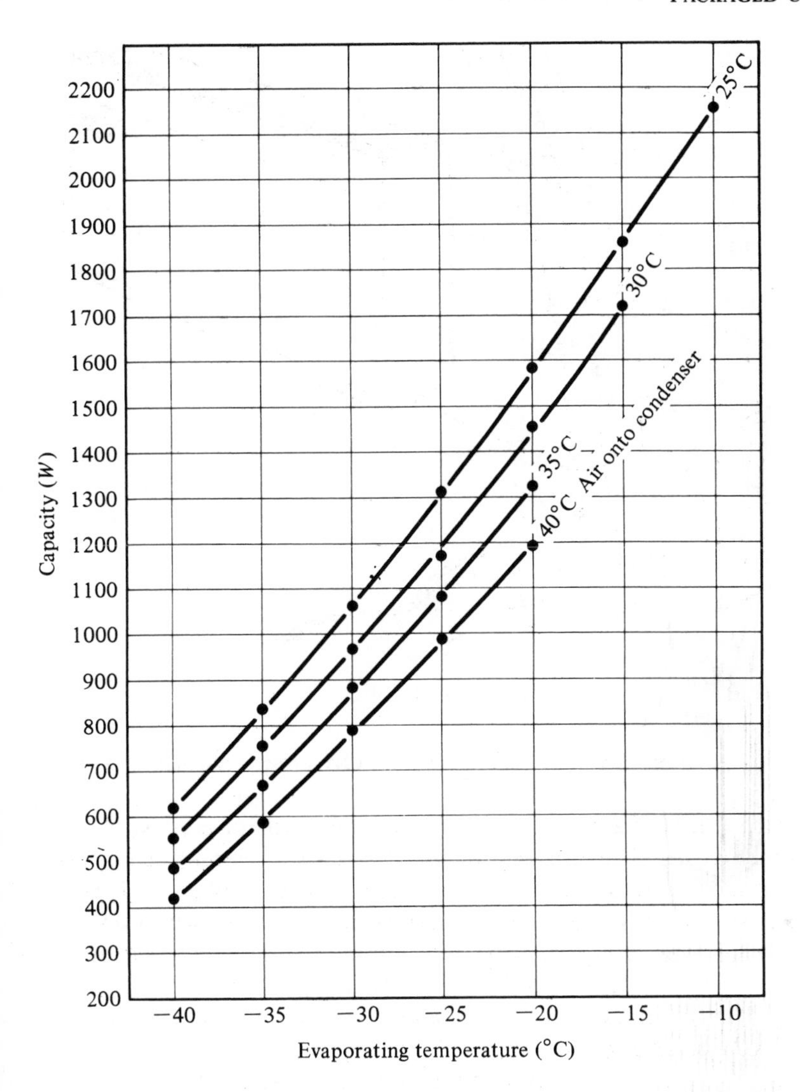

Figure 10-3 Capacity curves for AS75 condensing unit.

between the convenience of having the unit close to the load and the difficulty of obtaining condenser air or water, extra noise transmitted, or safety aspects.

10-4 SPLIT PACKAGES

To avoid the constraint of having all parts in one package, the evaporator set may be split from the condenser, the compressor going with either (see Fig. 10-4). The unit will be designed as a complete system but the two parts

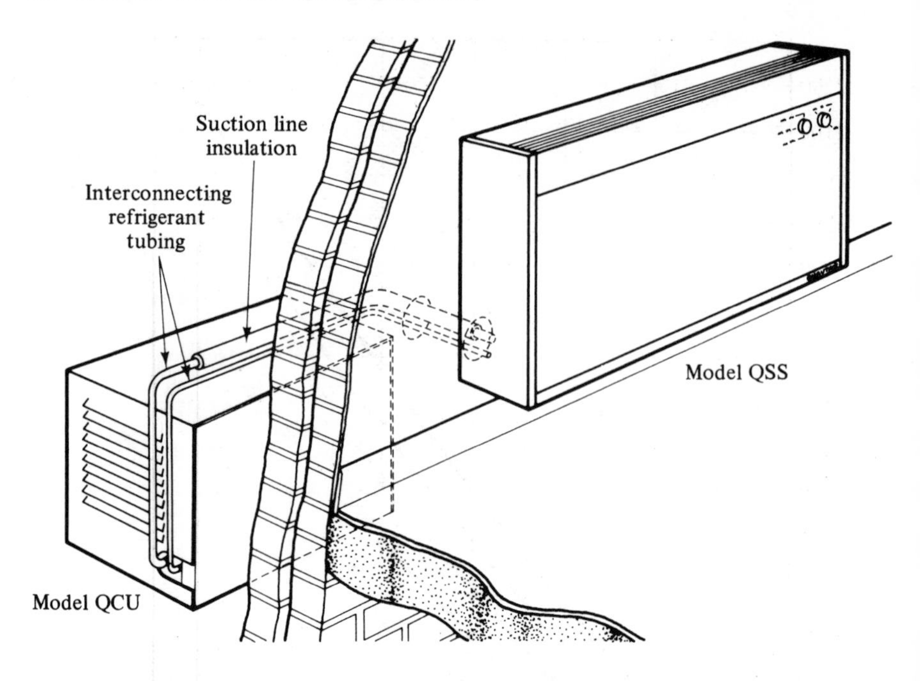

Figure 10-4 Split package unit. *(Courtesy Qualitair.)*

are located separately and connected on site. On some small units, flexible refrigerant piping may be provided.

If the system is of a range up to about 5 kW, coils of pre-charged soft copper tube, with self-sealing couplings, may be supplied for connection within a limited distance of 5 to 15 m. This facility enables full factory processing to be carried out to the standards of a one-piece unit. It is limited to the availability of suitable tubing, usually $\frac{5}{8}$ or $\frac{3}{4}$ in outside diameter. In such systems, the total charge is suitable for the final assembly, and pipes should not be extended beyond the factory-supplied length without prior consultation with the supplier.

Larger split packages must be piped on site by normal methods, and then processed and charged as an open plant, as in Chapter 9. Split unit evaporators should not be located more than 5 m higher than their condensers (see Chapter 9).

10-5 EVAPORATOR UNITS

Evaporator sets, as supplied as part of a split package or for application to a condensing unit, will be of three main types:

1. Air-conditioning, having the air-cooling coil with drip tray under, ex-

pansion valve, fan and motor, air filters, inlet and outlet grilles. They may also include dampers and duct connections for return and fresh air, heaters, humidifiers and various controls.
2. Cold-store evaporators having the coil with drip tray under, fans, and possibly the expansion valve.
3. Cold-store evaporators for use below +2°C must also have some means of defrosting the coil. If this is to be by electric heat, the elements will be inbuilt.

10-6 APPLICATION DATA

Comprehensive application data should be made available for all marketed packaged units, to allow the designer or sales engineer to make a correct selection for his purpose. However, it should be borne in mind that manufacturers or sales outlets are frequently not aware of all the parameters of an installation, and the interpretation of catalogue data has many pitfalls.

Errors in application stem mainly from a lack of understanding of the requirement and a tendency to buy at the lowest price without the protection of a clear specification. Once the application is fully understood and assessed, a specification needs to be drawn up, and the possibility of error and dispute is reduced (see also Chapter 25).

10-7 TESTING OF PACKAGED UNITS

Manufacturers' test procedures for packaged units take several forms:

1. Rating test data on the prototype, which forms the basis for the published capacity and application leaflets.
2. Rating check tests on a proportion of production units, to verify that standards are being maintained.
3. Function tests on all production units, to verify correct operation of components.
4. A short running test at normal conditions to check for reliability of operation plus, possibly, an approximate capacity check. This will not be possible on some types.
5. Safety tests at maximum operating conditions—usual on all or a high proportion of production units.

These test schedules are costly, requiring expensive equipment, and are reflected in the anticipated high quality.

Factory records will be kept of all such tests and, in the case of larger units, manufacturers will, if asked, provide a certified copy of the test on the equipment supplied.

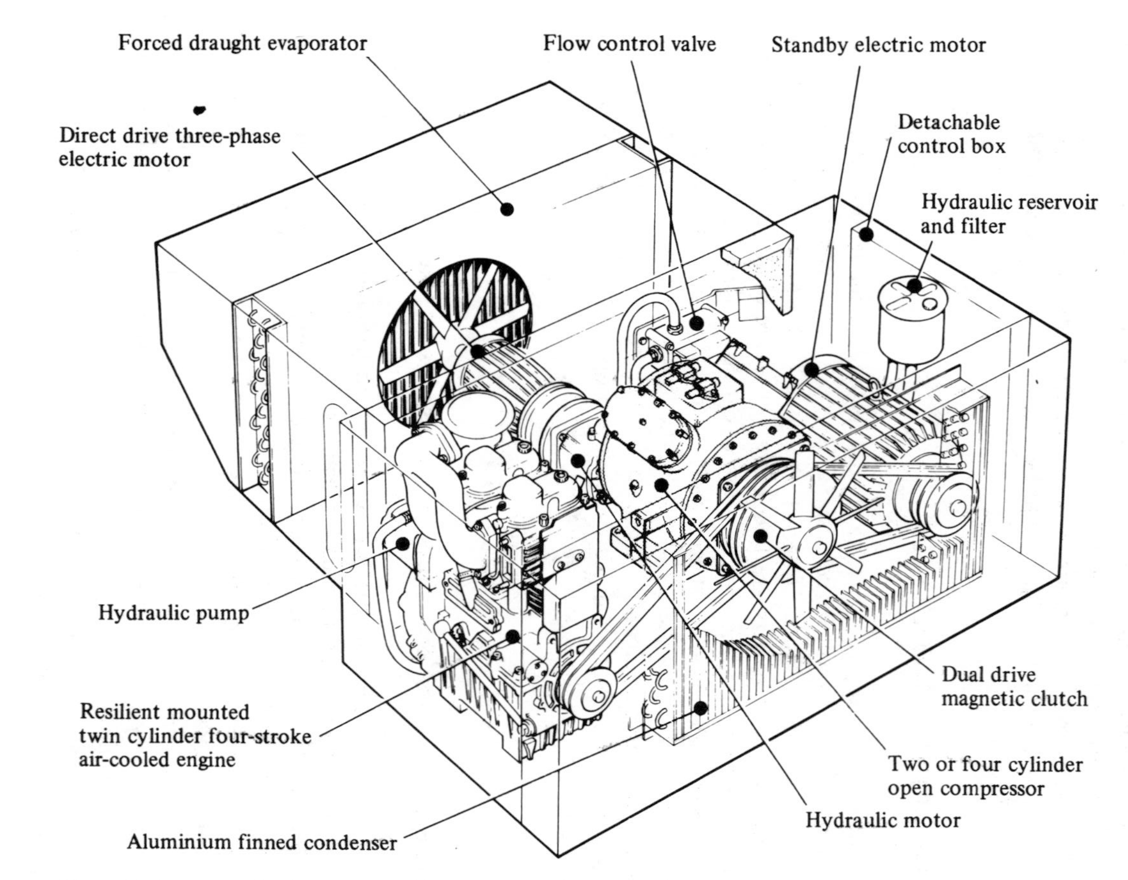

Figure 10-5 Packaged vehicle cooler. *(Courtesy Transfrig Limited.)*

10-8 MOBILE APPLICATION UNITS

The requirement for transport air-conditioning or refrigeration is for an air-cooling, air-cooled unit with reliable availability of service.

For long-distance travel, the prime mover is usually a built-in petrol or diesel motor, driving the compressor through belts and a clutch. An electric motor may also be provided which can be connected to mains supply when the vehicle is not moving. Other prime movers used are direct drives from the vehicle engine or indirect through hydraulic piping. The evaporator fan is usually electric for convenience, running off a 24 V d.c. feed with an auxiliary dynamo on the cooling unit (see Fig. 10-5).

Most of these methods allow essential maintenance and repairs to be carried out under workshop conditions without taking the vehicle off the road, providing a spare unit is available. A dominant feature of this market has become the wide availability of spares and service wherever such vehicles may go.

10-9 OTHER PACKAGES

A very large variety of self-contained refrigeration and air-conditioning packages are made, mainly for the consumer durable market and small domestic applications. They include:

1. The domestic refrigerator and freezer.
2. Ice-cream conservators.
3. Retail display cold and freezer cabinets and counters.
4. Cooling trays for bottles (beer, soft drinks, wines).
5. Instantaneous draught beer coolers. These usually comprise a tank of constantly chilled water, through which the beverage flows in stainless steel piping.
6. Ice makers—cubes and flakes.
7. Cooled vending machines.
8. Soft ice-cream freezers.
9. Dehumidifiers, in which air is passed first over the evaporator to remove moisture, and then over the condenser to re-heat and lower the humidity (see Fig. 13-2).
10. Drinking water chillers.

With the advantages of factory-built packaged cooling devices, this list cannot be exhaustive.

CHAPTER

ELEVEN

BRINES AND ICE

11-1 GENERAL

The cooling effect of a central refrigerating system can be distributed by a heat-transferring liquid or *secondary refrigerant*.

Where the working temperatures are always above 0°C, such as in air-conditioning, water is commonly used. At temperatures below this, non-freezing liquids are used, and these are collectively termed *brines*.

Brines may be, as the name suggests, solutions of inorganic salts in water, and the two in general use are sodium chloride and calcium chloride. Of these, the former is compatible with most foodstuffs and can be used in direct contact or in circumstances where the brine may come into contact with the product. Calcium chloride has an unpleasant taste and cannot be permitted to contaminate foods.

11-2 PHYSICAL PROPERTIES

With any solution, there will be one concentration which remains liquid until it reaches a freezing point, and then will freeze solid. This is the *eutectic* mixture, and its freezing point is the *eutectic point* of the solute (see Fig. 11-1). At all other concentrations, as the solution is cooled it will reach a temperature where the excess water or solute will crystallize out, to form a *slushy suspension* of the solid in the liquid, until the eutectic point is reached, when it will all freeze solid.

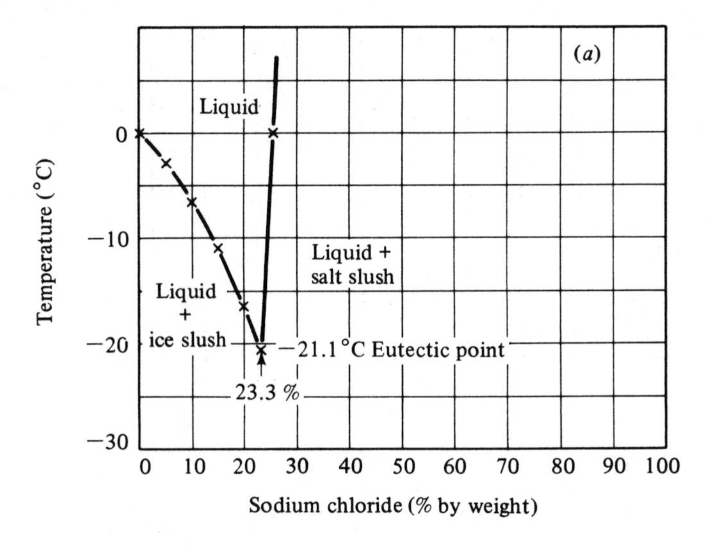

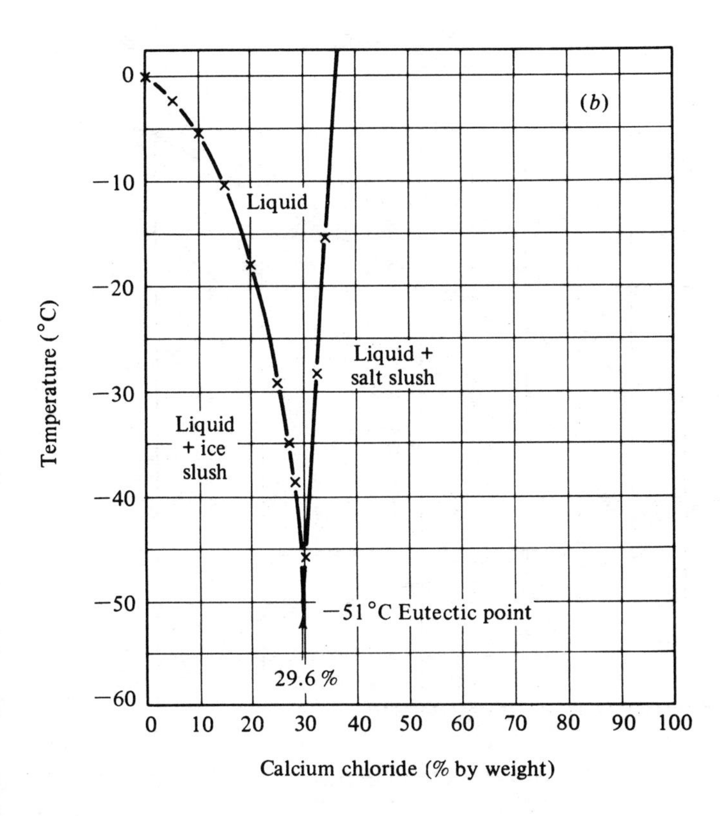

Figure 11-1 Eutectic curves: (*a*) sodium chloride in water, (*b*) calcium chloride in water.

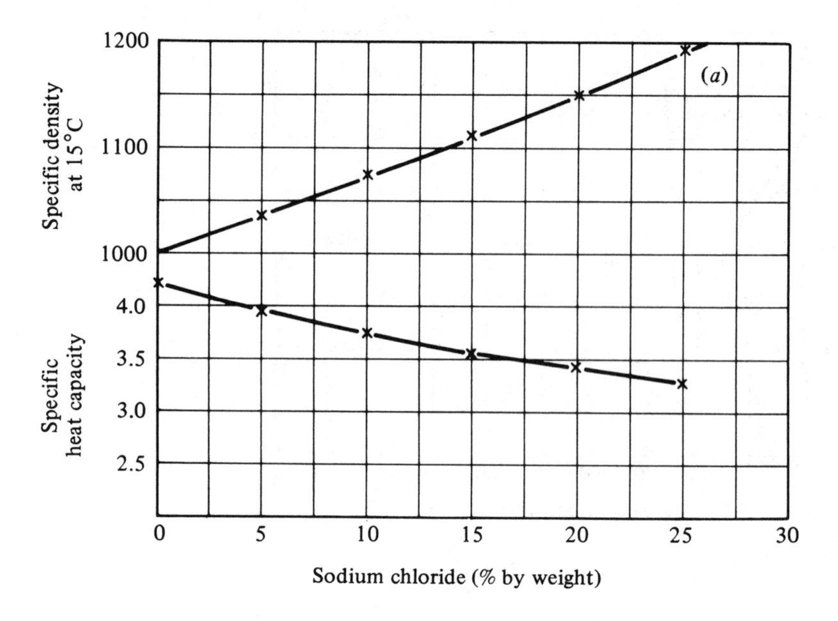

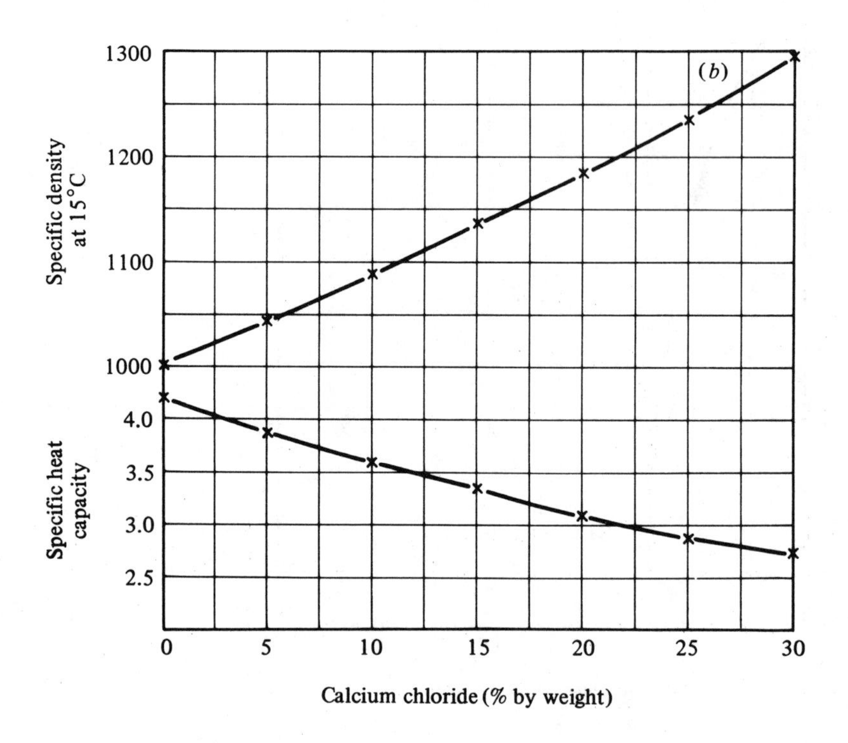

Figure 11-2 Density and specific heat capacity: (*a*) sodium chloride, (*b*) calcium chloride.

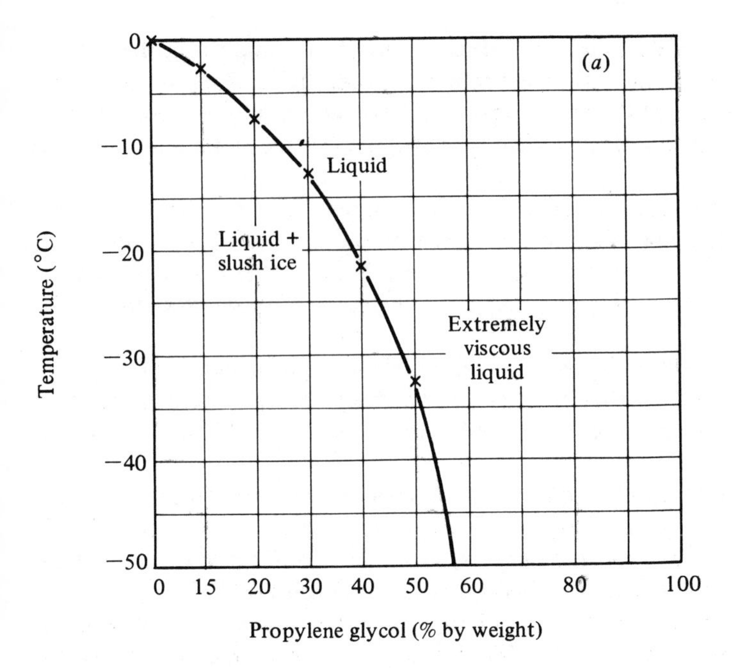

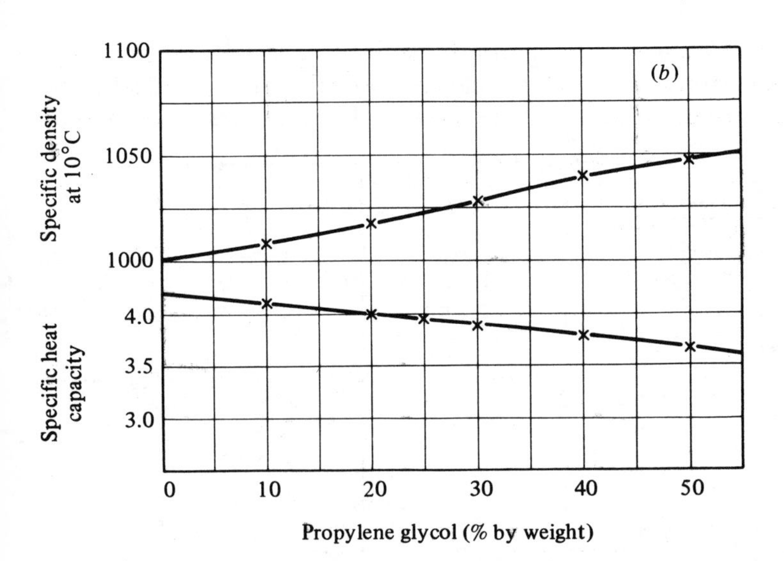

Figure 11-3 Propylene glycol in water: (*a*) eutectic curve, (*b*) density and specific heat capacity.

For economy of cost, and to reduce the viscosity (and so improve heat transfer), solutions weaker than eutectic are normally used, provided there is no risk of freezing at the evaporator. Non-eutectic mixtures have a safety factor in that a slush will form first, and the risk of freeze damage is much less.

In salt brines, the water may be considered as the heat transfer medium, since the specific heat capacity of the salt content is low (see Fig. 11-2). The specific heat capacity of the brine will therefore decrease as the salt concentration increases. At the same time, the specific mass will increase.

Non-freezing solutions can also be based on organic fluids, principally the glycols, of which ethylene and propylene glycol are in general use. Where contact with food is possible, propylene glycol (see Fig. 11-3) should be used.

The concentration of a solute has a considerable effect on the viscosity of the fluid and so on the surface convective resistance to heat flow. There is little published data on these effects, so applications need to be checked from basic principles. Industrial alcohol (comprising ethyl alcohol with a statutory addition of methyl alcohol to render it poisonous) may be used as a secondary refrigerant, either 100 per cent concentration or mixed with water. The fluid has a low viscosity and good heat transfer, but is now little used on account of its toxicity and the fire risk in high concentrations.

11-3 BRINE CIRCUITS

Where a brine system serves a multiple-temperature installation such as a range of food stores, the coolant may be too cold for some conditions, causing excessive dehydration of the product. In such cases, to cool these

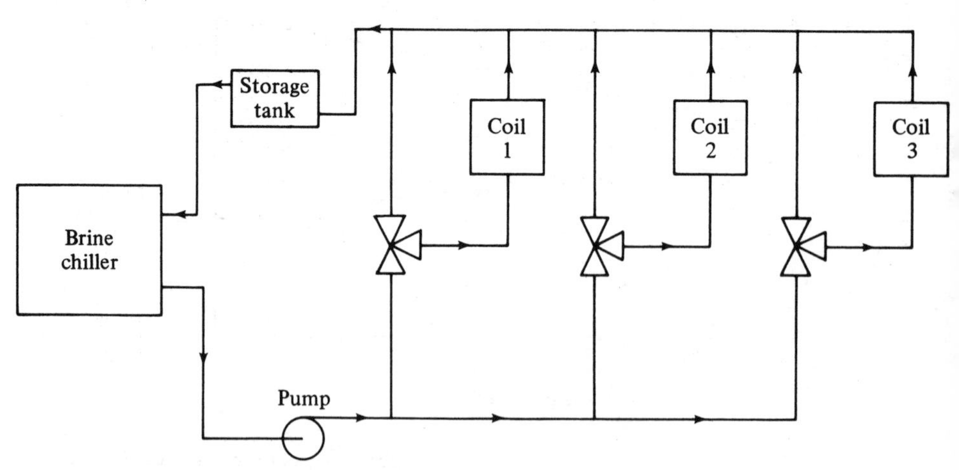

Figure 11-4 Brine circuit for separate rooms.

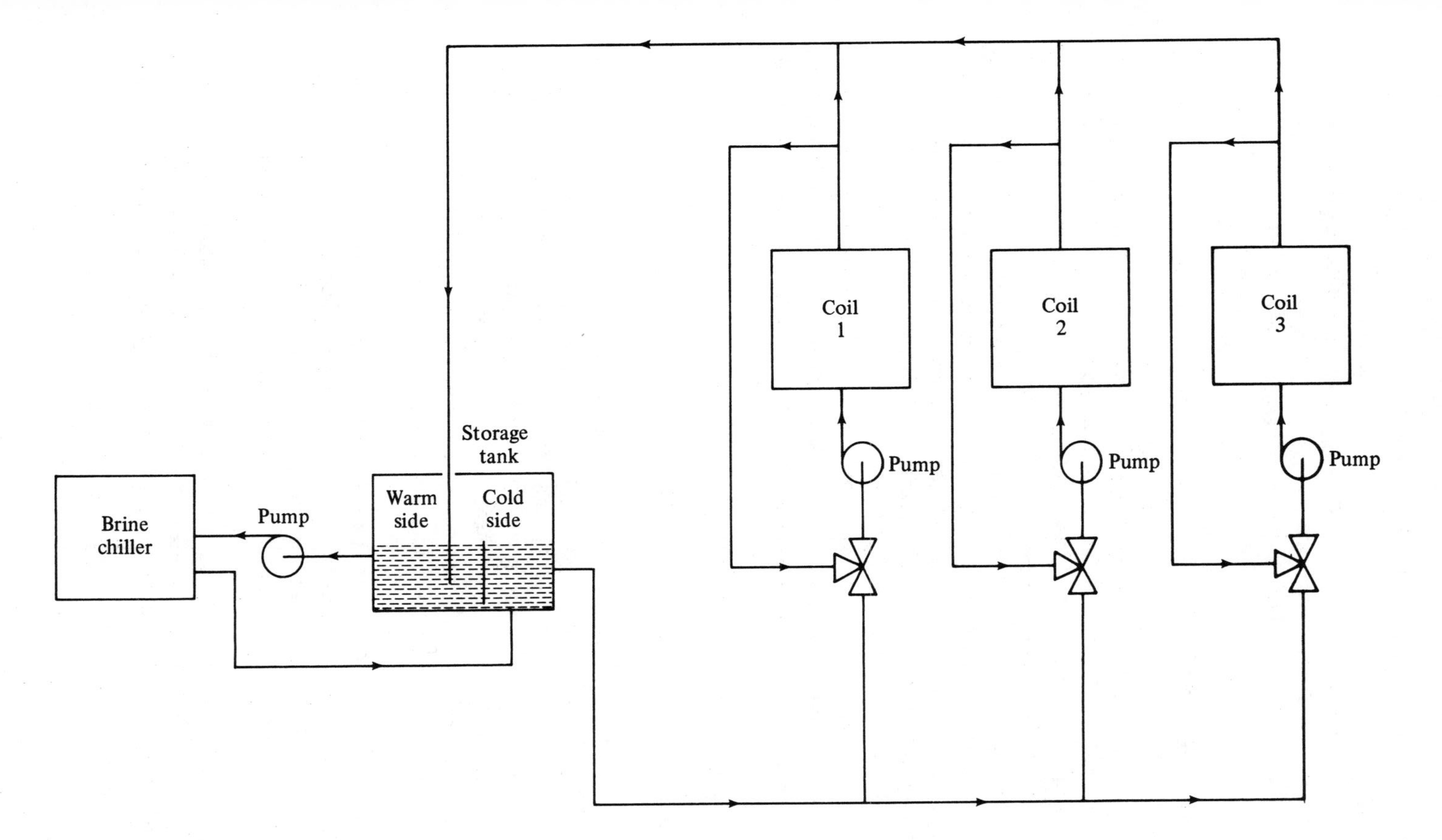

Figure 11-5 Brine circuit for rooms at different temperatures.

rooms the brine must be blended. A separate three-way blending valve and pump will be required for each room.

Figures 11-4 and 11-5 show two different brine circuits.

11-4 CORROSION

If brine circuits are open to the atmosphere, air may be entrained with consequent oxidation, and the solution will become acid. This will promote corrosion and should be prevented as far as possible by ensuring that return pipes discharge below the tank surface.

To reduce the effects of corrosion, inhibitors are added, typically sodium chromate in the salt brines and sodium phosphate in the glycols. These are alkaline salts and help to counteract the effects of oxidization, but periodic checks should be taken, and borax or similar alkali added if the pH value falls below 7.0 or 7.5.[1]

Brines are hygroscopic and will weaken by absorbing atmospheric moisture. Checks should be made on the strength of the solution and more salt or glycol added as necessary, to keep the freezing point down to the required value.

11-5 FROZEN BRINES AND ICE

Variations in cooling load can be provided from the latent heat of melting of ice or a frozen eutectic. Ice can be formed by allowing it to build up on the outside of evaporator coils in a tank. Brines are more normally held in closed tanks or plates, again with evaporator coils inside, the outside of the tank forming the secondary heat exchange surface. Eutectics can be formulated according to the temperature required (see Fig. 6-8).

This latent heat is used for two specific applications:

1. To handle a sudden peak load with a small refrigeration plant. Typically to make ice over a period of several hours and then use ice water for the cooling of a batch of milk on a dairy farm.
2. As holdover plates in road vehicles (see Sec. 6-5 and Fig. 6-8*d*).

Eutectic solutions in closed containers can be frozen in a domestic refrigerator or freezer and used to keep foods cool while away from the house. The use of ice cubes to cool beverages by contact or immersion will be well recognized.

CHAPTER

TWELVE

REFRIGERATION LOAD ESTIMATION

12-1 LOAD SOURCES

Refrigeration loads are of two types:

1. To cool something down, i.e., reduce its enthalpy
2. To keep something cool, i.e., remove incoming and internally generated heat

The components of the total cooling load will be:

1. Removal of heat, sensible or latent, from a product
2. Heat conducted in through the surfaces of the room, tank, pipe, etc., from warmer surroundings
3. Radiant heat from outside
4. Heat convected from outside (air infiltration or ventilation), both sensible and latent
5. Internal sources of heat—lights, fan motors, machinery, personnel, etc.—and heat generated by the product

Some of these can be calculated fairly accurately from known data. Others have unknown parameters, so estimates are based on a combination of available data and practical experience.

12-2 PRODUCT COOLING

Removal of sensible and latent heat is proportional to the mass, and can be summarized in the formula

$$Q = \frac{m[(c \times \Delta T) + h_s]}{t}$$

where Q = heat flow
c = specific heat capacity of product
ΔT = temperature decrease
h_s = latent heat
t = time

Example 12-1 What is the cooling duty to freeze water from 15°C to ice at 0°C, at the rate of 20 t/d?

$$Q = \frac{20\,000[(4.187 \times 15) + 335]}{24 \times 3600}$$

$$= 92 \text{ kW}$$

Example 12-2 What duty is required to cool 8 t lean meat (specific heat capacity 3.1 kJ/kg K) in 14 h from 22 to 1°C?

$$Q = \frac{8000[3.1 \times (22 - 1)]}{14 \times 3600}$$

$$= 10.3 \text{ kW}$$

There may be several unknown quantities in an estimate. For example, a dairy farm may produce 2400 litre/d (a rate of 100 litre/h), but this will come from two milkings, possibly 1400 litre in the morning and 1000 litre in the afternoon, and the milk must be cooled in 2 h, so the peak rate is 700 litre/h.

The entering temperature of a product may be uncertain, being warmer in the summer or after a long journey.

The dwell time within the cooling system may vary, beer leaving an instantaneous cooler at 4°C when first tapped, but at 12°C if drawn off continuously.

The exact product may not be known—a general foodstuffs cold store might contain bacon (sensible heat capacity 2.4) or poultry (sensible heat capacity 3.3).

Observations may need to be taken of the operation, to form an estimate of unknown figures, or the process analysed to decide representative rates. Assumptions should be stated and agreed by the parties concerned.

12-3 CONDUCTED HEAT

Conducted heat is that going in through cold-store surfaces, tank sides, pipe insulation, etc. It is normally assumed to be constant and the outside temperature an average summer temperature, probably 25 to 27°C, unless some other figure is known. Cold-room surfaces are measured on the outside dimensions and it is usual to calculate on the heat flow through the insulation only, ignoring other construction materials, since their thermal resistance is small.

Example 12-3 A cold room measures 35 m long by 16 m wide and is 5 m high inside. Insulation is 125 mm to walls and ceiling and 75 mm under the floor, of polystyrene having a thermal conductivity of 0.035 W/m K. Inside it is at −10°C, ambient 27°C, and the ground temperature is 12°C. What is the heat flow inwards?

$$\text{Area of walls} = 5.2 \times 2(35.25 + 16.25)$$

$$= 535.6 \text{ m}^2$$

$$\text{Area of ceiling} = 35.25 \times 16.25$$

$$= 572.8 \text{ m}^2$$

$$\text{Area of floor} = 572.8 \text{ m}^2$$

$$\text{Heat flow, walls} = 535.6 \times \frac{0.035}{0.125} \times [27 - (-10)] = 5\,549 \text{ W}$$

$$\text{ceiling} = 572.8 \times \frac{0.035}{0.125} \times [27 - (-10)] = 5\,935 \text{ W}$$

$$\text{floor} = 572.8 \times \frac{0.035}{0.075} \times [12 - (-10)] = 5\,881 \text{ W}$$

$$Q = 17\,365 \text{ W}$$

say 17.5 kW

Solar radiation may fall on outside walls or roofs, raising the skin temperature, and this must be taken into effect. Most new cold stores are built with an air space to avoid this. Heat load must be estimated through all surfaces including piping, ducts, fan casings, tank walls, etc., where heat flows inwards towards the cooled system.

Radiant heat is not a serious factor in commercial or industrial refrigeration systems, being confined to sunshine through refrigerated display windows (which should have blinds) and radiation into open shop display cabinets from lighting. (See also Chapter 19.)

12-4 CONVECTED HEAT

Warm air will enter from outside mainly during the opening of doors for the passage of goods. This must be estimated on the basis of the possible use of the doors, and such figures are based on observed practice. The parameters are the size of the store, the enthalpy difference between inside and outside air, and the usage of the doors. The latter is affected by the existence of air locks and curtains.

Standard textbooks give data on which to base an estimate, and this can be summed up as

$$Q_f = (0.7V + 2)\,\Delta T$$

where Q_f = heat flow
V = volume in cubic metres
ΔT = temperature difference between room and ambient

This is for cold rooms up to 100 m^3 with normal service. For heavy service, i.e., a great deal of traffic through the doors, this figure can be increased by 20 to 35 per cent.

Rooms above 100 m^3 tend to be used for long-term storage, and are probably fitted with curtains (air or plastic, see Chapter 15). For such rooms, the service heat gain by convection may be taken as

$$Q_f = (0.125V + 27)\,\Delta T$$

Example 12-4 Estimate the infiltration air heat gain for the cold room in Example 12-3.

$$\text{Volume} = 35 \times 16 \times 5$$
$$= 2800 \text{ m}^3$$
$$\Delta T = 27 - (-10)$$
$$= 37 \text{ K}$$
$$Q_f = (0.125 \times 2800 + 27) \times 37$$
$$= 13\,950 \text{ W, or } 14 \text{ kW} \quad \text{say (compare 13.9 kW)}^1$$

12-5 INTERNAL HEAT SOURCES

The main sources of internal heat are fan motors and circulating pumps. Where the motor itself is within the cooled space, the gross energy input to the motor is liberated as heat which must be removed. Where the motor is outside, only the shaft power is taken.

Other motors and prime movers may be present—conveyors, lifts, fork-lift trucks, stirrers, injection pumps, packaging machines, etc. The gross power input to these machines may be read from their nameplates or found from the manufacturers.

Personnel will give off about 120 W each.

All lighting within the space must be included on the basis of the gross input. The usual 80 W lighting tube takes about 100 W gross.

Where coolers are fitted with defrosting devices, the heat input from this source must be determined.

Example 12-5 The cold room in Example 12-3 has 80 fluorescent tubes labelled 80 W. The four evaporators each have three fan motors of 660 W gross per fan and 18 kW defrost heaters which operate alternately for 15 min twice a day. The fork-lift truck is rated 80 A at 24 V and will be in the store 20 min each hour during the 8-h working day. Two packers will be present for 10 min each hour. Estimate the average and peak loads.

	Average over 24 h	Peak
Lighting, 80 × 100 W gross, 8 h/d	2.67	8.00
Fan motors, 12 × 660 W	7.60	7.76
Defrost heaters, 72 kW, $\frac{1}{2}$ h/d	1.50	18.00
Fork-lift, 1.92 kW, $\frac{1}{3} \times 8$ h	0.21	1.92
Fork-lift driver, 120 W, $\frac{1}{3} \times 8$ h	—	0.12
Packers, 240 W, $\frac{1}{6} \times 8$ h	—	0.24
	11.98	36.04

This example shows that the greatest load is the fan motors, since these run all the time, except during defrosting, and the lighting. There are several unknowns. For example, it is assumed that the defrosting of the evaporators will not coincide, but this may occur if badly timed, and cause a peak load which may raise the store temperature for a time. The last two items can be ignored, making the load 12 kW average.

Certain stored foodstuffs are living organisms and give off heat as their sugar or starch reserves are slowly consumed. This is known as the heat of respiration since the products consume oxygen for the process. The heat of respiration varies with the sugar or starch content of the product, the variety, and its temperature, and is between 9 and 120 W/t at storage temperatures. Typical figures are shown in Table 12-1. These figures increase with temperature, roughly doubling for every 10 K, so that fruits and many vegetables deteriorate very rapidly if they are warm, using up their food reserves and then decaying.[30, 31]

Table 12-1

Product	Temperature, °C	Heat of respiration, W/t
Apples	2	12
Pears	1	16
Bananas	13	48
Strawberries	0	45
Potatoes	1.5	9

12-6 ESTIMATE ANALYSIS

It is frequently the case that very little definite information is available on which to base a heat load estimate. In these circumstances, the probable minimum and maximum should be calculated from the best available data and an average decided and agreed with the user.

Example 12-6 A dockside frozen meat store has a capacity of 1000 t, stored at −12°C, and leaving the store at a maximum rate of 50 t/d. Meat may arrive from a local abattoir at 2°C or from ships in batches of 300 t at −10°C. Estimate a product cooling load.

Case 1
Meat goes out at the rate of 350 t/week and may arrive from local supplies. There is possibly a four-day week, allowing for odd holidays, so may be 90 t/d from the abattoir. Cooling load is 90 t/d, from 2 to −12°C. Tables give the following:

$$\text{Specific heat capacity above } -1°\text{C} = 3.2 \text{ kJ/kg K}$$

$$\text{Freezing point of meat, average} = -1.0°\text{C}$$

$$\text{Latent heat of freezing} = 225 \text{ kJ/kg}$$

$$\text{Specific heat of frozen meat} = 1.63 \text{ kJ/kg K}$$

$$Q_f = \frac{90\,000}{24 \times 3600}[(3.2 \times 3) + 225 + (1.63 \times 11)] = 263 \text{ kW}$$

Case 2
Shipments may come in on consecutive days (unlikely, but possible if store is almost empty):

$$Q_f = \frac{300\,000}{24 \times 3600}(1.63 \times 2) = 11 \text{ kW}$$

These show a wide variation. Since meat will keep for several days at 2°C, rework case 1 on the basis of a steady input of 50 t/d, all coming from the abattoir.

Case 3

$$Q_f = \frac{50\,000}{24 \times 3600}[(3.2 \times 3) + 225 + (1.63 \times 11)] = 146 \text{ kW}$$

It would seem, then, that the minimum safe cooling capacity required is 146 kW, with the possible risk of 263 kW for a day or so. Most of the time the load will be much less.

A practical approach would be to install plant having a maximum continuous running capacity of 146 kW (to which must be added the other load components of heat leakage, internal heat, and service). Having formed an estimate of the total cooling load, this must be converted into a refrigeration plant capacity.

General practice, having calculated the average load over a period of 24 hours is to take the absolute maximum, or allow 50 per cent over the average, i.e., a plant running time of 16 h in the 24. This general rule must be assessed for the particular application.

Example 12-7 The milk-cooling requirement (above) of 700 litre/h is a maximum rate. There is no need to allow for any more than this, but it cannot be any less. Alternatively, this could be cooled using an ice bank, in which case the total load of 2400 litre could be spread over 16 h of running time. With an allowance for water tank insulation heat gains and an ice water pump, the load might be reduced to a refrigeration plant one-third the size.

Example 12-8 The meat-cooling load in Example 12-2 is probably a daily batch from an abattoir and the duty will be less at night, once the meat is cooled. The maximum capacity will therefore be 10.3 kW, plus the fans and other room losses, and the plant will run continuously while the meat is being chilled only.

All assumptions regarding the load, and estimated cooling duty, should be recorded as the design parameters of the system, and agreed with the user.

Exercise 12-1 The cold store in Example 12-3 is now to be located in an ambient of 35°C, and to have the internal load of Example 12-5 and the product load of case 3 of Example 12-6. Include for infiltration and estimate plant capacity.

Answer to Exercise 12-1

Product cooling load, from Example 12-6, case 3. Allow another 5 per cent for higher ambient, in case meat warms up in transit from abattoir = 153 kW

Heat gain through walls, etc., as Example 12-3 but corrected for 35°C:

walls	6749 W
ceiling	7218 W
floor	7733 W

$$21\,700 \times \frac{24}{16} = 33\text{ kW}$$

Service or convection gains:

$$[(0.125 \times 2800) + 27]45 \times \frac{1}{1000} \times \frac{24}{16} = 25\text{ kW}$$

Internal heat gains as Example 12-5:

$$12 \times \frac{24}{16} = 18\text{ kW}$$

Total plant capacity = 229 kW

CHAPTER

THIRTEEN

APPLICATIONS OF REFRIGERATION

Almost every human activity is affected in some way by the use of mechanical cooling. This chapter illustrates some of the more common, apart from the domestic refrigerator and freezer and air-conditioning.

13-1 COOLING TUNNELS

Air is used as the secondary refrigerant in cold rooms, to cool the product and remove heat gains. Air coolers of suitable types are shown in Chapter 6.

Product cooling on a continuous factory basis must be quicker, and part of the manufacturing process. Such coolers are arranged in the form of tunnels (Fig. 13-1) with the product passing through on a conveyor belt and having cold air blasted through or over the belt.

The evaporators are of the finned tube type, the fin spacing depending on the temperature and interval between defrosts. Blast freezers for quick-frozen foods probably defrost every 5, 6, or 8 h, giving a short, non-operating break between shifts. Such tunnels are used for:

1. Quick-freezing of fruits, vegetables, fish, and meats
2. Hardening of ice-cream
3. Pre-cooling of fresh-baked bread for slicing
4. Cooling of cooked meats such as pies
5. Cooling of chocolates, after the centre has been coated

The heat put in by the air circulation fans is a high proportion of the total cooling load, and modern designs minimize this input of energy.[32]

13-2 DEHUMIDIFIERS AND AIR DRYING

Dry air at atmospheric pressure is used for removing excess moisture from various end products. The air first passes over the evaporator to condense out part of the water vapour, and then over the condenser to reheat (see Figs 13-2 and 17-13). Because of the self-contained nature of this system, with no connections apart from the electric supply and therefore no constraint on location, they are commonly made as unit packages (see also Chapter 10). Applications are as follows:

1. Maintaining dry atmospheres for the storage of steel, cardboard, books, timber; i.e., any product which is better preserved in low humidity.
2. Removal of moisture from newly constructed buildings, to expedite final decoration and occupation.
3. Removal of excess moisture from the atmosphere of indoor swimming pools.

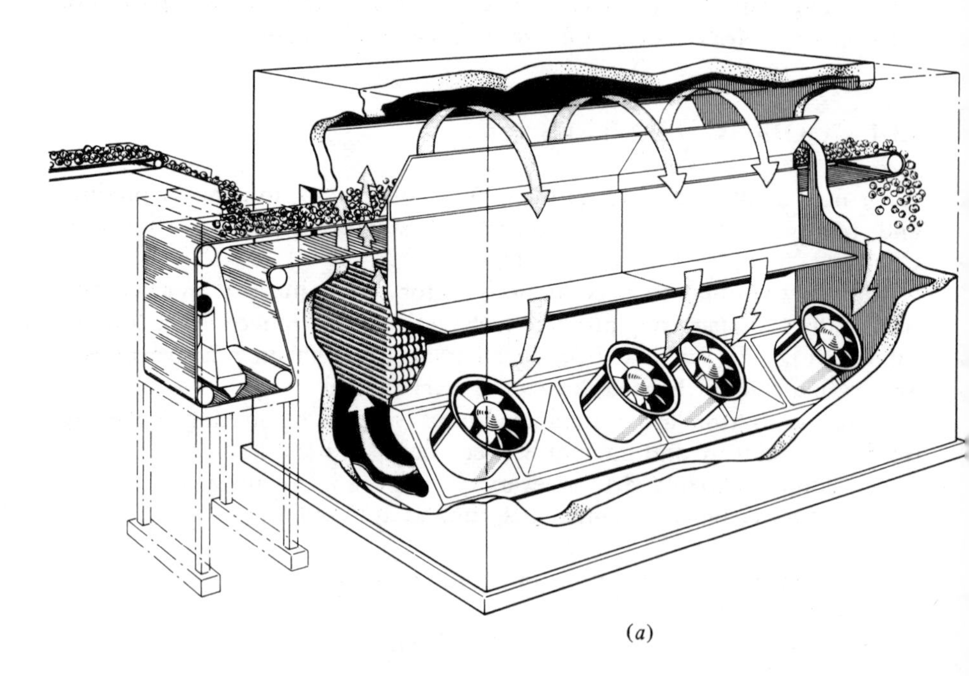

(*a*)

Figure 13-1 Freezing tunnels: (*a*) fluidized bed. *(Courtesy APV-Parafreeze Limited.)*

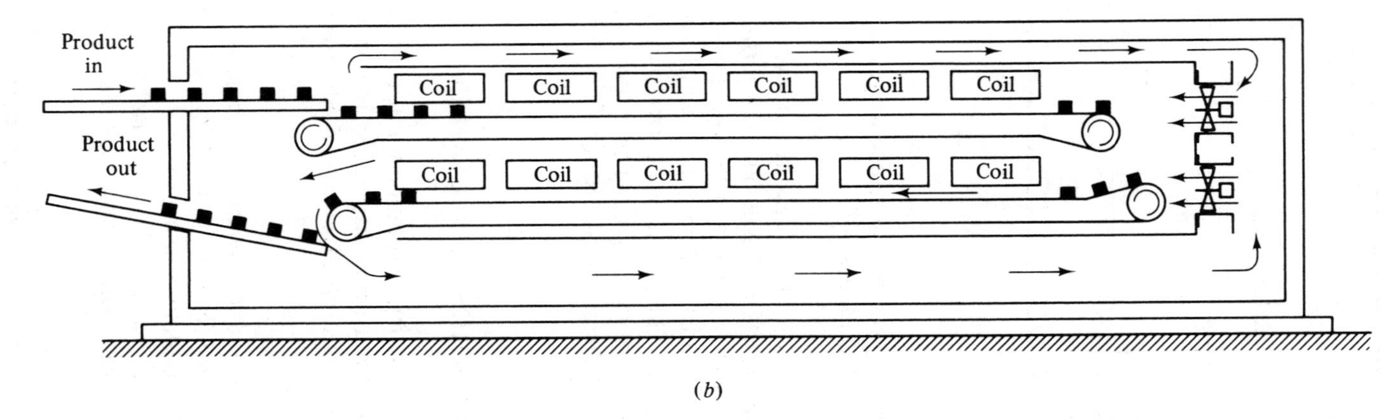

Figure 13-1 (*b*) belt (low fan energy). *(Courtesy S. Forbes Pearson.)*

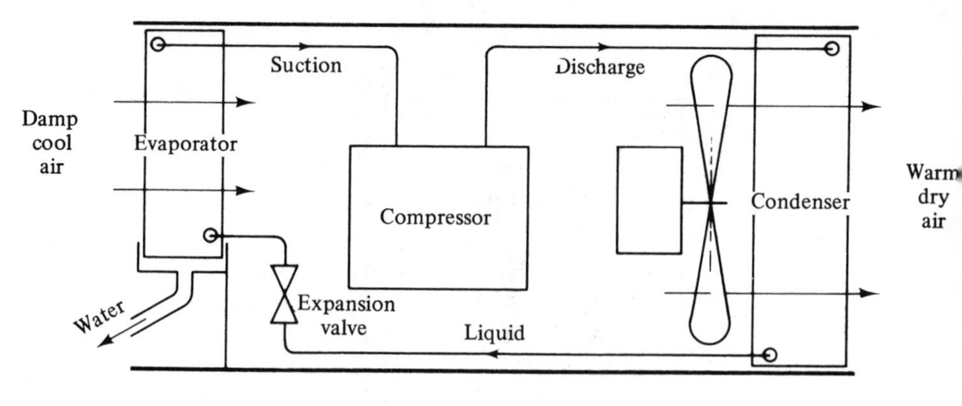

Figure 13-2 Dehumidifier.

4. Seasoning of timber. This is a high-temperature application and requires special refrigeration techniques.

Air-cooling evaporators for pressures above atmospheric will be designed as pressure vessels and will take the form of shell-and-tube or shell-and-coil types. Such coolers will be found on large compressed-air installations, to remove moisture from the air which would otherwise settle in distribution piping and valves, causing corrosion which is accelerated by the high partial pressure of the oxygen. These coolers must be designed to ensure easy drainage of the condensed water.

13-3 CHILLED LIQUIDS FOR COOLING

The use of water or a non-freeze solution for heat transfer is now replacing many applications where a direct expansion refrigerant has been used in the past. This method gives the advantage of employing packaged chillers. Uses are as follows:

1. Cooling milk, cream, and ice-cream mix, using chilled water
2. Cooling soft drinks or beer with water or brine
3. Multiple cold-room buildings, using brine
4. Cooling the moulds of plastic moulding machines with water
5. Manufacture of frozen confections such as lollipops by immersing moulds in a cold brine

The list of such applications is endless, since fresh uses arise with developing technologies.

13-4 ICE MANUFACTURE

Ice can be made as thin slivers on the surface of evaporator drums, and removed mechanically when the correct thickness has been formed. Either the drum or the scraper may rotate. This is a continuous process and the ice flakes fall directly onto the product or into a storage bin below the machine. Smaller units are made as packages with the bin integral and cooled by a few turns of the suction line or by a separate evaporator. Small pieces of ice can be formed in or on tubes or other prismatic shapes made as evaporator tubes, arranged vertically. Water is pumped over the surface to freeze to the thickness or shape required. The tube is then switched to 'defrost' and the moulded section thaws sufficiently to slide off, possibly being chopped into short pieces by a rotating cutter. The machine itself will be made as a package, and the smaller sizes will include the condensing unit.

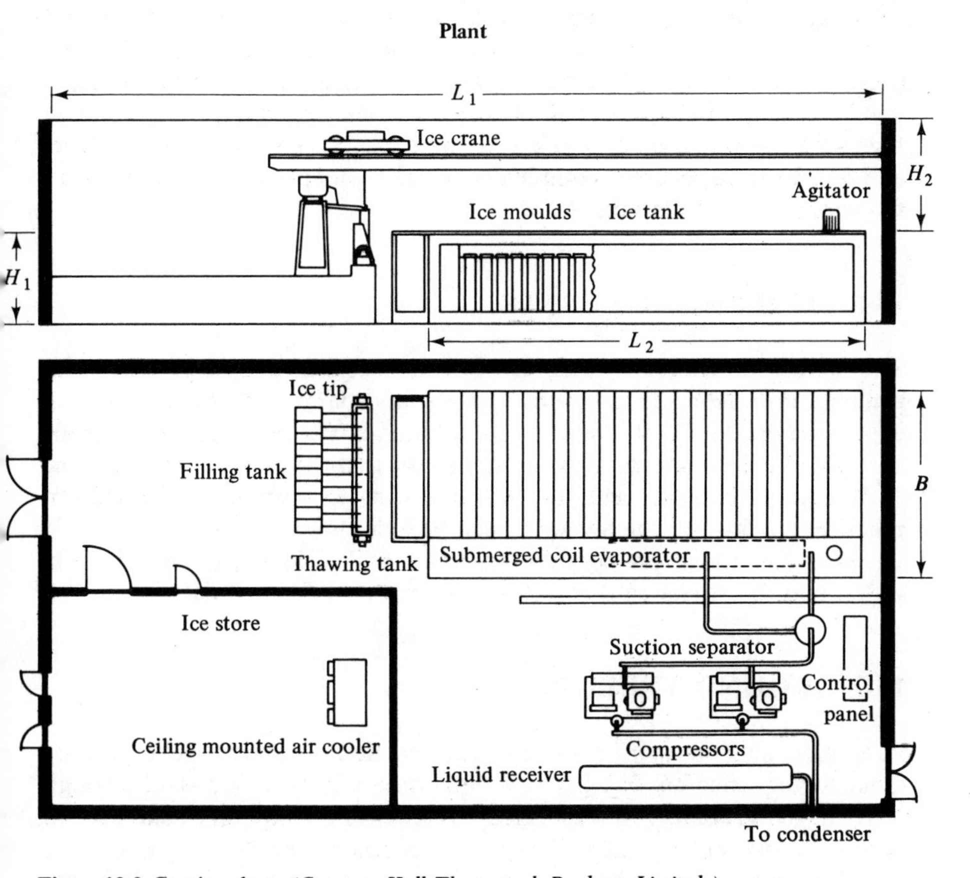

Figure 13-3 Can ice plant. *(Courtesy Hall-Thermotank Products Limited.)*

The manufacture of ice in large blocks is by the can method (see Fig. 13-3), where a number of mould cans, filled with water, are immersed to just below the rim in a tank of refrigerated brine. The smallest block made in this way is 25 kg and will freeze in 8 to 15 h, using brine at −11°C. Blocks up to 150 kg are made by this method. When frozen, the moulds are lifted from the tank and slightly warmed to release the ice block from the sides of the moulds, when they can be tipped out. Blocks may go into storage or for direct use.

Where the available water has a high proportion of solids, there are methods either of pre-treating the water or, by agitating the water in the centre of the block (which freezes last), of removing the concentrated dirty water before it becomes solid. The core is then refilled with fresh water.[30]

13-5 ICE RINKS

Artificial ice rinks are formed and maintained using a brine in tubes buried under the surface. Tubes may be steel or plastic for a permanent rink or plastic for a temporary installation. The brine temperature within the pipes will be about −11°C, and must be lower for rinks in the open air, owing to high solar radiation loads. Packaged liquid chillers are now generally used, and will be transportable, complete with brine pumps and other apparatus, for temporary installations.

13-6 COOLING CONCRETE

The setting of concrete is an exothermic reaction, and large masses of concrete in building foundations, bridges, and dams will heat up, causing expansion cracks if not checked. To counteract this heating, the materials are cooled before and as they are mixed, so that the concrete is laid some 15 K colder than ambient, so that it warms to ambient on setting. In practice, the final mix temperature can be held down to 10°C.

Methods are to pre-cool the aggregate with cold air, to chill the mix water, and to provide part of the mix water in the form of flake ice.[30, 33]

13-7 GROUND FREEZING

In mining and, more recently, the construction of underground storage tanks for liquefied natural gas, it is often necessary to sink a shaft through waterlogged ground. The requirement is to form a temporary cofferdam to permit excavation and the building of a permanent liner.

The general method is to drive in a ring of vertical pipes and pass chilled brine down through an inner pipe so that it flows up the annulus, to cool and eventually freeze the surrounding wet soil. This process is continued until the ice builds up a continuous wall around the proposed excavation. Depths of over 650 m have been excavated in this way. Calcium chloride brine, cooled by surface plant, is usual, but liquid nitrogen has been used on small shafts.[30, 34]

13-8 SOLVENT RECOVERY

Large quantities of solvent liquids are used in industrial and commercial processes and any loss of these creates an environmental hazard, apart from the cost of the material itself.

All these solvents are volatile liquids and will have a pressure–temperature characteristic (see Sec. 1-3), so can be condensed if cooled to their condensing temperature. Finned-tube evaporators are generally used, but the condensation may be at a high pressure requiring heat exchangers of the shell-and-tube type.

The size of equipment can vary from a 200 W unit for a commercial dry-cleaning machine to systems of megawatt size for synthetic fibre processes.

13-9 WET FISH TANKS

Freshly caught fish must be kept as cold as possible until it can be sold, frozen, or otherwise processed. The general practice is to put the fish in a tank, either in the hold or on the deck of the fishing or factory vessel. This tank is filled with sea water and kept down near 0°C by direct expansion coils. Original installations had the coils within the tank, but a remote evaporator is now more usual.

13-10 CHEMICAL INDUSTRY

Processes in the chemical industry require the control of temperatures of reactions where heat is liberated. Direct expansion refrigerant coils may constitute a hazard, and such heat exchangers generally use chilled water or brine. Coolers of this sort will be found in every branch of the chemical industry.

Since continuity of the process and safety are prime considerations, plant security will require duplication of all items of apparatus so that a temporary shut-down for repair or maintenance will not reduce the cooling capacity.

CHAPTER

FOURTEEN

COLD STORAGE

14-1 GENERAL

The preservation of foodstuffs and other perishable products is prolonged by lowering the temperature, since this slows chemical reactions and biological deterioration.

Cold storage can be grouped as follows:

1. Storage above freezing point
2. Freezing and storage below freezing point

The latter prevents movement of the liquid content of the product. However, it must be borne in mind that foodstuffs contain sugars and salts, and the ultimate freezing point of all parts of the solution may be below −20°C. For this reason, the freezing of a product may be shown as a range of temperatures, the latent heat being given up within this range.

14-2 PRE-STORAGE TREATMENT

Cooling, freezing, and cold storage cannot improve a product and the best that can be achieved is to keep it near to the condition in which it entered storage. Best-quality produce only should be stored, and this should be as fresh as possible on entry. (This general principal must, of course, be interpreted in the light of local conditions and needs. In some countries of the world, preservation in cold stores is essential to prevent wastage, regardless of the quality of the crop.)

All foods must be clean on entry. Some, such as fish, leaf vegetables, and some fruits may be washed and left wet. Fish will tend to dry out and lose its fresh appearance, so it is packed wet or given a sprinkling of ice chips to keep the surface moist.

Other products, especially the meats, must be dry, or bacteria will live on the moisture and make the skin slimy.

Potatoes will start to sprout after a long period in storage. This can be checked by spraying the freshly lifted tubers with a chemical sprout depressant.

Certain fruits, notably grapes and dates, may have some surface contamination or infestation when first picked, and they are fumigated with sulphur dioxide or other gas. They must, of course, then be thoroughly ventilated before going into storage.

Techniques of this processing will be known to the user or can be found in sources from the particular branch of the food industry.[30]

14-3 PRE-COOLING

If warm produce is taken into a cold store, moisture will evaporate from its surface and may condense on the cold produce already there. This will be of no consequence with wet products such as fish and leaf vegetables but cannot be permitted with meat or poultry. For these meats, pre-cooling is carried out in a separate room under controlled conditions so that the product is reduced to near-final storage temperature, the surface remaining dry all the time.[34]

A lot of pre-cooling can be gained by allowing produce to stand in ambient air, especially at night. For example, apples and pears picked in the daytime at 25°C may cool down to 12°C by the following morning, halving the final refrigerated cooling load.

Wet products can be pre-cooled in chilled water, or by the addition of flake ice. Ice is also used with fish and leaf vegetables to help maintain freshness in transit to storage.

14-4 FREEZING

Most products will keep longer and fresher in the frozen condition, and this process is used for those whose sale value will support the extra process cost.

The cells of animal and vegetable products contain a watery solution of salts and sugars. If this solution starts to freeze, surplus water will freeze out until the eutectic mixture is reached (see Sec. 11-2). If freezing is not carried out quickly, these ice crystals will grow and pierce the cell walls; then when the product thaws out the cells will leak and the texture will be spoiled. This is of no great consequence with the meats, whose texture is changed by cooking, but will not be suitable for fresh fruits or vegetables.

As a general rule, any product which will be eaten without cooking, or only a very brief cooking (such as green peas), should be quick-frozen in a blast-freezing tunnel or similar device. Other foodstuffs need not be frozen so quickly, and may be left in a cold room at a suitable temperature until frozen.

Frozen confections (ice-cream and lollipops) rely on speed of freezing to obtain a certain consistency and texture, and they require special treatment (see Sec. 6-4).

Once a product has been frozen, it must never be allowed to warm and then be re-cooled, or partial thawing may take place with slow re-freezing.

14-5 PACKING AND HANDLING

Cold storage packing must contain and protect the product, while allowing the passage of cooling air to keep an even temperature.

Packages generally will be small enough to be lifted by hand if required, and of a suitable shape to be stacked on pallets for mechanical handling by fork-lift trucks. Stacking on a pallet should allow the passage of air between the individual packs.

Fruits and vegetables which give off heat of respiration need to have perforated cases so that air may pass through the product.

Carcase meat does not lend itself to regular packages and, in any case, needs to be out of contact with other surfaces, including other carcases, or slime may form. These are hung from overhead rails on roller hooks so that they can travel along the rail system (see Fig. 14-1). Special cage pallets are also used for carcase meat.

Potatoes are cold stored in bulk or in large boxes of $\frac{1}{2}$ or 1 t capacity (also in sacks in some countries). They are always stored on their own, so special handling methods have evolved.

Pallets are now mostly standardized at 1.2×1.0 m and the total weight will be between $\frac{1}{2}$ and 1 t, depending on the product. Handling in small cold stores can be by hand pallet trucks or hand-steered electric drive trucks. These can transport but not lift one pallet onto another. The usual fork-lift truck is a ride-on vehicle, electric driven, and can lift to form a stack of two, three, four, or even five pallets high, according to the length of the telescopic lift and the skill of the driver.

Methods of arranging the product in the store will depend on the

Figure 14-1 Meat store with rails at Baxters (Butchers) Ltd. *(Courtesy of SIMON-IWEL Ltd.)*

number of varieties and the storage life (see Fig. 14-2). With short-term storage it may be necessary to get to any pallet, so access gangways will be required with only one row of pallets on each side. The gangway width for a high-lift truck is at least 2.6 m and may be as much as 3.7 m, so some 55 per cent of the available floor space will be gangways.

Some sacrifice of perfect accessibility is usually made in the interest of economy. Where storage life is long, pallets may be stacked as many as four rows deep, requiring one gangway for eight rows of pallets. In this case, a gangway of 3.7 m is required to shuffle pallets to get to those at the rear, and the usable floor space comes up to 68 per cent.

Racking can be installed to support the pallets above floor level, and permit a pallet to be removed without disturbing those above it. Alternatively, *post-pallets* having corner pillars to support upper pallets provide a firm stacking method. The height per pallet is about 1.4 m. (See Fig. 14-3).

Meat pallets, for hanging carcases and sides, will have about the same floor area but will be up to 2 m high, with cage sides.

Where the product is in rigid boxes or cartons, it is possible to stack pallets up to three high without auxiliary support, i.e., one on the other.

Fork-lift drivers need to be skilled, experienced, and safety-conscious, since the misplacement of a pallet on a high stack can present a serious hazard. They work in well-insulated protective clothing and in short spells within a shift if the room is much below 0°C. Extra staff are required if the

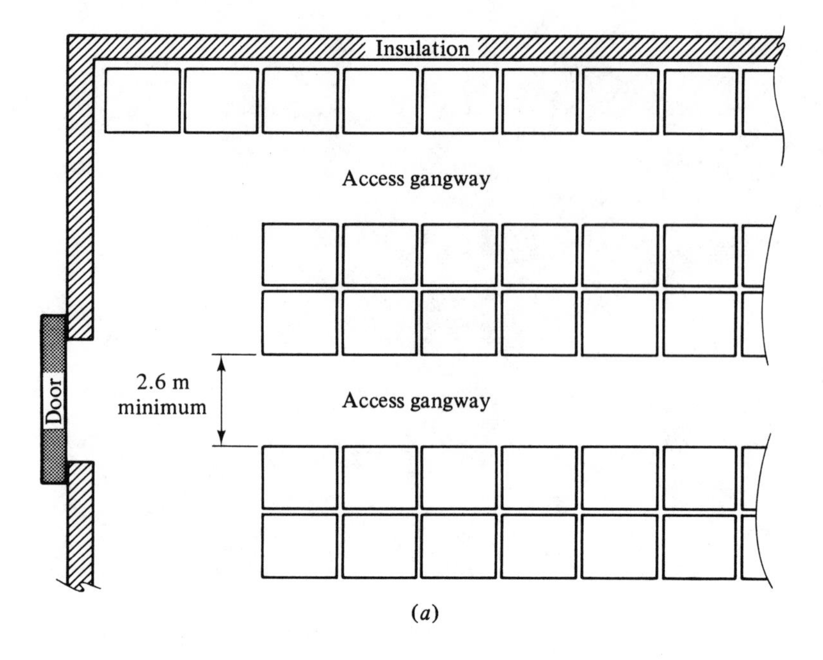

(*a*)

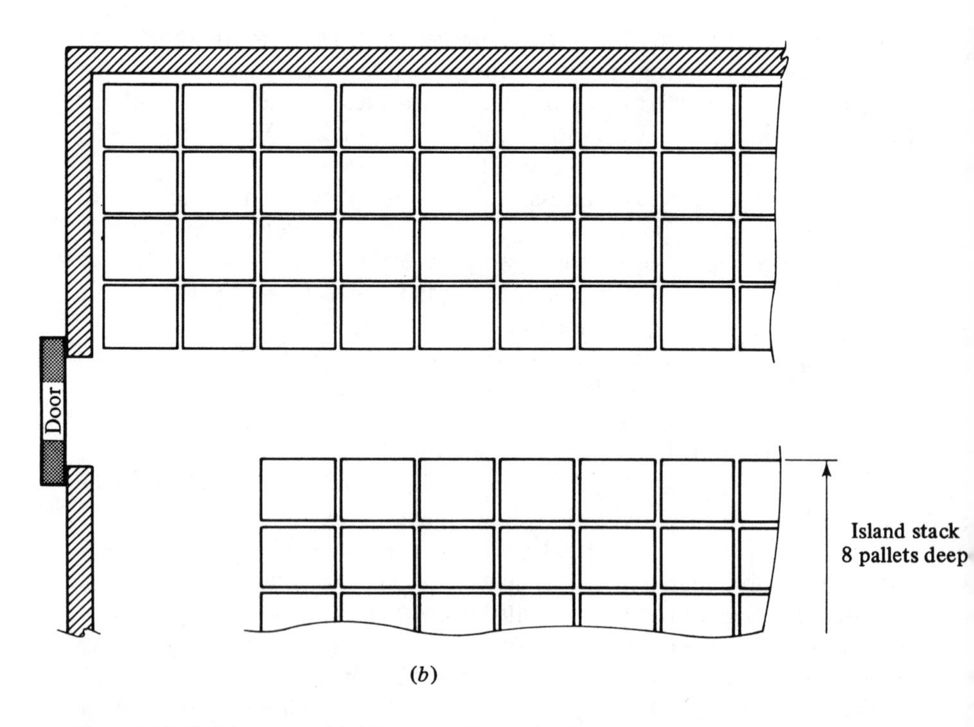

(*b*)

Figure 14-2 Pallet storage: (*a*) full access, (*b*) restricted access.

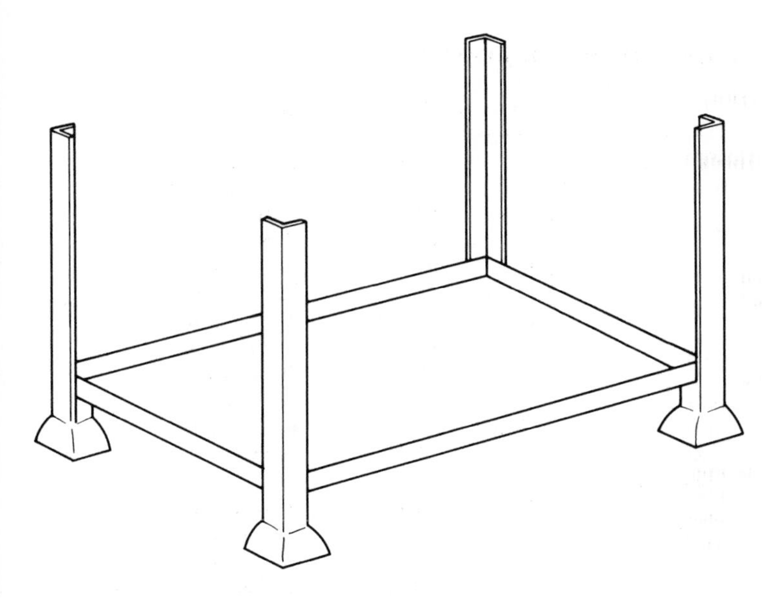

Figure 14-3 Post-pallet.

traffic is constant. Large stores will have a wide loading platform outside at floor level to permit fork-lift trucks to manoeuvre their loads onto vehicles.

Stock control must present a clear picture of the contents and location at any time, to ensure correct rotation of stocks.

14-6 GROUPING OF PRODUCTS

Most cold storage installations will have a wide variety of products to hold, with several different types in each chamber. Apart from obvious separation by storage temperature, some foodstuffs are not compatible with others, e.g.:

1. Wet fish will impart its smell to butter, cheese, eggs, and fresh meat.
2. Citrus fruits such as lemons do the same.
3. Onions must not be stored too long at the high humidity of other vegetables, or they will rot.
4. Frozen meat for medium-term storage at −10°C is the only product kept at this temperature.
5. Frozen foods below −20°C may be mixed with impunity.

14-7 STORAGE CONDITIONS

Table 14-1 shows recommended storage conditions for a few basic products. Comprehensive tables will be found in standard works of reference.[30, 35] In the event of a product being encountered which cannot

Table 14-1 Storage conditions for foodstuffs

Product	Temperature	Humidity	Life
Apples	1–4†	85–90	2–8 months
Bananas, green	12–14	90	10–20 days
ripe	14–16	90	5–10 days
Beer, barrel	2–12	65	3–6 months
Cabbage	0–1	95	3–5 weeks
Carrots, young	0–1	95	1–2 months
old	0–1	95	5–8 months
Celery	0–1	95	1–2 months
Cucumber	10–12	90–95	10–14 days
Dairy products, milk	0–1	—	2–4 months
cream	−23–(−28)	—	6–12 months
cheese	1–4	65–70	6–18 months
yoghourt			
Dried fruits	0–1	Low	6 months up
Eggs, shell	−1–0	80–85	5–6 months
Fish, wet	1–2	90–95	5–15 days
Fruit, soft (berries)	0–1	90–95	5–7 days
Grapefruit	10–14	85–90	4–6 weeks
Grapes	0–1	90–95	2–5 months
Lemons, green	14–15	85–90	1–6 months
Lettuce	0–1	90–95	1–2 weeks
Meats, bacon	1–4	85	1–3 months
beef	−1–(+1)	85–90	1–6 weeks
ham, fresh	0–1	85–90	7–14 days
lamb, mutton	0–1	85–90	5–14 days
pork, fresh	0–1	85–90	3–7 days
poultry	−1–0	85–90	1 week
frozen	−12	90–95	2–8 months
frozen	−18	—	4–12 months
Melons	4–10‡	85–90	1–4 weeks
Mushrooms	0	90	1–4 days
Onions	0–1	65–70	1–8 months
Oranges	0–9‡	85–90	3–12 weeks
Pears	−1–(+1)†	90–95	2–6 months
Pineapples	7–10	90	2–4 weeks
Plums	0–1	85–90	2–8 weeks
Potatoes, new crop	10–12	85–95	3–6 weeks
main crop	1–3	90–95	6–10 months
Tomatoes, green	12–15	85–90	3–5 weeks
ripe	10	85–90	8–12 days
Wine, unfortified	8–10	—	Indefinite

† See also Sec. 14-9.

‡ Depending on variety, harvest time, and other factors.

be found in general references, information can often be obtained from agricultural and other research establishments.

It will be noted that fruits and vegetables, with the exception of dried fruit and onions, are stored in high humidity to prevent drying out through the skin. Meats generally must be in drier air, or slime may form with the growth of bacteria.

14-8 POST-STORAGE OPERATIONS

As a general principle, products leaving cold storage for ultimate consumption may be allowed to rise slightly in temperature but, if so, must not be re-cooled. It follows that they should be kept at the storage temperature as long as possible down the chain of delivery. This requires prompt handling and the use of cooled vehicles up to the final retail outlet.

Some products require special treatment, for which provision should be made, e.g.:

1. Frozen meat coming out of long-term storage to be sold chilled must be thawed out under controlled conditions. This is usually carried out by the retail butcher who will hang the carcase in a chill room (−1°C) for two or three days. On a large scale, thawing rooms use warmed air at a temperature below 10°C.
2. Potatoes and onions coming out of storage will collect condensation from the ambient air and must be left to dry or they will rot.

14-9 GAS STORAGE OF APPLES AND PEARS

All fruits respire oxygen and, in doing so, start to decay. If the oxygen concentration can be reduced, the rate of respiration will be slowed and the storage life may be extended. The maintenance of a low partial pressure of oxygen requires a gas-tight structure to prevent diffusion. Such controlled atmosphere stores are carefully constructed and sealed to achieve this, and are generally termed *gas stores*.

The fruits are loaded and the store sealed. Within a few days they consume a proportion of the available oxygen and respire carbon dioxide. Considerable research over the past 60 years, mainly in the United Kingdom,[36] has determined the correct balance of gases to prolong the storage life of the different varieties of apples and pears, both home grown and imported.

Apparatus is required to monitor the atmosphere within the store and keep the right proportions by chemical removal or controlled ventilation.

14-10 FREEZE DRYING

Certain products cannot be kept in the liquid form for an appreciable time and must be reduced to dry powders, which can then be kept at chill or ambient temperatures. The water must be removed to make them into powders, but any heating above ambient to boil off the water would lead to rapid deterioration. The water must therefore be removed at low temperature, requiring low pressures of the order of 125 Pa.

The process is carried out in a vacuum chamber fitted with refrigerated contact freezing plates, heaters, and a vacuum pump. Between the chamber and the pump may be a refrigerated separator to prevent too much of the moisture entering the pump. The product is placed in containers on the plates and frozen down to about −25°C, depending on the product, but may be as low as −50°C. The vacuum and, at the same time, a carefully controlled amount of heat is then applied, to provide the latent heat of sublimation (ice to vapour) without allowing the temperature to rise. As the water is driven off, the product collapses to a dry powder. This is extremely hygroscopic and must be packed in air-tight containers as quickly as possible on completion of the cycle.

This process was developed for the preservation of antibiotics, but is now in widespread use for other products such as 'instant' coffee, tea, soup, etc.

CHAPTER

FIFTEEN

COLD-STORE CONSTRUCTION

15-1 SIZE AND SHAPE

The purpose of a cold store is to provide an insulated and refrigerated enclosure suitable for the handling and storage of perishable goods, at predetermined temperatures. The net capacity of the cold-store chambers will be estimated on the basis of the quantity and type of goods to be stored, starting with the individual packages and working through to a gross volume.

Example 15-1 What will be the internal dimensions of a cold room to store 1800 t of boxed frozen meat if the box size is 700 × 450 × 150 mm and the net weight 60 kg?

$$\text{Number of boxes in store} = \frac{1\,800\,000}{60} = 30\,000 \text{ boxes}$$

These will be stacked on pallets 1200 × 1000 mm, the height per pallet being 1400 mm, say a box height of 1250 mm, allowing for the base of the pallet. Boxes could be loaded flat, three per layer and eight layers high, making 24 boxes per pallet:

$$\text{Number of pallets} = \frac{30\,000}{24} = 1250 \text{ pallets}$$

These pallets can be stacked three high without auxiliary support, and this will be the cheapest and most flexible arrangement, leaving the floor clear for any other type of produce if there is less than a full load in the store:

$$\text{Floor space required, pallets} = \frac{1250}{3} = 420$$

The net area will be $420 \times 1.2\ \text{m}^2$, plus a space of some 75 mm between pallets to allow clearance in handling, giving an occupied floor area of about $575\ \text{m}^2$.

Since it is a one-product store, it seems unlikely that any one pallet must be accessible, so stacking the pallets in rows three deep should be acceptable.

One or two sketch layouts can be tried based on these figures, unless the shape of the store is already decided by the available site. One such layout is shown in Fig. 15-1. It will hold a maximum of 1278 pallets and requires a floor area 34.5 m long by 27 m wide, a gross floor area of $932\ \text{m}^2$. The stacks will be 4.05 m high. To this must be added a clearance for lifting and the depth of the evaporators. If the latter are 1.1 m, the store internal height will be 5.25 m. The volume is $4890\ \text{m}^3$ and the storage density is $2.72\ \text{m}^3/\text{t}$ or $368\ \text{kg/m}^3$.

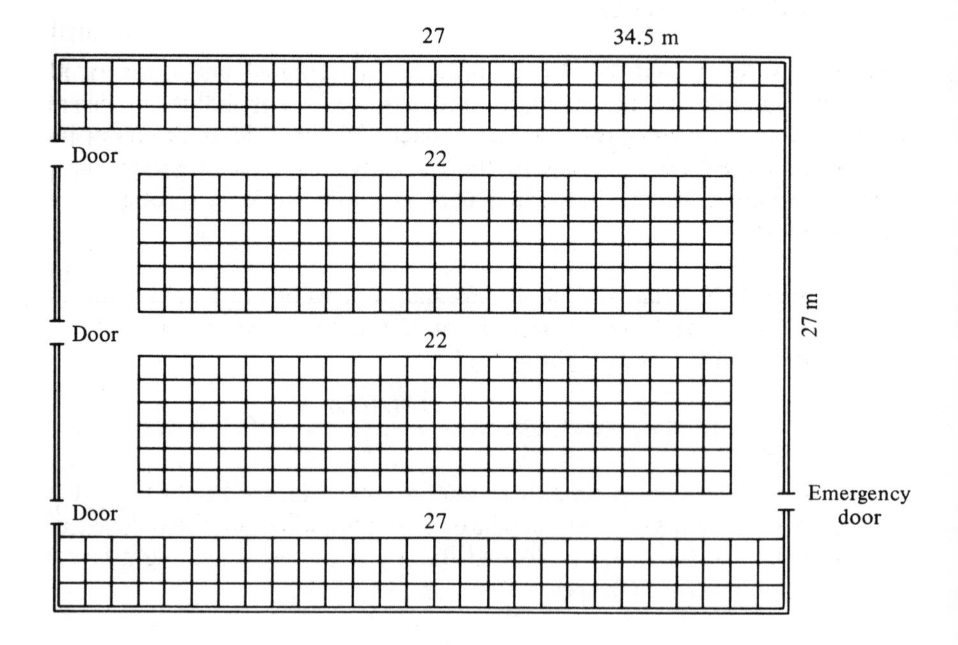

Figure 15-1 Schematic layout of 1800-t pallet store.

Example 15-2 What would be the volume of this store if the pallets were arranged in the same way, but on racks?

The solution to this would require constructional details of the proposed racking system. If the vertical struts are 100 mm, and 100 mm width clearance is required for positioning, each pallet now requires some 1.400 × 1.075 m. Height clearance must now be allowed for each tier—a total of 1.600 m per pallet.

The internal dimensions will now be approximately as follows:

$$\text{Length, } 27 \times 1.4 = 38 \text{ m}$$

$$\text{Width as before} = 27 \text{ m}$$

$$\text{Floor area} = 1\,026 \text{ m}^2$$

$$\text{Height, } 3 \times 1.6 + 1.2 = 6 \text{ m}$$

$$\text{Volume} = 6\,156 \text{ m}^3$$

$$\text{Density} = 292 \text{ kg/m}^3$$

A store of the same capacity, designed to give immediate access to any one pallet, would have a floor area of about 1460 m^2 and a volume of 8760 m^3 with a storage density of 205 kg/m^3.

Cold stores intended only for carcase meat will have the product hung from an overhead rail system (see Fig. 15-2). The meat is hung on hooks

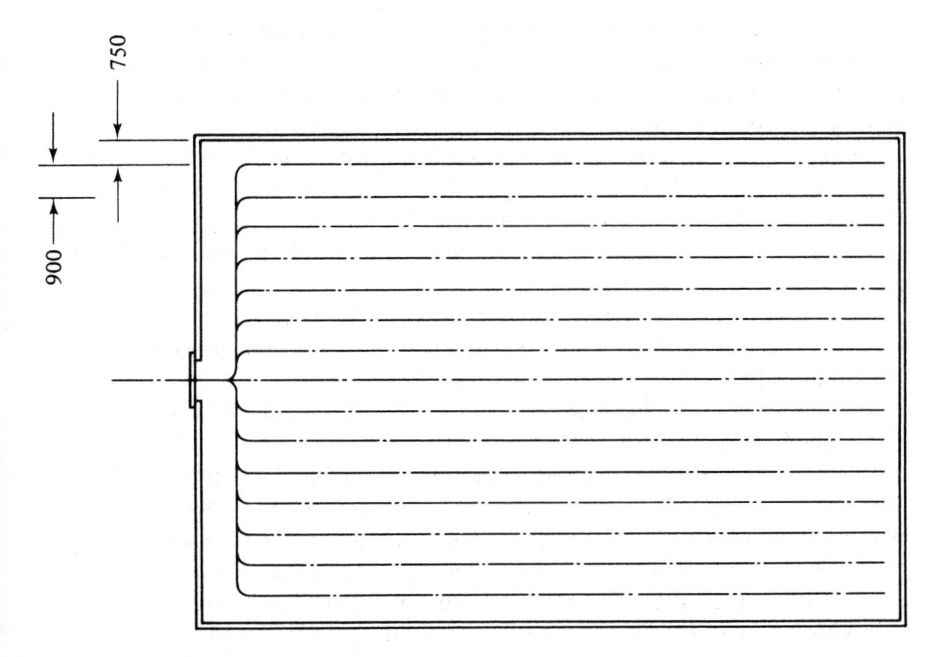

Figure 15-2 Schematic plan of rail layout.

on roller carriers—possibly one side of beef on a single hook or smaller carcases and cuts on multiple-hook carriers. Rails for beef sides will be spaced 900 to 1200 mm apart, and the length for a side is 450 mm of rail. The height to the top of the rail may be up to 3.35 m, but may be less. The average side weight in the United Kingdom is 140 kg, with a variation of 80 to 225 kg throughout the world; the local trend should be taken as a guide. Space is required for the service rails and junctions at the door end, possibly 1 m being taken up by this. Rails should be well clear of walls so that air can circulate and the meat cannot touch the wall.

Example 15-3 What will be the floor area of a railed cold room to take 500 head of beef in sides?

$$\text{Number of sides} = 500 \times 2 = 1000$$
$$\text{Rail length required} = 1000 \times 0.450 \text{ m} = 450 \text{ m}$$

A possible arrangement would be 15 rails, each 30 m long.

$$\text{Width, say 1 m spacing, } 15 \times 1 = 15 \text{ m}$$
$$\text{Extra wall clearance on each long side} = 0.5 \text{ m}$$
$$\text{Overall} \quad 15.5 \text{ m}$$
$$\text{Length, } 30 \text{ m} + \text{clearance for service rails} = 31 \text{ m}$$
$$\text{Giving a floor area of } 31 \times 15.5 = 480 \text{ m}^2$$

A room as large as this may require an access door at each end, with rail points and an exit rail, adding another 1 m to the length. This would be decided by the amount of traffic in and out and the direction, if part of a factory production line.

Small stores for miscellaneous products must allow for random stacking of a variety of different packages. The storage density will be between 150 and 450 kg/m^3 when the store is fully loaded, allowing for access passages. A closer estimate could be made if probable loadings and package shapes were known.[30]

More than one chamber may be required, based on the separation of products by type and by storage temperature. These will be sized on the probable individual contents. Where a low-temperature room is required as well as some at higher temperatures, it should be placed between them, to reduce heat gains.

Old stores, especially those occupying valuable land in city centres, were built several stories high. Access was by lifts and most of the handling was by hand or with hand pallet trucks. Such stores are occasionally built today, but the general use of mechanical handling has led to the single-storey building close by a main road but away from a city centre.

The majority of transport of frozen foods is by 12 m long articulated trailers. Access, turning, docking, and parking space is needed for such vehicles and the loading dock should be at the tailboard height, with adjustable ramps to allow for small differences in this. The loading platform usually runs across the full side or end of the store with doors opening onto it. The absolute minimum width is 3 m and many docks are as wide as 12 m. The check-in office will be on the dock and may have a weighbridge or rail scale for carcases. The refrigeration machine room should have separate access.

15-2 INSULATION

The purpose of insulation is to reduce heat transfer from the warmer ambient to the store interior. Many different materials have been used for this purpose but most construction is now with the following:

1. Cork, a natural material—the bark of the Mediterranean cork oak tree. It is largely air cells and the fibrous cell walls have a high resin content. When baked, the resin softens and welds the pieces of bark into a comparatively homogeneous mass, which is sliced into blocks, commonly 50, 75, and 100 mm thick.
2. Expanded polystyrene. The plastic is formed into beads containing an expanding agent. When placed in a mould and heated they swell and stick together.
3. Foamed polyurethane. The basic chemicals are mixed in the liquid state with foaming agents, and swell into a low-density foam which sets by polymerization into a flexible or rigid mass. The rigid board is used for cold-store insulation. It has strong adhesive properties and is mostly used as the core of a sandwich construction.

The value of an insulation to reduce heat flow is expressed as resistivity or its reciprocal conductivity. The units of the latter are watt metres per square metre kelvin or watts per metre kelvin (W m/m^2 K or W/m K). Values for these materials used are approximately as follows:

Corkboard	0.04 W/m K
Expanded polystyrene	0.034 W/m K
Foamed polyurethane	0.026 W/m K

Example 15-4 What is the heat conduction through a panel of foamed polyurethane 125 mm thick, 46.75 m long and 6 m high if the inside temperature is −25°C and the ambient 27°C?

$$\text{Area} = 46.75 \times 6 = 280.5 \text{ m}^2$$

$$\Delta T = 27 - (-25) = 52 \text{ K}$$

$$Q = 280.5 \times 52 \times \frac{1}{0.125} \times 0.026 = 3034 \text{ W}$$

This assumes that a wall of that size could be made of an unbroken sheet of the insulant. Since there will be some structural breaks, an allowance of some 5 per cent should be added, making the leakage 3.2 kW.

Insulation thicknesses used are 50, 75, 100, 125, and 150 mm, but they can be obtained in non-standard thicknesses for special applications. A general rule to determine the possible temperature difference for different thicknesses gives the following:

	Corkboard	Expanded polystyrene	Foamed polyurethane
50 mm	11 K	13 K	17 K
75 mm	16 K	19 K	25 K
100 mm	22 K	25 K	33 K
125 mm	27 K	32 K	42 K
150 mm	32 K	38 K upwards	50 K upwards
200 mm	43 K upwards		

In most cases, the insulation will be the greatest resistance to heat flow and other materials in the construction, and surface resistances are ignored in estimating heat gains through cold-store walls, ceilings, and floors.

Conductivity figures for other materials will be found in standard references.[2]

15-3 VAPOUR SEALS

When the evaporator begins to cool a cold store, surplus moisture in the air in the room will condense on the coil and, if cold enough, will freeze. This will continue until the water vapour pressure inside the room approaches the saturation pressure at the coil fin temperature, e.g., with a coil temperature of −20°C the vapour pressure would be 0.001 bar. Since this is lower than the vapour pressure of the ambient air, water vapour will try to diffuse from the hot side to the cold, through the wall (see Fig. 15-3). At the same time, heat is passing through the wall, and the temperature at any point within the insulation will be proportional to the distance through it.

At some point through the wall, the temperature will be equal to the saturation temperature of any water vapour passing through it, and this vapour will condense into liquid water within the insulation. This process will continue and the water travel inwards until it reaches that part of the insulation where the temperature is 0°C, where it will freeze. The effect of water is to fill the air spaces in the material and increase its conductivity. Ice, if formed, will expand and split the insulant.

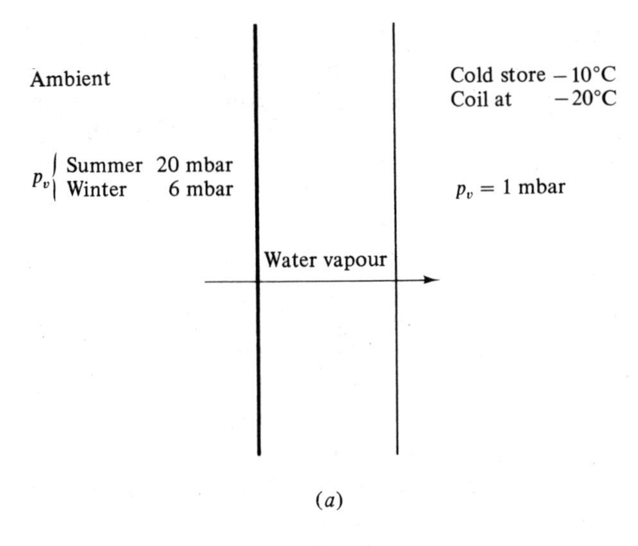

(*a*)

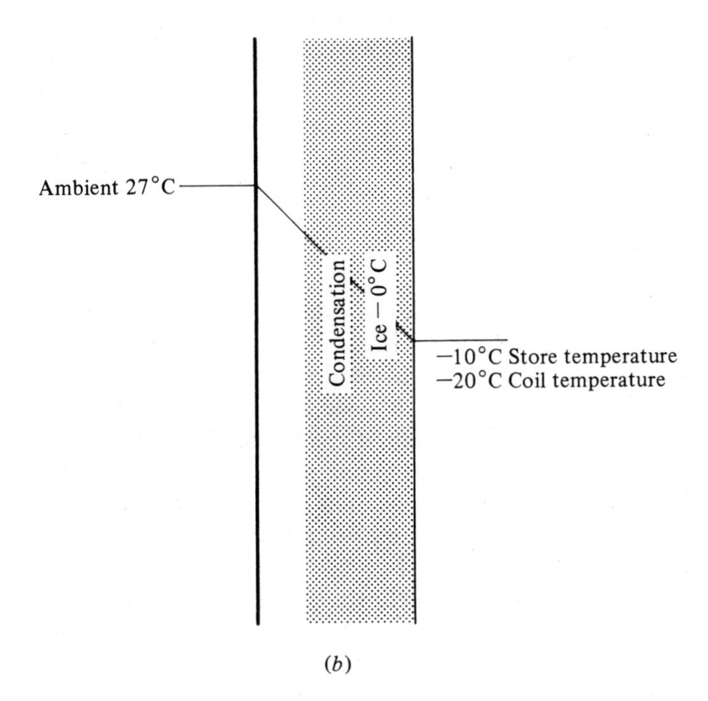

(*b*)

Figure 15-3 Section through coldroom insulation: (*a*) vapour diffusion, (*b*) thermal gradient.

To prevent this deterioration of the insulation, a vapour seal is required across the warm face. This seal must be continuous and offer the best possible barrier to the transmission of water vapour. The traditional vapour barrier was bituminous emulsion or hot bitumen, applied in two or more layers. More recent materials are heavy-gauge polythene sheet, metal foil, and metal sheet. It is sometimes thought that the plastic insulants, since they do not easily absorb moisture, are vapour barriers. This is not so, and no reliance should be placed on the small resistance to vapour transmission which they may have.

Any small amount of vapour which might enter through faults in the vapour seal should be encouraged to pass through the inner (cold side) skin of the structure to the coil, rather than be trapped within the insulation. It follows that, if the vapour seal is at all suspect, the inner wall coating should be more porous. In traditional construction, this was provided by an inner lining of cement plaster or asbestos cement sheet, both of which transmit vapour. The modern use of impervious materials on both skins required meticulous attention to the sealing of any joints.

Any conductive material, such as masonry and metal structural members or refrigerant pipes, which must pass through the insulation will conduct heat, and the outer part may become cold enough to collect condensation and ice. Such heat bridges must be insulated for some distance, either inside or outside the main skin, to prevent this happening. If outside, the vapour seal must, of course, be continuous with the main skin vapour seal.

15-4 SECTIONAL COLDROOMS

Small cold rooms can be made as a series of interlocking and fitting sections, for assembly on site on a flat floor (see Fig. 15-4). Standard ranges are made up to about 70 m^3, but larger stores can be made on this principle. The floor section(s) is placed on a flat floor and the sides erected on this, located, sealed, and pulled up together. The roof sections then bridge across the walls. Such packages are supplied complete with all fittings. They can be dismantled and moved to another location if required. Specialist site work is restricted to cutting necessary holes for pipework and fitting the cooling equipment.

Stores of this size can be built, using standard size factory-made sandwich panels—all cutting, fitting, jointing, and sealing being carried out on site. This form of construction is prone to fitting errors, with subsequent failure of the insulation, if not carried out by skilled and experienced craftsmen.

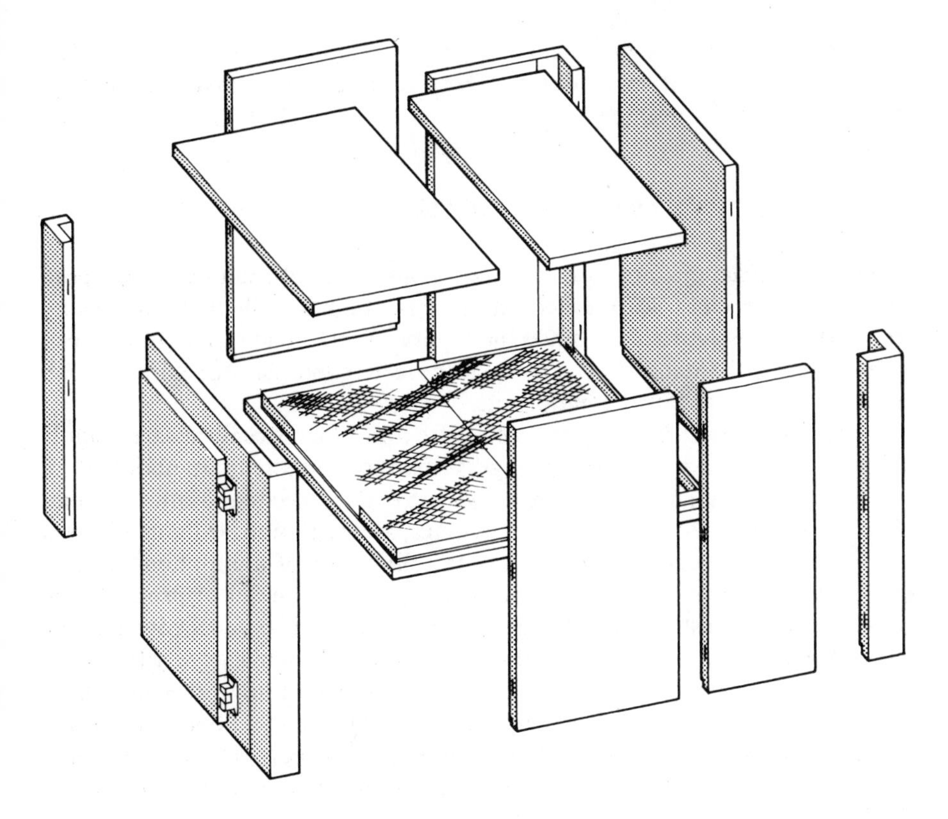

Figure 15-4 Assembly of section coldroom. [*Courtesy Hemsec (Construction) Limited.*]

15-5 INBUILT CONSTRUCTION

Traditional cold-store construction was to build an insulated lining within a masonry shell. The outer skin would be erected in brick and concrete, and rendered as smooth as possible inside with cement plaster, to take the insulation. When the surface was dry, it would have several coats of bitumen applied as a vapour seal and slabs of insulation material stuck to this with hot bitumen. This was normally carried out in two or more layers so that joints did not pass right through the insulant, but were staggered. The inner skin would be finished with cement plaster, reinforced with wire mesh. The usual insulant was slab cork.

Any columns passing through cold rooms would be insulated, at least partially, to reduce conduction along the heat bridge and the build-up of condensation and ice. Floors would have a layer of hard concrete on the floor insulation. Ceilings were stuck to a concrete ceiling or fixed to a false timber ceiling.

This form of construction is seen to be quite sound, and there are still many such stores in service which were built 50 and more years ago. The method is still used in countries where cork is cheap and craft labour available at an economic price.

15-6 FACTORY PANEL SYSTEMS

The plastic insulants are rigid, homogeneous materials, suitable as the core of sandwich panels. Such a method of fabrication is facilitated when using foamed rigid polyurethane, since the liquids can be made to foam between the inner and outer panel skins and have a good natural adhesion, so making a stiff structural component.

Panels made in this way for cold-store and other structures are usually 1.2 m wide and can be made in lengths of up to a maximum of about 14 m. The manufacture incorporates interlocking edging pieces and other fittings. Such panels are used for walls and ceilings, although not for floors above a certain store size. The inner and outer skins are of aluminium or rust-proofed steel sheet, usually finished white, and may be flat or profiled. The edge seals are plastic extrusions or similar material. The panel edge locking devices may be built in or applied on site. To build such a store, the floor is first prepared (see Sec. 15-5), bringing the vapour seal up at the outer face. Wall sections are erected on end on the edge of the floor and locked together, making the interpanel seal at edges and corners. Ceiling panels are fitted over the tops of the walls and sealed at the warm face of the junction.

Since the panels must be rigid enough to support their own weight, thickness cannot be reduced below a minimum, and this is usually 100 mm, although less insulation might suffice for the purpose. For a large store, panels will be 125 or possibly 150 mm thick.

It is not customary to rely on the outer surface of the insulated panel as the weather shell of the building, and such panels are erected within a protective frame building. Advantage is taken of this outer structure to steady long vertical panels. Ceiling panels with more than a 6 m span are supported by suspension wires or rods from the roof framework of the outer structure (see Fig. 15-5).

15-7 FLOORS

Heavy floor loadings and the use of ride-on electric trucks demand a strong, hard-working floor surface, which must be within the insulation envelope.

Floor construction starts with a firm foundation slab about 200 to 250 mm below the final floor level. This is covered with the vapour seal,

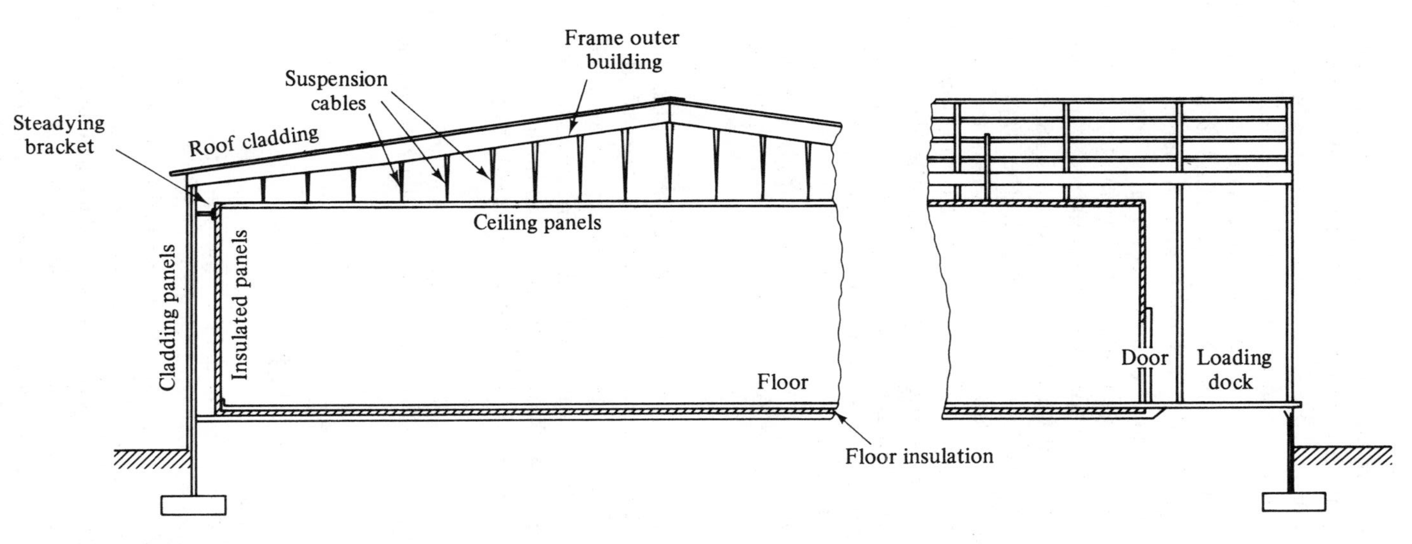

Figure 15-5 Panel construction.

probably of overlapping layers of heavy-gauge polythene sheet. On this is placed the insulation board in two layers with staggered joints, which is fitted as tightly as possible. The upper joints may be covered with strips of plastic to prevent concrete running in, but a continuous layer of vapour-tight sheet must not be used on this cold side of the insulation. The concrete floor is made with granite aggregate, laid to the final level, as dry as possible, reinforced with steel mesh and in panels not more than 10 m square, to allow for contraction on cooling. Where fork-lift trucks are in use, it is best to lay these panels with no gap, to minimize cracking of the edges under load. If the floor will be wet in use, a finite gap is left, and filled with mastic to prevent water getting into the insulation.

If floors are laid on wet ground, the vapour pressure gradient (Fig. 15-3) will force water vapour up towards the vapour seal. Given a ground temperature of 13°C in the United Kingdom, the underside slab may become as cold as 0°C after many months of store operation, and any moisture condensed under the floor insulation will freeze and, in freezing, expand. In time this layer of ice under the floor slab, unable to expand downwards, will lift the floor (*frost-heave*).

The prevention of frost-heave is to supply low-intensity heat to the underside of the insulation, to keep it above freezing point. This may take several forms:

1. Low-voltage electric resistance heater tapes placed under the vapour seal during construction, thermostatically controlled.
2. Pipes buried in the base slab, carrying warm condenser water or a non-freeze solution heated from the condenser system.
3. Air vent pipes to allow a current of ambient air through the ground under the base slab. This is not very suitable in cold climates.

On very damp or uncertain ground, it might be advisable to raise the cold store completely off the soil, giving full air circulation below.

15-8 DOORS AND SAFETY EXITS

Cold-store doors must combine the functions of door and insulation. Small doors will be hinged and have an arrangement of double gaskets to reduce the transmission of convected heat (air leakage) and consequent ice accumulations at the door edges. Such doors are normally wood-framed to reduce conduction, but may now have plastic moulded frames. Insulation is by one of the foam plastics, and the face panels are sheet metal or g.r.p. In order to keep the seals in good alignment throughout the life of the door, hinges will be made adjustable. The closing latch will have a cam or lever action to compress the large gasket area and give a tight seal.

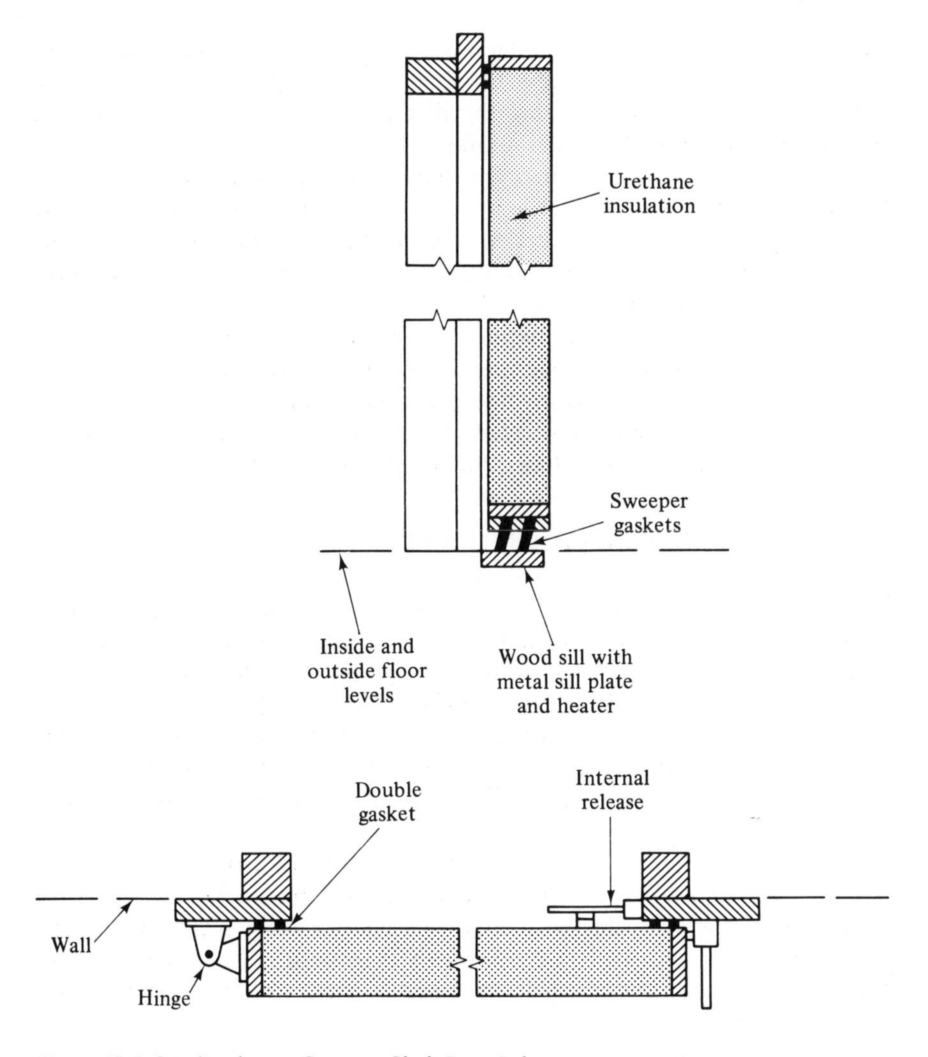

Figure 15-6 Overlap door. *(Courtesy Clark Door Ltd.)*

Where a flush door sill is required, the gaskets on the lower edge will be in the form of two or three flexible blades which just brush the floor.

A simpler and more adaptable method of sealing is a face-fitting or *overlap* door (Fig. 15-6). The door itself overlaps the opening by some 150 mm all round, and two or three soft gaskets seal the overlapping surfaces. This type of door is general in rooms operating below 0°C, and may have warming tapes embedded in the wall face to prevent freezing of any vapour which penetrates. The smaller sizes, and the rebated doors, are hand operated. Larger doors, especially those to take fork-lift trucks, must be mechanically operated for speed and convenience, and because the doors should never be left open too long. For most purposes, horizontal sliding doors are used, closing onto face gaskets in the same way as the overlap doors. The slide system is generally arranged so that the door moves out from the wall during the first part of its travel, so as to free the gaskets and make for easier sliding.

Various electric and pneumatic mechanisms are used, and the switches for opening and closing are controlled by toggle ropes hanging down where the fork-lift driver can reach them without dismounting. Protection posts each side reduce the risk of damage to the door frame or wall if the truck collides with them.

All mechanical doors are required by law to be capable of hand operation in the event of power failure, and doors of all types must have fastenings which can be opened from either side in case an operator is shut in the store. Larger rooms must have an escape door or breakout hatch or panel at the end remote from the doors, for use in an emergency.

15-9 INTERIOR FINISH AND FITTINGS

The interior surface finish, to comply with EEC and other health standards, must be rustproof, cleanable, and free from any crevices which can hold dirt. Bare timber in any form is not permitted. Most liners are now aluminium or galvanized steel sheet, finished white with a synthetic enamel or plastic coating. G.r.p. liners are also in use. Floors are of hard concrete or tiles. Very heavy working floors may have metal grids let into the concrete surface. Floor concrete is coved up at the base of the walls to form a protective curb.

In the past, timber dunnage battens were fixed around the walls to protect the surface from collision damage and ensure an air space for circulation of the air from the evaporators. Since timber is no longer used, dunnage may be provided in the form of metal rails. The provision of the floor curb at the walls will ensure that pallets cannot be stacked to prevent air circulation.

Lighting is normally by fluorescent tubes, fixed to the ceiling and having starters which will work in low temperatures. The design of efficient lighting systems merits close attention, since all energy put into the store for lighting must be removed again. Control switches are of the waterproof type and may be outside, but some emergency inside light must be provided, both mains and automatic battery standby operated, to give minimum illumination inside the store at all times.

15-10 EVAPORATORS

In small cold stores, the coolers will be fixed to the walls, probably blowing the air downwards or, to the ceiling, blowing sideways (see Fig. 6-2).

Larger evaporators (see Fig. 15-7) will also be mounted at high level if

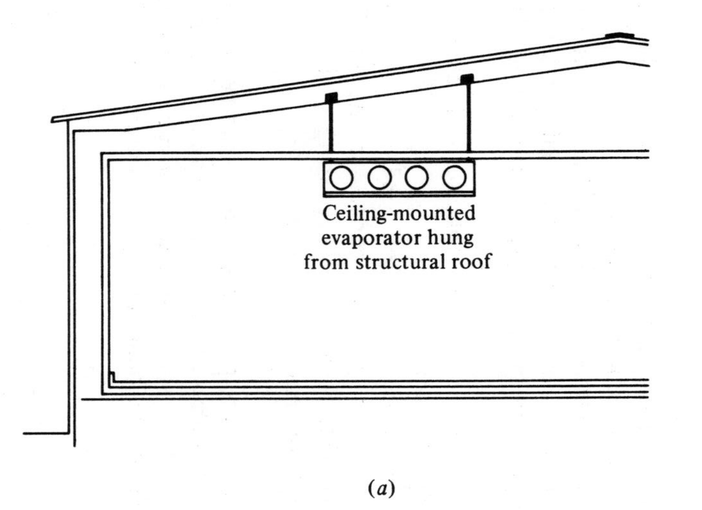

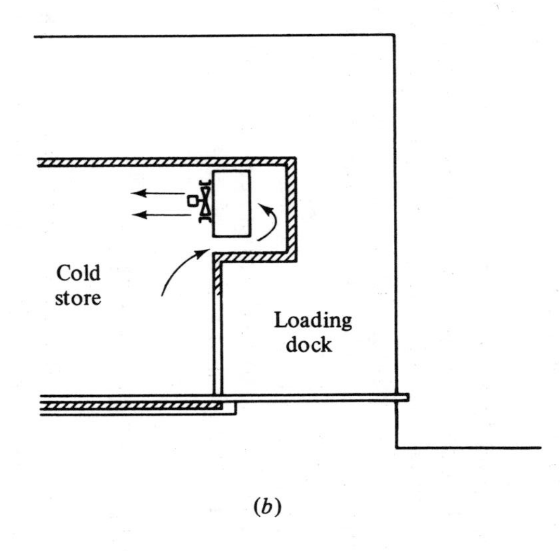

Figure 15-7 Coldroom evaporators: (*a*) ceiling hung, (*b*) above loading bay.

possible, to save useful floor space. Owing to the weight, they must be supported from the outer structural roof by tie-rods passing through the insulation. Access gangways are needed in the roof void to facilitate maintenance and inspection of piping, valves, and insulation. Some stores have the coolers mounted in a recess above the loading bay, providing a maintenance platform. This can only be done where the fans can cover the full length or width of the chamber.

Ducting is sometimes used to distribute cold air from the evaporators, but is very prone to damage and is then difficult to repair.

15-11 SECURITY OF OPERATION

The value of the produce in a large cold store may be several times the cost of the store itself, and every effort should be made to maintain the refrigeration service at all times, even if plant may be inoperative for inspection, overhaul, or repair. The principle of plant security is that there should be sufficient pieces of each item of plant and that they should have enough capacity for conditions to be held as required by the produce, regardless of any item which might be stopped.[24]

Usual arrangements can be summarized as follows:

1. At least two compressors, either of which can keep the store at temperature. It may run continuously to hold this.
2. Two condensers, or a condenser assembly having two separate refrigerant circuits and permitting repair to one circuit while the other is working. If there is one assembly with forced convection, there are at least two fans.
3. All circulating pumps to be in duplicate, with changeover valves to permit immediate operation.
4. At least two evaporators, to maintain conditions if one is not working.
5. Where two compressors and two condensers are installed as independent circuits, provide changeover valves so that either compressor can work with either condenser or evaporator.

Before installation, the planned system should be analysed in terms of possible component failures to ensure that it can operate as required. Commissioning running tests should include simulated trials of plant failure, and operatives should be made aware of failure drills to keep the plant running.

CHAPTER

SIXTEEN

AIR AND WATER VAPOUR MIXTURES

16-1 GENERAL

The atmosphere consists of a mixture of *dry air* and *water vapour*. Air is itself a mixture of several elemental gases, mainly oxygen and nitrogen, but the proportions of these are consistent throughout the atmosphere and it is convenient to consider air as one gas. This has a *molecular mass* of 28.97 and the standard atmospheric pressure is 1 013.25 bar or 101 325 Pa.

Water may be present in air in the liquid form, as rain or mist, or as a solid (snow, hail). However, in general ambient and indoor conditions the water present in the air will be in the vapour form, i.e., as superheated low-pressure steam.

16-2 CALCULATION OF PROPERTIES

If air and water are present together in a confined space, a balance condition will be reached where the air has become saturated with water vapour. If the temperature of the mixture is known, then the pressure of the water vapour will be the pressure of steam at this temperature (see also Sec. 1-3), viz.

Temperature, °C	Vapour pressure, mbar
0	6.10
10	12.27
15	17.04
20	23.37
25	31.66

Dalton's Law of partial pressures (see also Sec. 1-7) states that the total pressure of a mixture of gases is equal to the sum of the individual pressures of the constituent gases, taken at the same temperature and occupying the same volume. Since the water vapour pressure over water will remain constant, depending on temperature and not on volume, this pressure can be obtained from steam tables as above. The partial pressure exerted by the dry air must therefore be the remainder.

Thus, for an air–water vapour mixture at 25°C:

$$\begin{aligned} \text{Total (standard) pressure} &= 1\,013.25 \text{ mbar} \\ \text{Partial pressure of saturated vapour} &= \underline{\quad 31.66} \text{ mbar} \\ \text{Partial pressure of dry air} &= 971.59 \text{ mbar} \end{aligned}$$

This calculation of the proportions by partial presssure can be converted to proportions by weight, by multiplying each pressure by the molecular mass (Avogadro's hypothesis), to give:

$$\begin{aligned} \text{Proportion by mass of water} &= 31.66 \times 18.016 = 570.4 \\ \text{Proportion by mass of dry air} &= 971.59 \times 28.97 = 28\,146 \end{aligned}$$

$$\text{Proportion by weight of } \frac{\text{water}}{\text{dry air}} = \frac{570.4}{28\,146} = 0.020\,3 \text{ kg/kg}$$

Since neither dry air or water vapour is a perfect gas, there will be a slight difference between published tables (0.020 16) and this simplified calculation.

The *specific enthalpy* (or total heat) of the mixture can be taken from 0 K (−273.15°C) or from any convenient arbitrary zero. Since most air-conditioning processes take place above the freezing point of water, and we are concerned mostly with differences rather than absolute values, this is commonly taken as 0°C, dry air. For conditions of 25°C, saturated, the specific enthalpy of the mixture, per kilogram of dry air, is

$$\begin{aligned} \text{Sensible heat of dry air} &= 1.006 \times 25 &&= 25.15 \text{ kJ/kg} \\ \text{Sensible heat of water} &= 0.020\,16 \times 25 \times 4.187 &&= 2.11 \\ \text{Latent heat of water} &= 0.020\,16 \times 2440 &&= \underline{49.19} \\ & \text{Total} && 76.45 \text{ kJ/kg} \end{aligned}$$

(Again, there are some slight variations in these properties within the range considered, and the published figure[4] is 76.49 kJ/kg.)

The *specific volume* of the mixture can be obtained, taking either of the two gases at their respective partial pressures, and using the General Gas Law. Only basic SI values must be used, so the pressures must be expressed in pascals:

$$pV = mRT$$

$$\text{or} \quad V = mRTp^{-1}$$

$$\text{For the dry air } V_a = \frac{1 \times 287 \times (25 + 273.15)}{97\,159}$$

$$= 0.8807 \text{ m}^3$$

$$\text{For the water vapour } V_v = \frac{0.020\,16 \times 461 \times (25 + 273.15)}{3166}$$

$$= 0.8752 \text{ m}^3$$

(The published figure is 0.871 5 m^3/kg dry air.)

16-3 MOISTURE CONTENT, PERCENTAGE SATURATION, AND RELATIVE HUMIDITY

The moisture content in the example at 25°C, saturated, was given in standard tables as 0.020 16 kg/kg dry air. This is also termed its *specific humidity*.

Air will not always be saturated with water vapour in this way, but may contain a lower proportion of this figure, possibly half, or 50 per cent:

$$\frac{0.020\,16}{2} = 0.010\,08 \text{ kg/kg dry air}$$

This lower figure can be expressed as a percentage of the saturation quantity:

$$\text{Percentage saturation} = 100 \times \frac{g}{g_{ss}}$$

$$= 100 \times \frac{0.010\,08}{0.020\,16}$$

$$= 50\% \text{ sat.}$$

Properties for this new mixture can be calculated as above to obtain the specific enthalpy and specific volume.

The proportion of moisture can also be expressed as the ratio of the vapour pressures, and is then termed *relative humidity:*

$$\text{Relative humidity} = 100 \times \frac{p_s}{p_{ss}}$$

$$= 50.80\% \text{ relative humidity}$$

(for the example taken)

Since most air-conditioning calculations are based on weights of air and moisture, *percentage saturation* is usually employed, and moisture content is expressed as kilograms per kilogram of dry air. Much published data is still expressed in the original quantities of Willis H. Carrier, i.e., *grains per pound*, where 1 lb = 7000 grains.

16-4 DEW POINT

Saturated air at 25°C, having a water vapour content of 0.020 16 kg/kg can be shown as a point *A* on a graph of moisture content against temperature (Fig. 16-1). Air which is 50 per cent saturated at this temperature will contain 0.010 08 kg/kg and will appear on this graph as point *B*. If this 50 per cent saturation mixture is slowly cooled, the change of condition

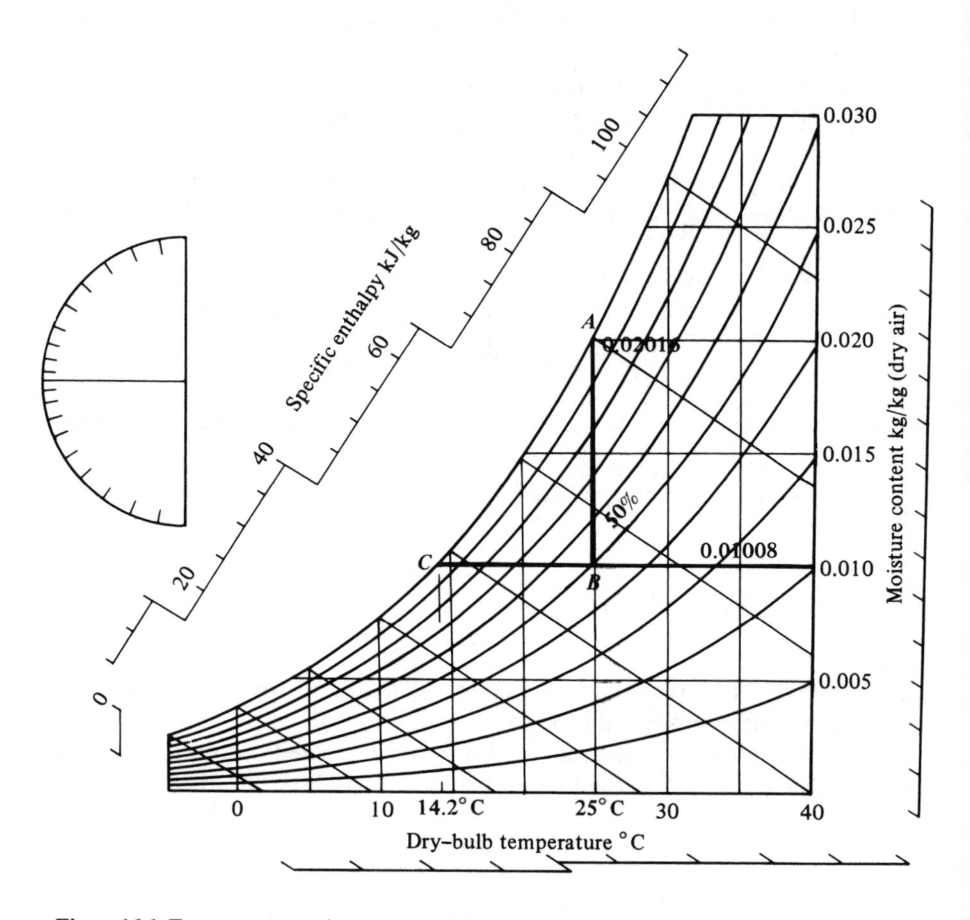

Figure 16-1 Temperature–moisture content graph.

will be along the line *BC*, with constant moisture content but decreasing temperature. It will eventually reach point *C* on the saturation line, where the maximum moisture it can hold is 0.01008 kg/kg (about 14.2°C). It cannot be cooled below this temperature and still hold this proportion of water vapour, so moisture will be precipitated as *dew*. The point *C* for the mixture originally at *B* is termed the *dew point* temperature.

16-5 WET BULB TEMPERATURE

If the percentage saturation of an air sample is less than 100, i.e., it is less than saturated, and it comes into contact with water at the same temperature, there will be a difference in vapour pressures. As a result, some of the water will evaporate. The latent heat required for this change of state will be drawn from the sensible heat of the water, which will be slightly cooled. This drop in the water temperature provides a temperature difference, and a thermal balance will be reached where the flow of sensible heat from the air to the water (Fig. 16-2) provides the latent heat to evaporate a part of it.

The effect can be observed and measured by using two similar thermometers (Fig. 16-3), one of which has its bulb enclosed in a wet wick.

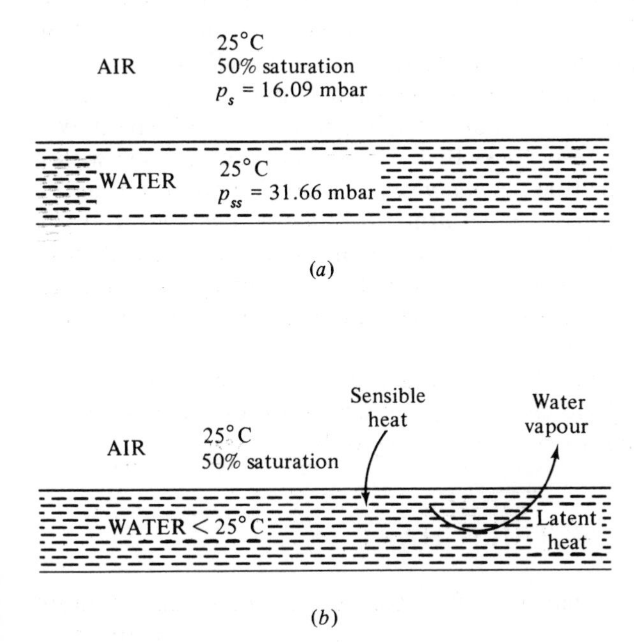

Figure 16-2 Exchange of sensible and latent heat at water–air surface.

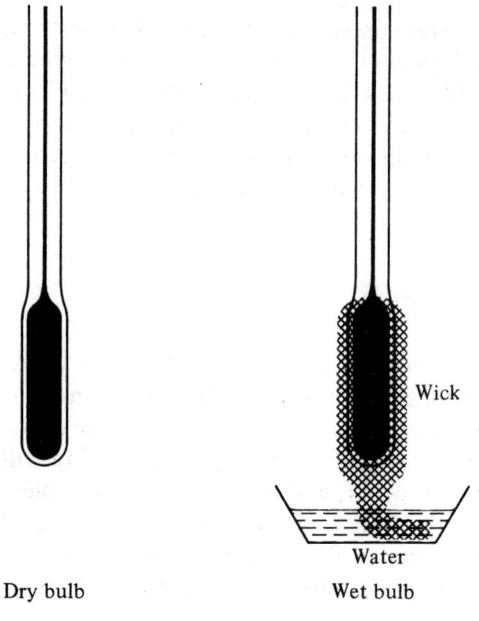

Figure 16-3 Thermometers, *dry bulb* and *wet bulb*.

The drier the air passing over them, the greater will be the rate of evaporation from the wick and the greater the difference between the two readings. In the case of air at 25°C, 50 per cent saturation, the difference will be about 6.5 K. The measurements are termed the *dry bulb* and *wet bulb* temperatures, and the difference the *wet bulb depression*.

In order that consistent conditions can be obtained, the air speed over the thermometers should be not less than 1 m/s. This can be done with a mechanical aspiration fan (the *Assmann psychrometer*) or by rotating the thermometers manually on a radius arm (the *sling psychrometer*). If the thermometers cannot be in a moving air stream, they are shielded from draughts by a perforated screen and rely only on natural convection. In this case the wet bulb depression will be less and the reading is termed the *screen wet bulb* (see Fig. 16-4).

It follows that the drier the air, the greater will be the difference between the dry bulb, wet bulb, and dew point temperatures and, conversely, at 100 per cent saturation these three will coincide.

16-6 THE PSYCHROMETRIC CHART

All the above properties may be tabulated, but can be displayed more effectively in graphical form. The basic properties to be shown are dry bulb temperature, moisture content, and specific enthalpy. Within the limits of

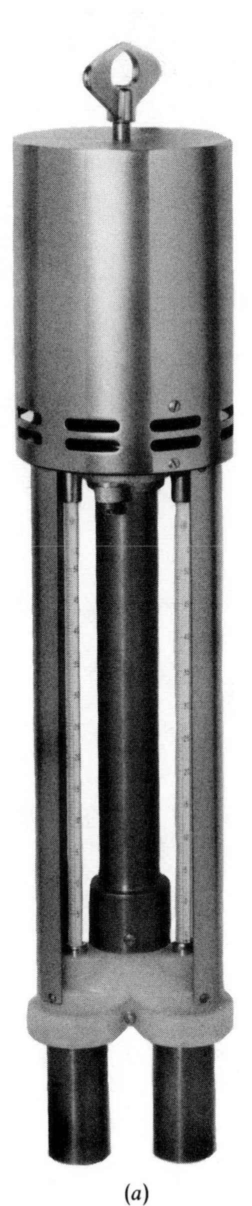

(*a*)

Figure 16-4 Psychrometers. *(Courtesy of C. F. Casella & Co. Ltd.)*
(*a*) Assmann.

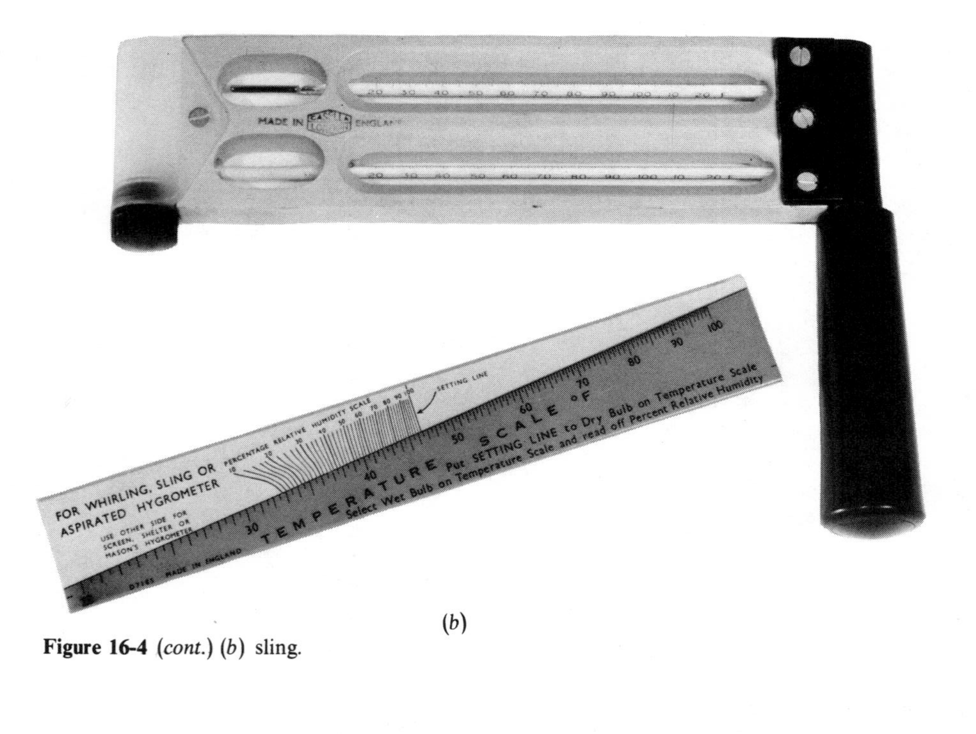

(*b*)

Figure 16-4 (*cont.*) (*b*) sling.

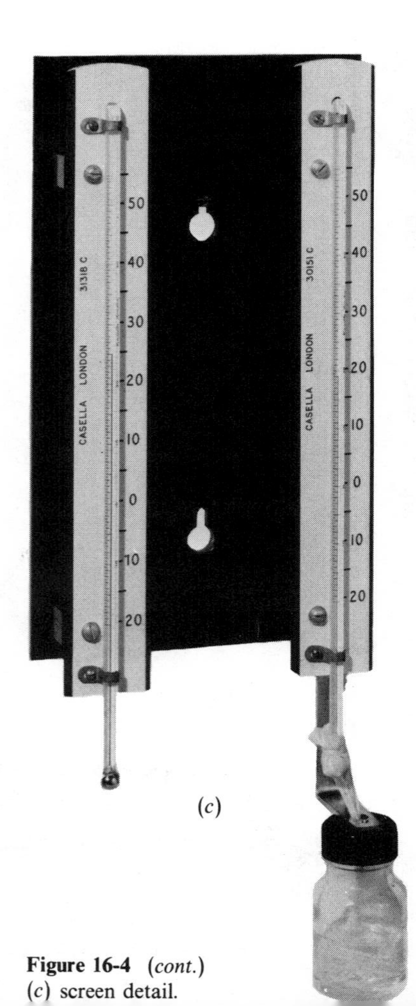

(*c*)

Figure 16-4 (*cont.*)
(*c*) screen detail.

Figure 16-4 (*cont.*) (*d*) screen housing.

the graph required for ordinary air-conditioning processes, the grid lines can be assumed as parallel and form the basis of the *psychrometric chart* (Fig. 16-5). (It will be seen from the full chart, Fig. 16-6, that the dry bulb lines are slightly divergent. The moisture content and enthalpy grids are parallel.)

On this chart, the wet bulb temperatures appear as diagonal lines, coinciding with the dry bulb at the saturation line. The specific enthalpy will increase with dry bulb (sensible heat of the air) and moisture content (sensible and latent heat of the water). The adiabatic (isoenthalpic) lines for an air–water vapour mixture are almost parallel with the wet bulb lines so, to avoid any confusion, the enthalpy scale is placed outside the body of the chart, and readings must be taken using a straight-edge.

A further property which is shown on the psychrometric chart is the specific volume of the mixture, measured in cubic metres per kilogram. This appears as a series of diagonal lines, commonly at intervals of 0.01 m^3.

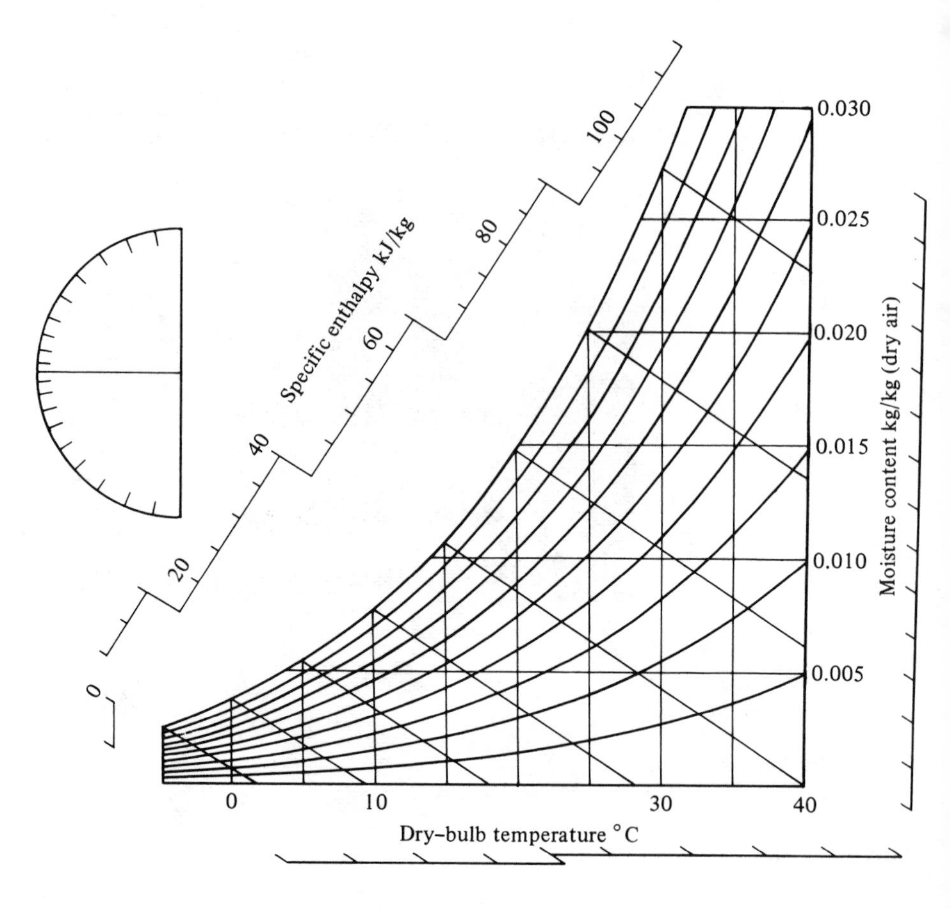

Figure 16-5 Basic CIBS psychrometric chart. *(Courtesy of the Chartered Institution of Building Services.)*

16-7 EFFECTS ON HUMAN COMFORT

The human body takes in chemical energy as food and drink, and oxygen, and consumes these to provide the energy of the metabolism. Some mechanical work may be done, but the greater proportion is liberated as heat, at a rate between 90 W when resting and 440 W when doing heavy work.

A little of this is lost by radiation if the surrounding surfaces are cold and some as sensible heat, by convection from the skin. The remainder is taken up as latent heat of moisture from the respiratory tissues and perspiration from the skin (see Table 16-1). Radiant loss will be very small if the subject is clothed, and is ignored in this table.

Convective heat loss will depend on the area of skin exposed, the air speed, and the temperature difference between the skin and the ambient. As

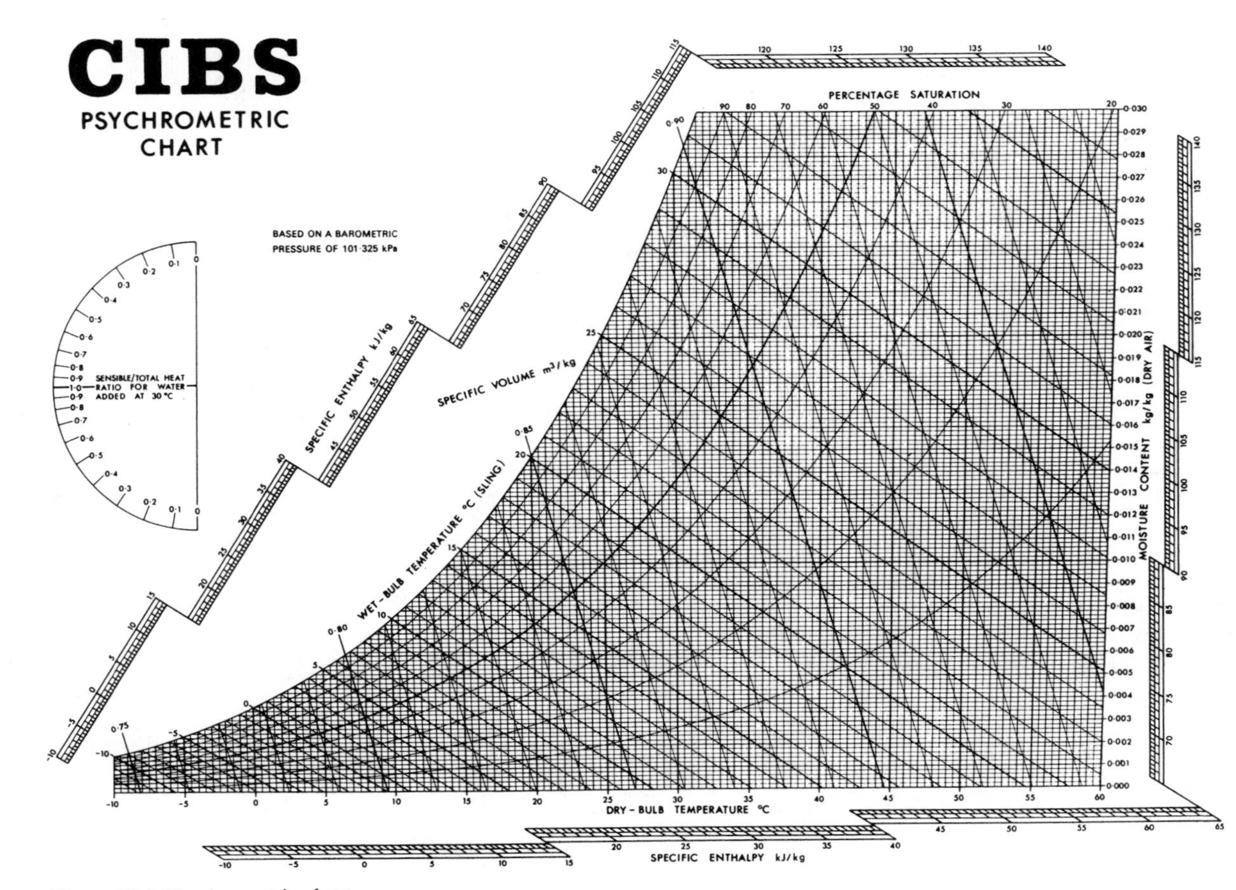

Figure 16-6 Psychrometric chart.

Table 16-1 Heat emission from the human body (adult male, body surface area 2 m²) (*From CIBS Guidebook A*)

Application			Sensible (s) and latent (I) heat emissions, W, at the stated dry bulb temperatures, °C									
			15		20		22		24		26	
Degree of activity	Typical	Total	(s)	(I)	(s)	(I)	(s)	(I)	(s)	(I)	(s)	(I)
Seated at rest	Theatre, hotel lounge	115	100	15	90	25	80	35	75	40	65	50
Light work	Office, restaurant†	140	110	30	100	40	90	50	80	60	70	70
Walking slowly	Store, bank	160	120	40	110	50	100	60	85	75	75	85
Light bench work	Factory	235	150	85	130	105	115	120	100	135	80	155
Medium work	Factory, dance hall	265	160	105	140	125	125	140	105	160	90	175
Heavy work	Factory	440	220	220	190	250	165	275	135	305	105	335

† For restaurants serving hot meals, add 10 W sensible and 10 W latent for food.

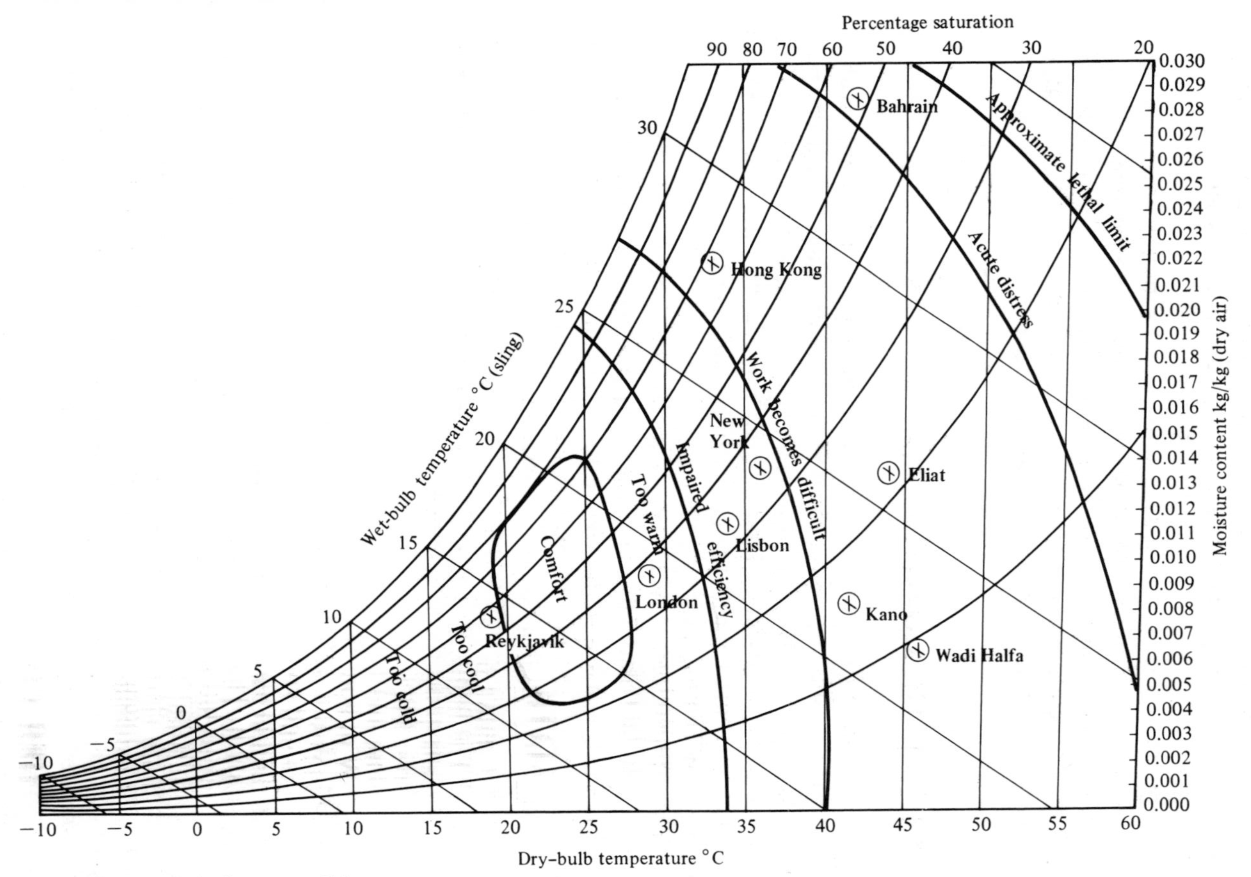

Figure 16-7 Typical climate conditions.

the dry bulb approaches body temperature (36.9°C) the possible convective loss will diminish to zero. At the same time, loss by latent heat must increase to keep the body cooled. This, too, must diminish to zero when the wet bulb reaches 36.9°C.

In practice, the human body can exist in dry bulb temperatures well above blood temperature, providing the wet bulb is low enough to permit evaporation. The limiting factor is therefore one of wet bulb rather than dry bulb temperature and, the closer the upper limits are approached, the less heat can be rejected so the less work is done.

16-8 CLIMATIC CONDITIONS

Figure 16-7 shows the maximum climatic conditions in different areas of the world. The humid tropical zones have high humidities but the dry bulb rarely exceeds 35°C. The deserts have an arid climate, with higher dry bulb temperatures. Approximate limits for human activities are related to the enthalpy lines and indicate the ability of the ambient air to carry away the 90 to 440 W of body heat.

The opposite effect will take place at the colder end of the scale. Evaporative and convective loss will take place much more easily and the loss by radiation may become significant, removing heat faster than the body can generate it. The rate of heat production can be increased by greater bodily activity, but this cannot be sustained, so losses must be prevented by thicker insulation against convective loss and reduced skin exposure in the form of more clothing. The body itself can compensate by closing sweat pores and reducing the skin temperature.

16-9 OTHER COMFORT FACTORS

A total assessment of bodily comfort must take into account changes in convective heat transfer arising from air velocity, and the effects of radiant heat gain or loss. These effects have been quantified in several objective formulas, to give *equivalent*, *corrected effective*, *globe*, *dry resultant*, and *environmental* temperatures, all of which give fairly close agreement. This more complex approach is required where air speeds may be high, there is exposure to hot or cold surfaces, or where other special conditions call for particular care.

For comfort under normal office or residential occupation, with percentage saturations between 35 and 70 per cent, control of the dry bulb will result in comfort conditions for most persons. Feelings of personal comfort are as variable as human nature and at any one time, 10 per cent of the occupants of a space may feel too hot and 10 per cent too cold, while the 80 per cent majority are comfortable. Such variations frequently arise from lack or excess of local air movement, or proximity to cold windows, rather than an extreme of temperature or moisture content.

16-10 FRESH AIR

Occupied spaces need a supply of outside air to provide oxygen, remove respired carbon dioxide, and dilute body odours and tobacco smoke. The quantities are laid down by local regulations and commonly call for 6 to 8 litre/s per occupant. Such buildings are usually required also to have mechanical extract ventilation from toilets and some service areas, so the fresh air supply must make up for this loss, together with a small excess to pressurize the building against ingress of dirt.[3]

CHAPTER

SEVENTEEN

AIR TREATMENT CYCLES

17-1 WINTER HEATING

Buildings lose heat in winter by conduction out through the fabric, convection of cold air, and some radiation. The air from the conditioning system must be blown into the spaces warmer than the required internal condition, to provide the heat to counteract this loss.

Heating methods are as follows:

1. Hot water or steam coils
2. Direct fired—gas and sometimes oil
3. Electric resistance elements
4. Refrigerant condenser coils of heat pump or heat reclaim systems

Figure 17-1 shows the sensible heating of air.

Example 17-1 Air circulates at the rate of 68 kg/s and is to be heated from 16 to 34°C. Calculate the heat input and the water mass flow for an air heater coil having hot water entering at 85°C and leaving at 74°C.

$$Q = 68 \times 1.006 \times (34 - 16) = 1231 \text{ kW}$$

$$m_w = \frac{1231}{4.187 \times (85 - 74)} = 27 \text{ kg/s}$$

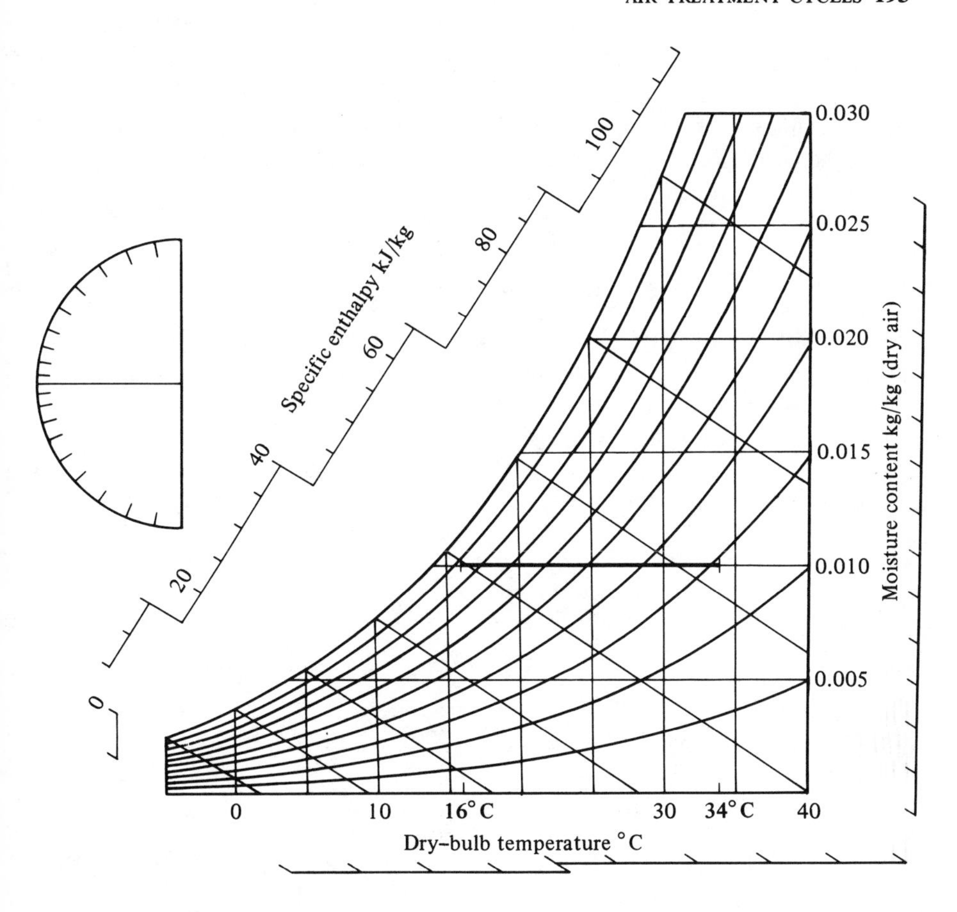

Figure 17-1 Sensible heating of air.

Example 17-2 A building requires 500 kW of heating. Air enters the heater coil at 19°C at the rate of 26 kg/s. What is the air-supply temperature?

$$t = 19 + \frac{500}{26 \times 1.006} = 19 + 19.1$$
$$= 38.1°\text{C}$$

If the cycle is being traced out on a psychrometric chart, the enthalpy can be read off for the coil inlet and outlet conditions. In Example 17-1, the enthalpy increase as measured on the chart is 18.1 kJ/kg dry air (taken at any value of humidity), giving

$$68 \times 18.1 = 1230 \text{ kW}$$

17-2 MIXING OF AIRSTREAMS

Air entering the conditioning plant will probably be a mixture of return air from the conditioned space, with a proportion of outside air. Since no heat or moisture is gained or lost in mixing,

$$\text{Sensible heat before} = \text{sensible heat after}$$

and

$$\text{Latent heat before} = \text{latent heat after}$$

The conditions after mixing can be calculated, but can also be shown graphically by a mix line joining the conditions A and B (see Fig. 17-2). The position C along the line will be such that

$$AC \times m_a = CB \times m_b$$

This straight-line proportioning holds good to close limits of accuracy. The

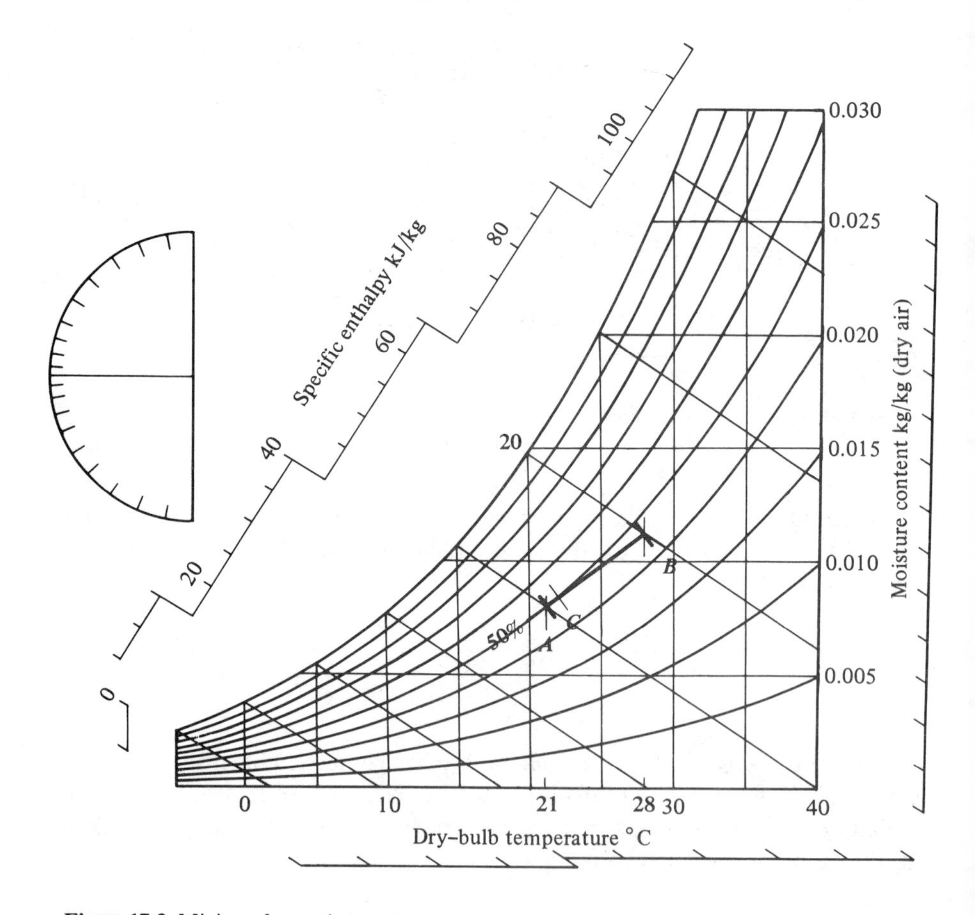

Figure 17-2 Mixing of two airstreams.

horizontal divisions of dry bulb temperature are almost evenly spaced, so indicating units of sensible heat. The vertical intervals of moisture content indicate units of latent heat.

Example 17-3 Return air from a conditioned space at 21°C, 50 per cent saturation, and a mass flow of 20 kg/s mixes with outside air at 28°C dry bulb and 20°C wet bulb, flowing at 3 kg/s. What is the condition of the mixture?

Method (*a*) Construct on the psychrometric chart as shown in Fig. 17-2 and measure off:

Answer = 22°C dry bulb, 49% sat.

Method (*b*) By calculation, using dry bulb temperatures along the horizontal component, and moisture content along the vertical. For the dry bulb, using

$$AC \times m_a = CB \times m_b$$

$$(t_c - 21) \times 20 = (28 - t_c) \times 3$$

giving $$t_c = 21.9°\text{C}$$

The moisture-content figures, from the chart or from tables, are 0.0079 and 0.0111 kg/kg at the return and outside conditions, so

$$(g_c - 0.0079) \times 20 = (0.0111 - g_c) \times 3$$

giving $$g_c = 0.0083 \text{ kg/kg}$$

If only enthalpy is required, this can be obtained from the same formula in a single equation, viz.

$$(h_c - h_a) \times m_a = (h_b - h_c) \times m_b$$

$$(h_c - 41.8) \times 20 = (56.6 - h_c) \times 3$$

giving $$h_c = 43.7 \text{ kJ/kg dry air}$$

Readers will recognize that the calculation methods lend themselves to computing.

17-3 SENSIBLE COOLING

If air at 21°C dry bulb, 50 per cent saturation, is brought into contact with a surface at 12°C, it will give up some of its heat by convection. The cold surface is warmer than the dew point, so no condensation will take place, and cooling will be sensible only (Fig. 17-3).

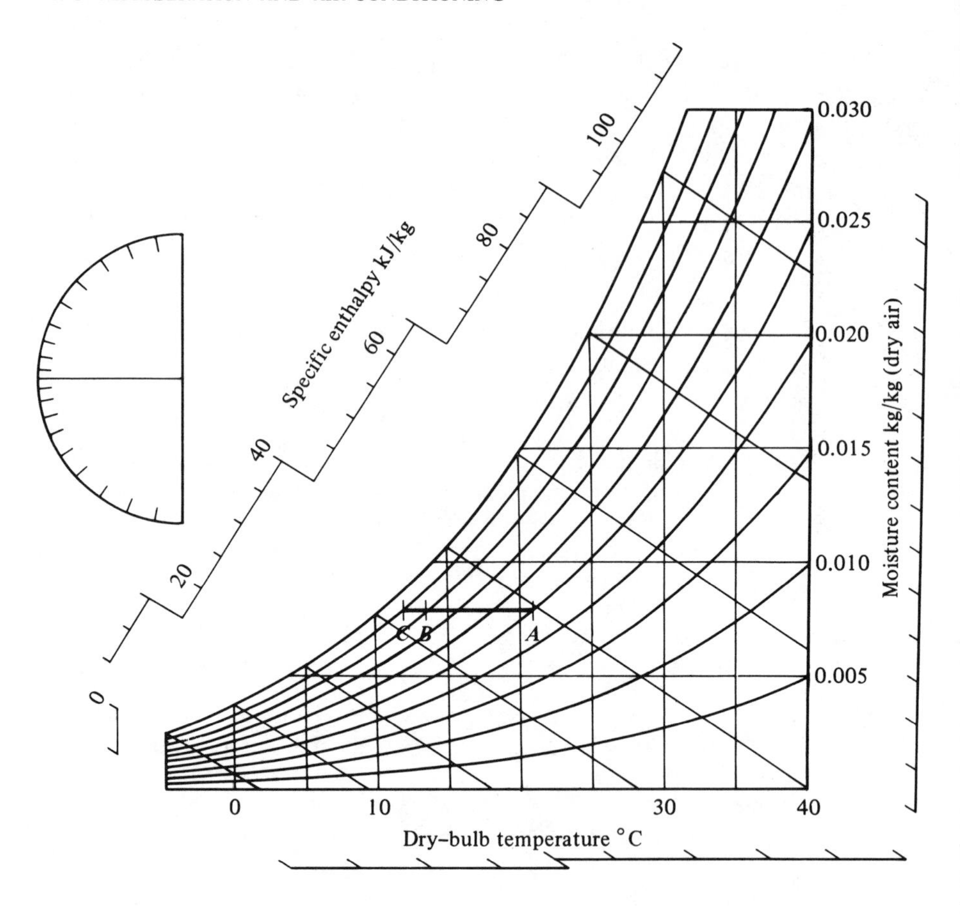

Figure 17-3 Sensible cooling of air.

This process is shown as a horizontal line on the chart, since there is no change in the moisture content. The loss of sensible heat can be read off the chart in terms of enthalpy, or calculated from the dry bulb reduction, considering the drop in the sensible heat of both the dry air and the water vapour in it.

17-4 WATER SPRAY (ADIABATIC SATURATION)

The effect of spraying water into an airstream will be as shown in Fig. 16-2, assuming that the air is not already saturated. Evaporation will take place and the water will draw its latent heat from the air, reducing the sensible heat and therefore the dry bulb temperature of the air (Fig. 17-4).

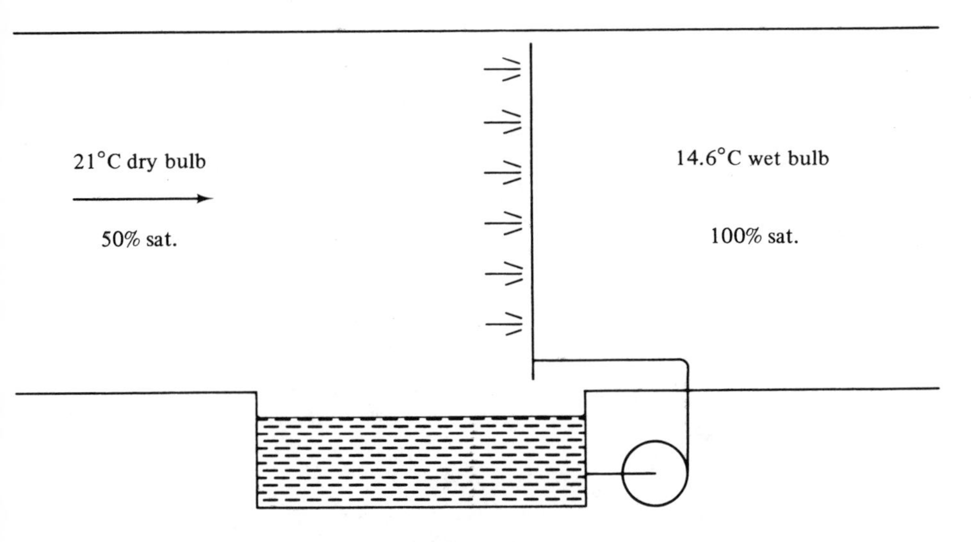

Figure 17-4 Adiabatic saturation to ultimate condition.

Example 17-4 Water is sprayed into an airstream at 21°C dry bulb, 50 per cent saturation. What would be the ultimate condition of the mixture?

No heat is being added or removed in this process, so the enthalpy must remain constant, and the process is shown as a movement along the line of constant enthalpy. Latent heat will be taken in by the water, from the sensible heat of the air until the mixture reaches saturation, when no more water can be evaporated.

$$\text{Initial enthalpy of air} = 41.08 \text{ kJ/kg}$$

$$\text{Final enthalpy of air} = 41.08 \text{ kJ/kg}$$

Final condition, 14.6°C dry bulb, 14.6°C wet bulb, 14.6°C dew point, 100% saturated.

It should be noted that this ultimate condition is difficult to reach, and the final condition in a practical process would fall somewhat short of saturation, possibly to point C in Fig. 17-5. The proportion AC/AB is termed the *effectiveness* of the spray system.

The adiabatic (constant enthalpy) line AC is almost parallel to the line of constant wet bulb. Had the latter been used, the final error would have been about 0.2 K, and it is sometimes convenient and quicker to calculate on the basis of constant wet bulb. (This correlation applies only to the mixture of dry air and water vapour, and not to other gas mixtures.)

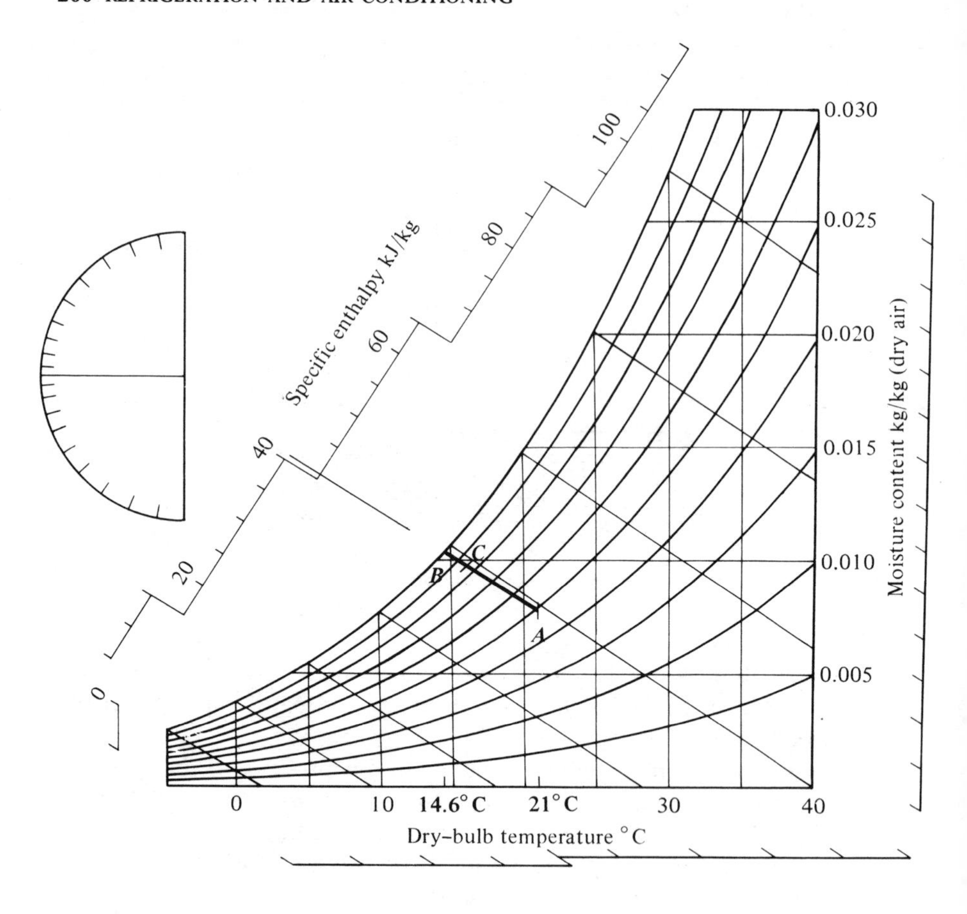

Figure 17-5 Adiabatic saturation—process line.

17-5 STEAM INJECTION

Moisture can be added to air by injecting steam, i.e., water which is already in vapour form and does not require the addition of latent heat (Fig. 17-6). Under these conditions, the air will not be cooled and will stay at about the same dry bulb. The steam will be at 100°C when released to the atmosphere (or may be slightly superheated), so raises the final temperature of the mixture.

Example 17-5 Steam at 100°C is blown into an airstream at 21°C dry bulb, 50 per cent saturation, at the rate of 1 kg steam/150 kg dry air. What is the final condition?

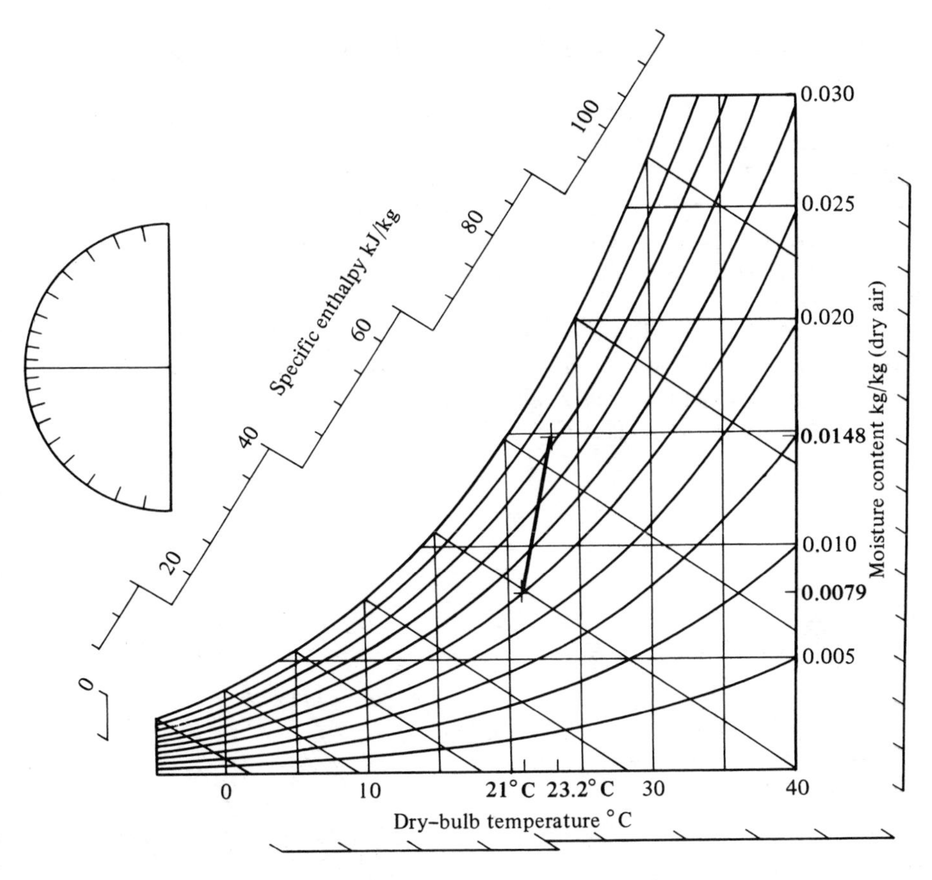

Figure 17-6 Addition of steam to air.

$$
\begin{aligned}
\text{Moisture content of air before} &= 0.0079 \text{ kg/kg} \\
\text{Moisture added, 1 kg/150 kg} &= \underline{0.0067 \text{ kg/kg}} \\
\text{Final moisture content} &= 0.0148 \text{ kg/kg}
\end{aligned}
$$

An approximate figure for the final dry bulb temperature can be obtained, using the specific heat capacity of the steam through the range 20 to 100°C, which is about 1.972 kJ/kg. This gives

$$\text{Heat lost by steam} = \text{heat gained by air}$$

$$0.0067 \times 1.972\,(100 - t) = 1.006\,(t - 21)$$

giving

$$t = 22.2\text{°C}$$

Where steam is used to raise the humidity slightly, the increase in dry bulb can usually be ignored.

17-6 AIR WASHER WITH CHILLED WATER

The process of adiabatic saturation in Sec. 17-4 assumed that the spray water temperature had no effect on the final air condition. If, however, a large mass of water is used in comparison with the mass of air, the final condition will approach the water temperature. If this water is chilled below the dew point of the entering air, moisture will condense *out* of the air, and it will leave the washer with a lower moisture content (see Fig. 17-7).

The ultimate condition will be at the initial water temperature *B*. Practical saturation efficiencies (the ratio AC/AB) will be about 50 to 80 per cent for air washers having a single bank of sprays and 80 to 95 per cent for double spray banks (see Fig. 17-8).

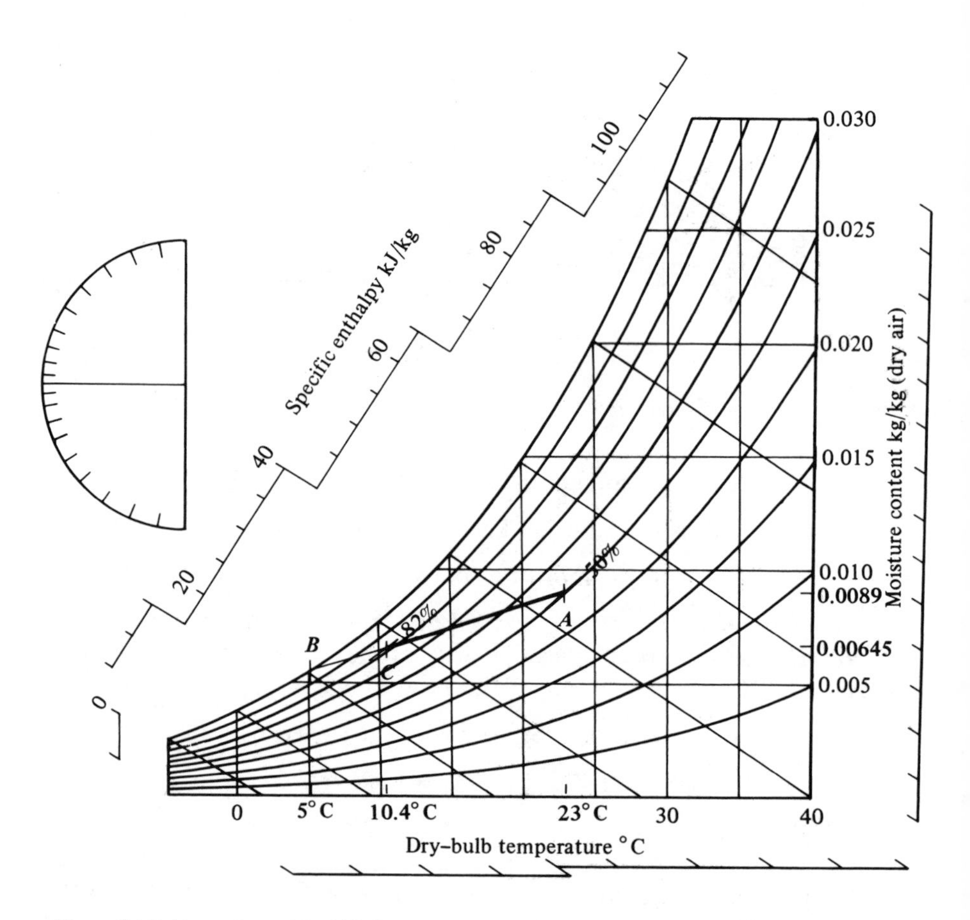

Figure 17-7 Air washer with chilled water.

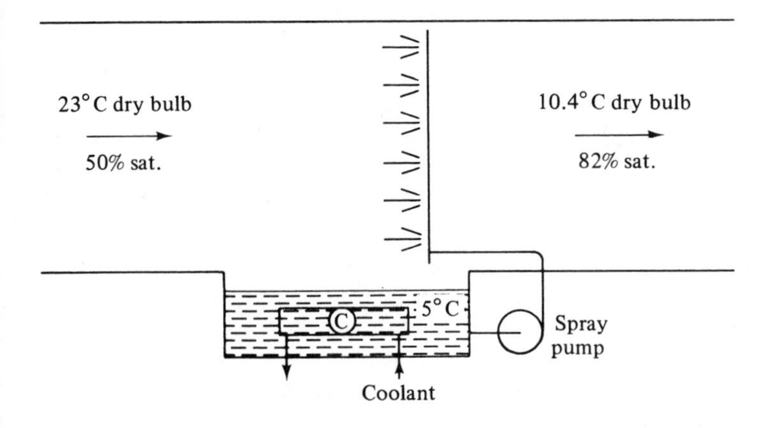

Figure 17-8 Chilled water spray.

Example 17-6 Air at 23°C dry bulb, 50 per cent saturation, enters a single-bank air washer having a saturation efficiency of 70 per cent and is sprayed with water at 5°C. What is the final condition?

(*a*) By construction on the chart (Fig. 17-7), the final condition is 10°C dry bulb, 82 per cent saturation.

(*b*) By proportion:
Dry bulb is 70 per cent of the way from 23°C down to 5°C

$$23 - [0.7(23 - 5)] = 10.4°\text{C}$$

Moisture content is 70 per cent down from 0.008 9 to 0.005 4 kg/kg

$$0.008\,9 - [0.7(0.008\,9 - 0.005\,4] = 0.006\,45 \text{ kg/kg}$$

Example 17-7 In Example 17-6, water is sprayed at the rate of 4 kg water for every 1 kg air. What is the water temperature rise?

$$\text{Enthalpy of air before} = 45.79 \text{ kJ/kg dry air}$$

$$\text{Enthalpy of air after} = \underline{26.7 \text{ kJ/kg}}$$

$$\text{Heat lost per kilogram air} = 19.09 \text{ kJ}$$

$$\text{Heat gain per kilogram water} = 19.09/4$$

$$= 4.77 \text{ kJ}$$

$$\text{Temperature rise of water} = \frac{4.77}{4.187}$$

$$= 1.1 \text{ K}$$

17-7 COOLING AND DEHUMIDIFYING COIL

In the previous process, air was cooled by close contact with a water spray. No water was evaporated, in fact some was condensed, because the water was colder than the *dew point* of the entering air.

A similar effect occurs if the air is brought into contact with a solid surface, maintained at a temperature below its dew point. Sensible heat will be transferred to the surface by convection and condensation of water vapour will take place at the same time. Both the sensible and latent heats must be conducted through the solid and removed. The surface used is that of a finned pipe coil, and the heat is carried away by refrigerant or a chilled fluid within the pipes. This coolant must be colder than the fin surface to transfer the heat inwards through the metal.

The process is shown on the chart in Fig. 17-9, taking point *B* as an

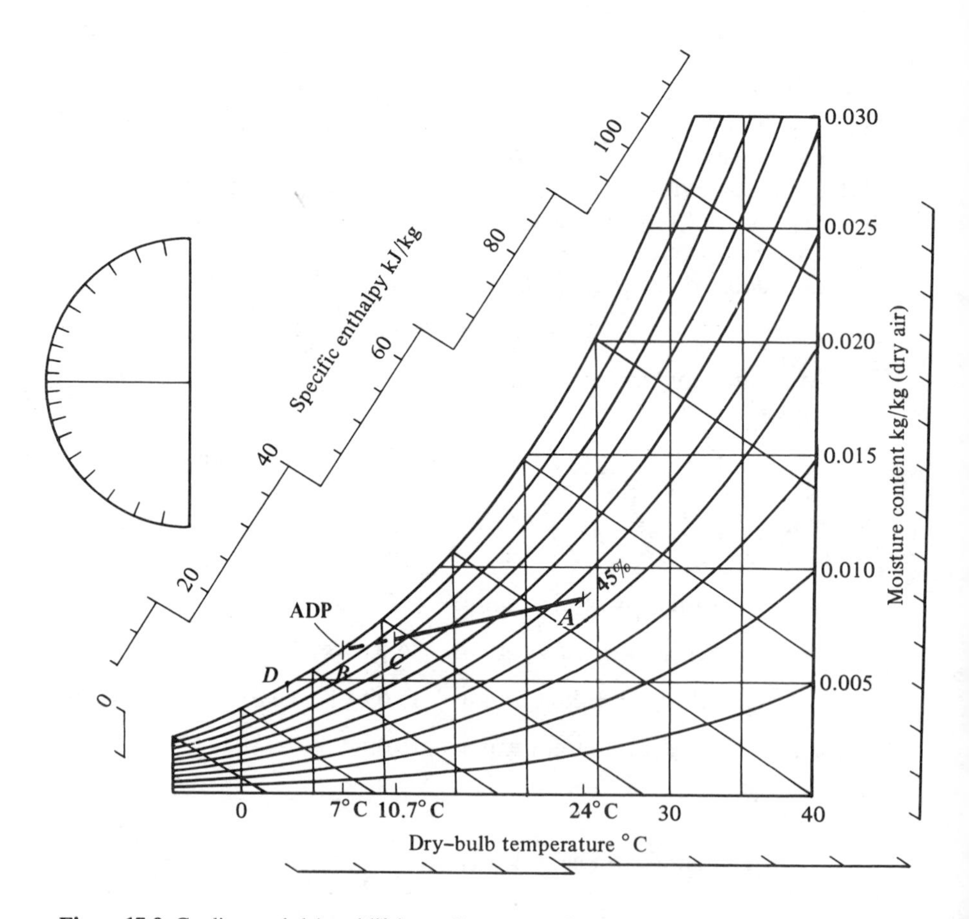

Figure 17-9 Cooling and dehumidifying coil—process line.

average of the fin surface temperature. This will vary with the fin height from the tube wall, the material, and any changes in the coolant temperature, so is not constant. Since this would be the ultimate dew point temperature of the air for an infinitely sized coil, the point B is termed the *apparatus dew point* (ADP). The average coolant temperature will be at some lower point D, and the temperature difference $B - D$ will be a function of the conductivity of the coil.

The process line AB is assumed to be straight for convenience of working, but may be curved, depending on the internal coolant temperatures and other smaller variations. The true shape of the line AB, if known, will not greatly affect the position C, showing the condition leaving the coil.

The proportion AC/AB is termed the coil *contact factor*. The proportion CB/AB is sometimes used, and is termed the *bypass factor*.

Example 17-8 Air at 24°C dry bulb, 45 per cent saturation, passes through a coil having an ADP of 7°C and a contact factor of 78 per cent. What is the off-coil condition?

(*a*) By construction on the chart (Fig. 17-9), 11.0°C dry bulb, 85 per cent saturation.

(*b*) By calculation, the dry bulb will drop 78 per cent of 24 to 7°C:

$$24 - [0.78 \times (24 - 7)] = 10.7°\text{C}$$

and the enthalpy will drop 78 per cent of 45.85 to 22.72 kJ/kg:

$$45.85 - [0.78 \times (45.85 - 22.72)] = 27.81 \text{ kJ/kg}$$

The two results obtained here can be compared with tabulated figures for saturation and give about 84 per cent saturation.

Example 17-9 Air is to be cooled by a chilled water coil from 27°C dry bulb, 52 per cent saturation, to 15°C dry bulb, 80 per cent saturation. What is the ADP?

This must be done by construction on the chart, and gives an ADP of 9°C. The intersection of the process and saturation lines can also be computed. Again, it has been assumed that the process line is straight, and there may be an error here which is not known.

17-8 SENSIBLE/LATENT RATIO

In all cases the horizontal component of the process line is a close indication of the sensible heat, and the vertical component gives the latent heat. It follows that the slope of the line shows the ratio between them and the

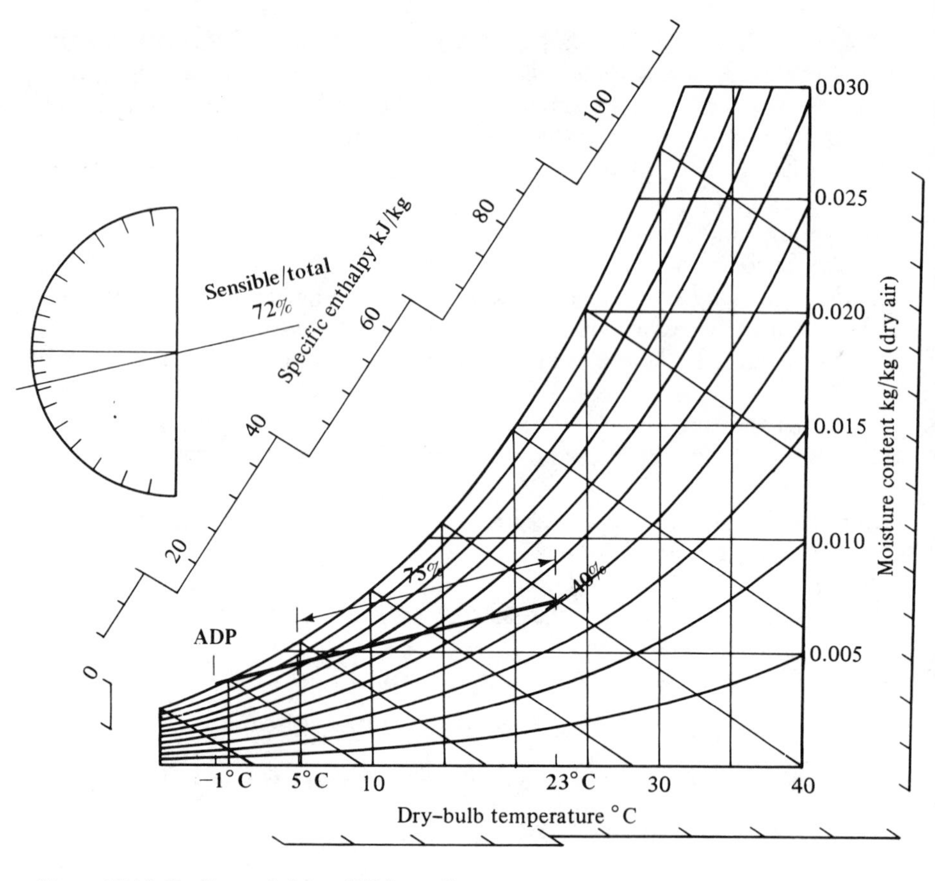

Figure 17-10 Cooling and dehumidifying coil.

angles, if measured, can be used to give the ratios of sensible to latent to total heat. On the psychrometric chart in general use (Fig. 16-5), the ratio of sensible to total heat is indicated as angles in a segment to one side of the chart. This can be used as a guide to coil and plant selection.

Example 17-10 Air enters a coil at 23°C dry bulb, 40 per cent saturation. The sensible heat to be removed is 36 kW and the latent 14 kW. What are the ADP and the coil contact factor if air is to leave the coil at 5°C?

Plotting on the chart (Fig. 17-10) from 23/40 per cent and using the ratio

$$\frac{\text{Sensible heat}}{\text{Total heat}} = \frac{36}{36 + 14} = \frac{36}{50} = 0.72$$

The process line meets the saturation curve at −1°C, giving the

ADP (which means that condensate will collect on the fins as frost).

Taking the 'off' condition at 5°C dry bulb and measuring the proportion along the process line gives a coil contact factor of 75 per cent.

17-9 MULTISTEP PROCESSES

Some air treatment processes cannot be made in a single operation, and the air must pass through two or more consecutive steps to obtain the required leaving condition.

Example 17-11 If air is to be cooled and dehumidified, as in Example 17-10, it may be found that the process line joining the inlet and outlet conditions does not meet the saturation line, e.g., in cooling air from

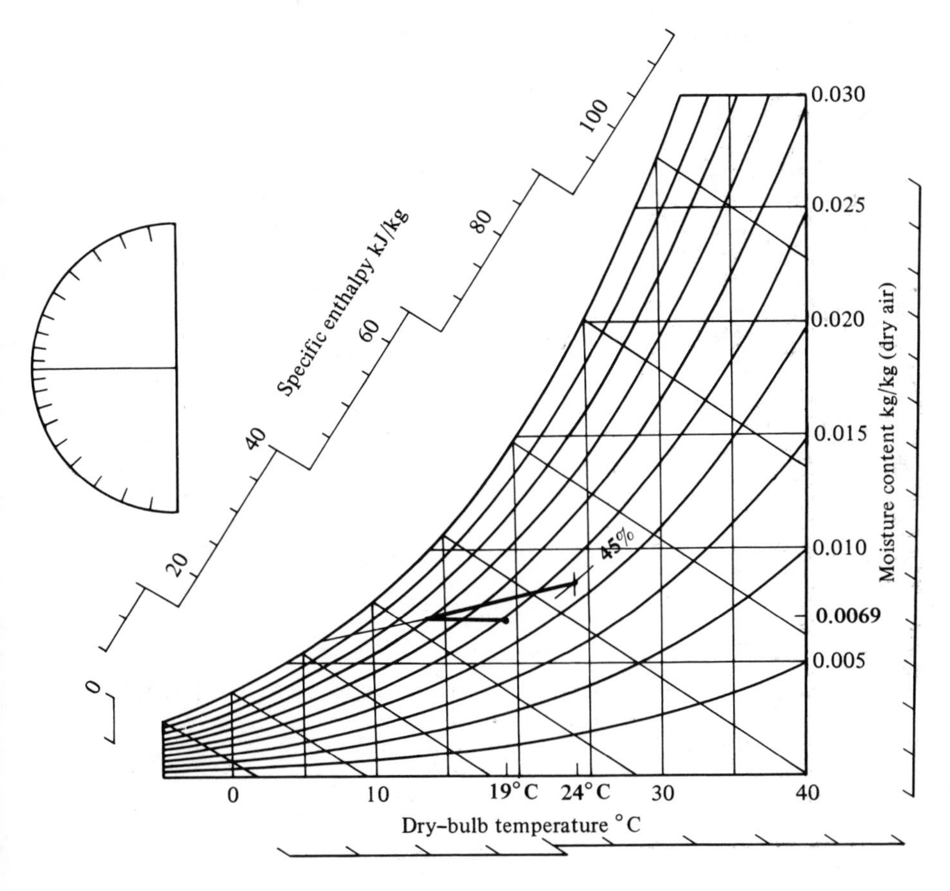

Figure 17-11 Cooling with dehumidifying, followed by reheat—process lines.

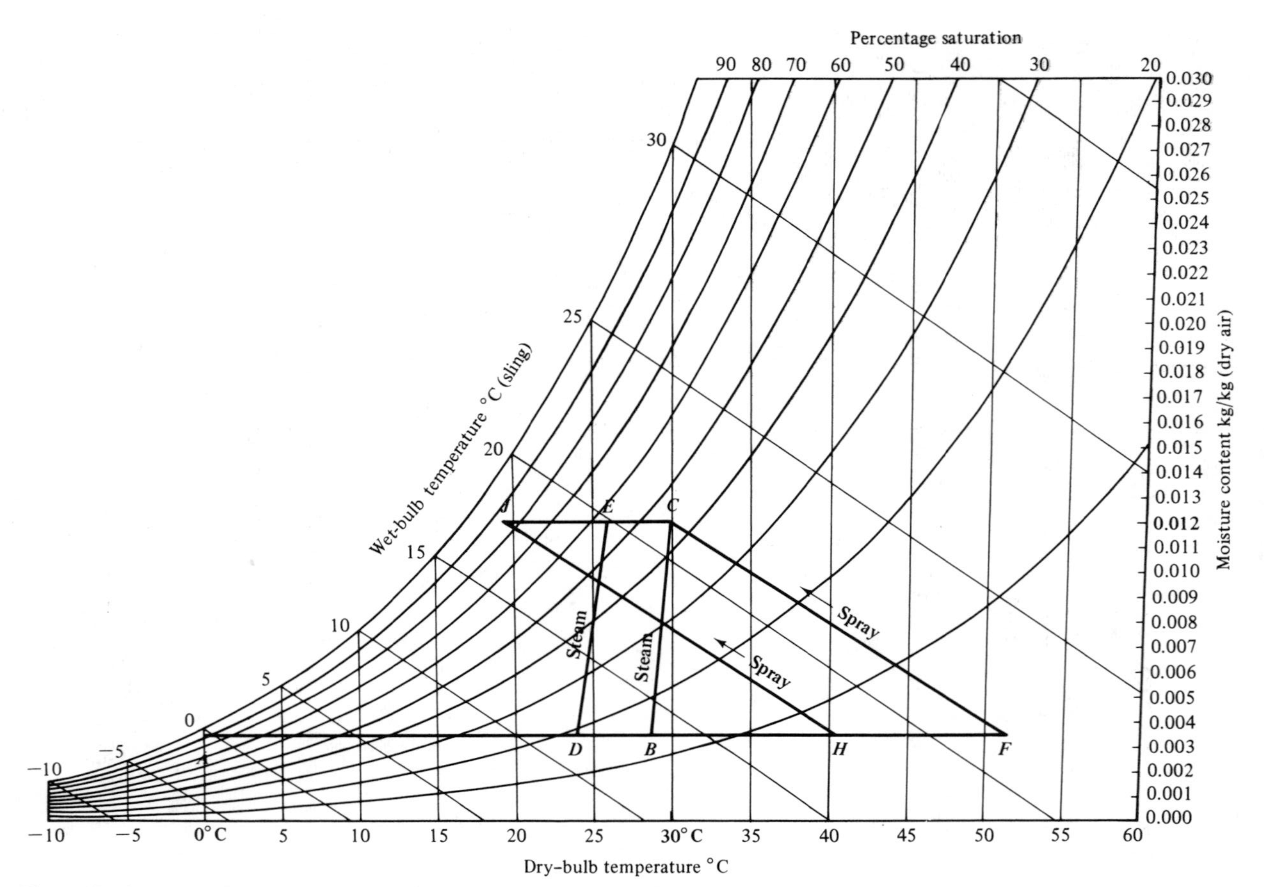

Figure 17-12 Pre-heating and humidification in winter—process lines.

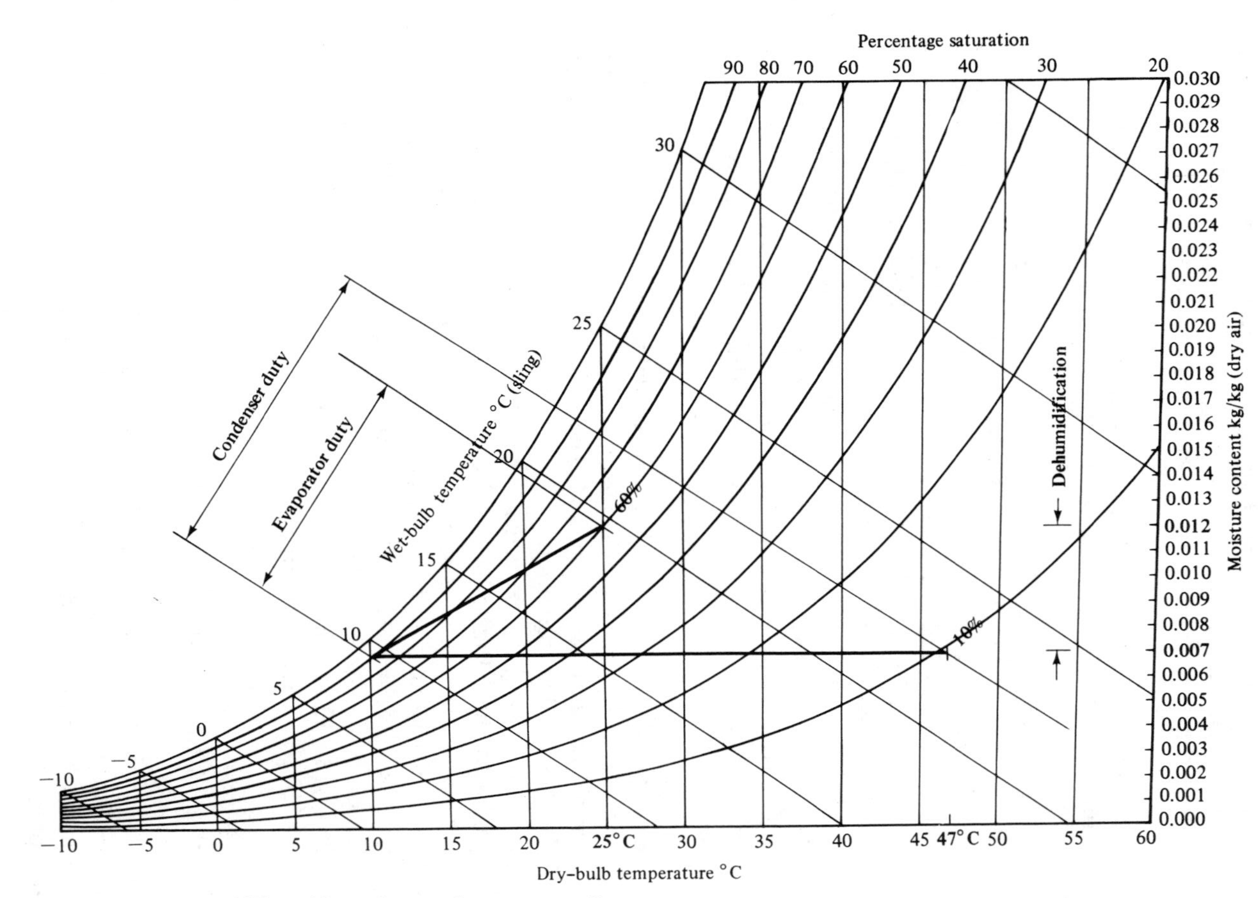

Figure 17-13 Dehumidifier with condenser reheat—process lines.

24°C dry bulb, 45 per cent saturation, to 19°C dry bulb, 50 per cent saturation, the process line shows this to be impossible in one step (Fig. 17-11). The air must first be cooled and dehumidified to reach the right moisture level of 0.0069 kg/kg and then reheated to get it back to 19°C. The first part is identical to that in Example 17-8, and the second step is the addition of sensible heat in a reheat coil.

Example 17-12 Winter outside air enters at 0°C dry bulb, 90 per cent saturation, and is to be heated to 30°C, with a moisture content of 0.012 kg/kg.

This can be done in several ways, depending on the method of adding the moisture and final dry bulb control (see Fig. 17-12). If by steam injection, the air can be pre-heated to just below 30°C and the steam blown in, line *ABC*. To give better control of the final temperature, the steam may be blown in at a lower condition, with final re-heat to get to the right point, line *ADEC*.

If by water spray or washer, the necessary heat must be put into the air first to provide the latent heat of evaporation. This can be done in two stages, *A* to *F* to *C*, or three stages *A* to *H* to *J* to *C*, if re-heat is required to get the exact final temperature. The latter is easier to control.

Example 17-13 Air enters a packaged dehumidifier (see Sec. 13-2) at 25°C dry bulb and 60 per cent saturation. It is cooled to 10°C dry bulb and 90 per cent saturation, and then re-heated by its own condenser. What is the final condition?

All of the heat extracted from the air, both sensible and latent, passes to the refrigerant and is given up at the condenser to re-heat, together with the energy supplied to the compressor and the fan motor (since the latter is in the airstream). Figures for this electrical energy will have to be determined and assessed in terms of kilojoules per kilogram of air passing through the apparatus. A typical cycle is shown in Fig. 17-13 and indicates a final condition of about 47°C dry bulb and 10 per cent saturation.

17-10 CYCLE ANALYSIS

The last three examples indicate the importance of analysis of the required air treatment cycle on the psychrometric chart, as a guide to the methods which can be adopted and those which are not possible. This analysis can also provide optimization of energy flows for a process.

Direct desk calculations would have indicated the overall energy flows between the inlet and outlet states, but may not have shown the cycles.

CHAPTER

EIGHTEEN

PRACTICAL CYCLES

18-1 HEATING

The majority of air-conditioned buildings are offices or for similar indoor activities, and are occupied intermittently. The heating system must bring them up to comfortable working conditions by the time work is due to start, so the heating must come into operation earlier to warm up the building.

A large part of the heating load when operating in daytime will be for fresh or outside air, which is not needed before occupation, and the heat-up time will be reduced if the fresh air supply can remain inoperative for this time.

The required warm-up time will vary with ambient conditions, being longer in cold weather and least in warm. Optimum-start controllers are now coming into general use which are programmed for the building warm-up characteristics and sense the inside and ambient conditions. They then transduce the required start-up period and set the heating plant going only when needed. This, and the previous scheme, will save fuel.

Air cooling systems commonly have a mass flow of 0.065 kg/s kW of cooling load. The normal heating load will be less than the cooling load for most of the time and, if this full air flow is maintained, the air inlet temperature will be of the order of 30 to 32°C. This is below body temperature and may give the effect of a cold draught, although it is heating. Where possible, the winter air flow should be reduced to give warmer inlet air. This is particularly so with packaged air-conditioners of all sizes, which may have to be located for convenience rather than for the best air-flow pattern.

The addition of moisture to the winter air in the United Kingdom is not usually necessary, except for systems using all outside air, or where persons with severe respiratory trouble are accommodated. With a winter ambient of 0°C dry bulb, 90 per cent saturation, outside air pre-heated to 25°C will then be 17 per cent saturation, which could itself cause discomfort. However, this is diluted with the return air, and it is unlikely that indoor humidities will fall below 35 per cent saturation. Humidification of this to 60 per cent saturation would permit a slightly lower dry bulb (0.5 K less) to give a similar degree of comfort, thus slightly reducing the conduction losses from the building fabric. However, this is at the cost of the latent heat to evaporate this moisture and a higher dew point (13°C instead of 5°C) with increased condensation on cold building surfaces and greater deterioration (see Fig. 18-1).

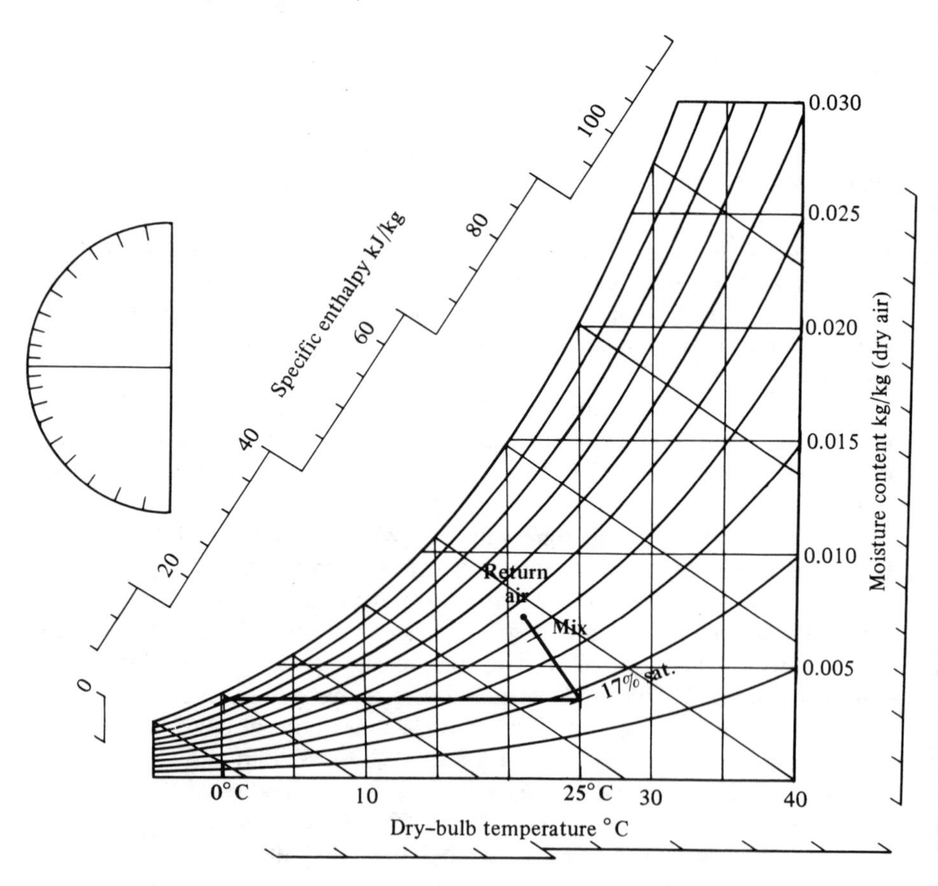

Figure 18-1 Pre-heating of outside air and mixing with return air—process lines.

18-2 ADDITION OF MOISTURE

Methods of adding moisture to the airstream (see Secs. 17-4, 17-5, and 17-9) are difficult to control, since a lot of water remains in the apparatus at the moment of switching off humidification. For this reason, the heat–humidify–re-heat cycle as shown in Fig. 17-12 and Fig. 18-2 is to be preferred, as the final heater control can compensate for overshoot.

Air washers require water treatment and bleed-off, since they concentrate salts in the tank. Steam will be free from such impurities, but the boiler will need attention to remove accumulations of hardness.

Mist and spray humidifiers, unless the water is pure, will leave a powder deposit of these salts in the conditioned space.

The use of standard factory packaged air-conditioners to hold close humidities, together with a humidifier to correct for overdrying, is a common source of energy wastage, since both may operate at the same time. Packaged units, unless specifically built for the duty, will pull down to 45 per cent saturation or lower under UK conditions. Humidity tolerances for process conditioning such as computer and standards rooms can usually be 45 to 55 per cent saturation, and this differential gap should be wide enough to prevent simultaneous operation of both humidifying and dehumidifying plant.

18-3 OUTSIDE AIR PROPORTION

The high internal heat load of many modern buildings means that comfort cooling may be needed even when the ambient is down to 10°C or lower. Under these conditions, a high proportion of outside air can remove building heat and save refrigeration energy. This presupposes that:

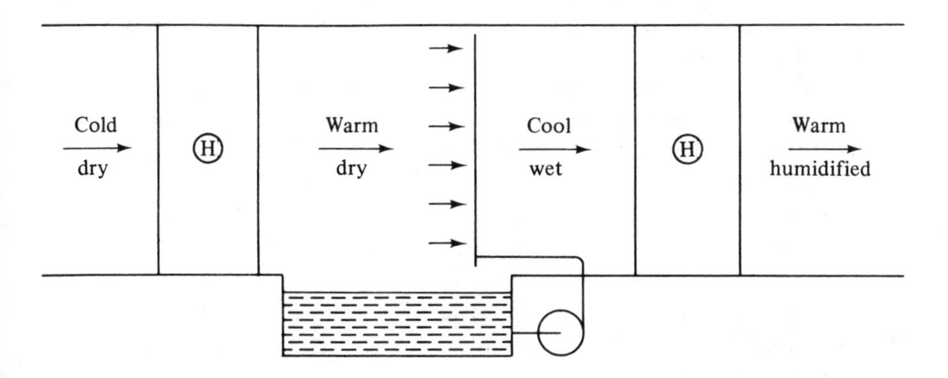

Figure 18-2 Pre-heat, humidify, re-heat cycle—apparatus.

1. The fresh air ducting and fan can provide more air.
2. This outside air can be filtered.
3. There are adequate automatic controls to admit this extra air only when wanted.
4. Surplus air in the building can be extracted.

18-4 COOLING AND DEHUMIDIFICATION

The cooling load will always be greatest in the early afternoon, so no extra start-up capacity is required. The general practice of using a single coil for cooling and dehumidification without reheat for comfort cooling will give design balance conditions only at full load conditions. Slightly different conditions must be accepted at other times. Closer control can be obtained by variation of the coolant temperature and air mass flow over the coil, but such systems can easily be thrown out of calibration, and measures should be taken to avoid unauthorized persons changing the control settings or energy will be wasted with no benefit in the final conditions.

18-5 EVAPORATIVE COOLERS

Many of the warmer climates have a dry atmosphere (see Fig. 16-6). In such areas, considerable dry bulb temperature reduction can be gained by the adiabatic saturation cycle (Sec. 17-4). The apparatus draws air over a wetted pad and discharges it into the conditioned space. It is termed an *evaporative* or *desert* cooler (Fig. 18-3).

Example 18-1 Air at 37°C dry bulb, 28 per cent saturation, is drawn through a desert cooler having an adiabatic saturation efficiency of 75 per cent. What is the final dry bulb, and how much water is required?

The entering enthalpy is 62.67 kJ/kg, and this remains constant through the process.

By construction on the chart, or from tables, the ultimate saturation condition would be 21.5°C and 75 per cent of the drop from 37 to 21.5°C gives a final dry bulb of 25.4°C.

The water requirement can be calculated from the average latent heat of water over the working range, which is 2425 kJ/kg. The amount of water to be evaporated is $1/2425 = 0.4 \times 10^{-3}$ kg/s kW.

A two-stage evaporative cooler (Fig. 18-4) uses the cooled water from the first stage to pre-cool the air entering the second stage. The two air systems are separate.

Example 18-2 Taking the first stage as Example 18-1, the water would be cooled to 25.4°C and could be used in a coil of 80 per cent contact factor to pre-cool outside air to

$$37 - 0.8(37 - 25.4) = 27.7°\text{C (point } D\text{, Fig. 18-4}a\text{)}$$

The wet bulb is now 18.9°C and the enthalpy is 53 kJ/kg. A second-stage evaporative cooler with an efficiency of 75 per cent will bring this down to 21°C dry bulb (point *F*).

The evaporative cooler has no refrigeration system and only requires electric power for fans and water pumps plus, of course, an adequate supply of water. No moisture can be removed from the air. It is very widely used in arid climates.

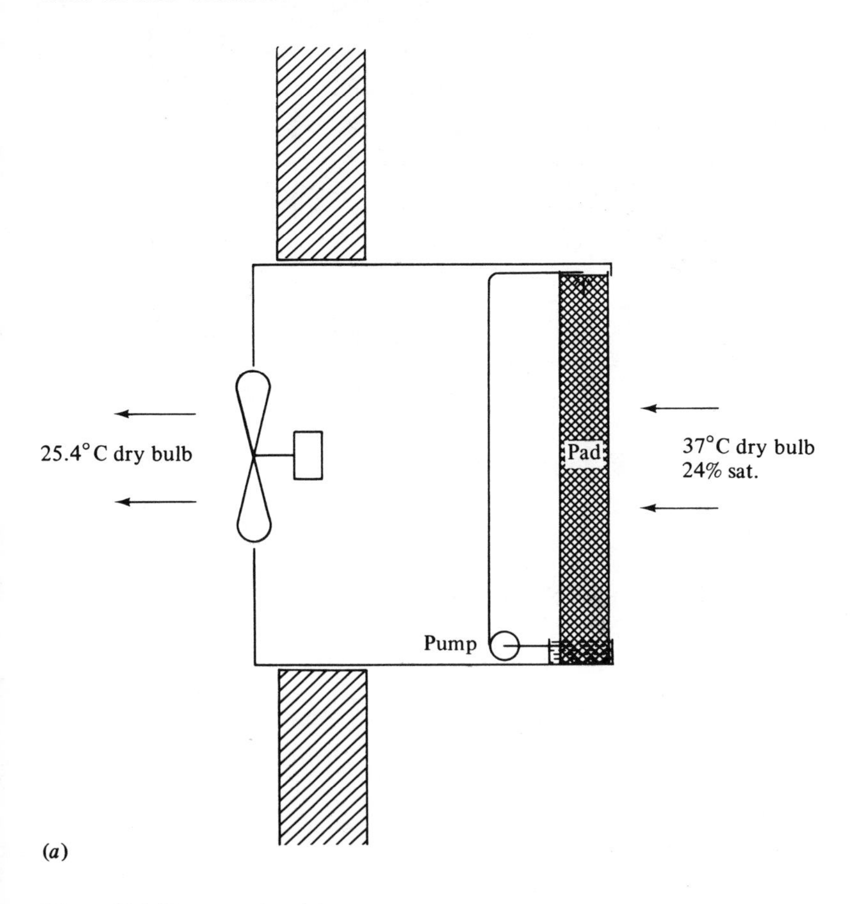

Figure 18-3 Desert cooler: (*a*) apparatus.

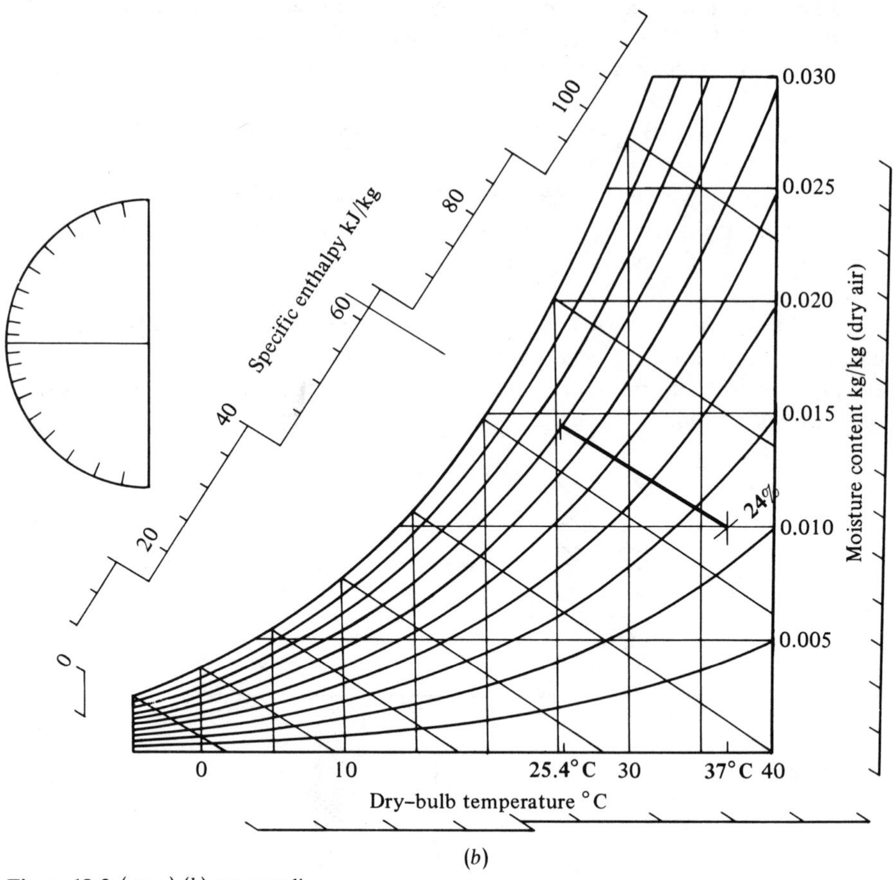

(*b*)

Figure 18-3 (*cont.*) (*b*) process line.

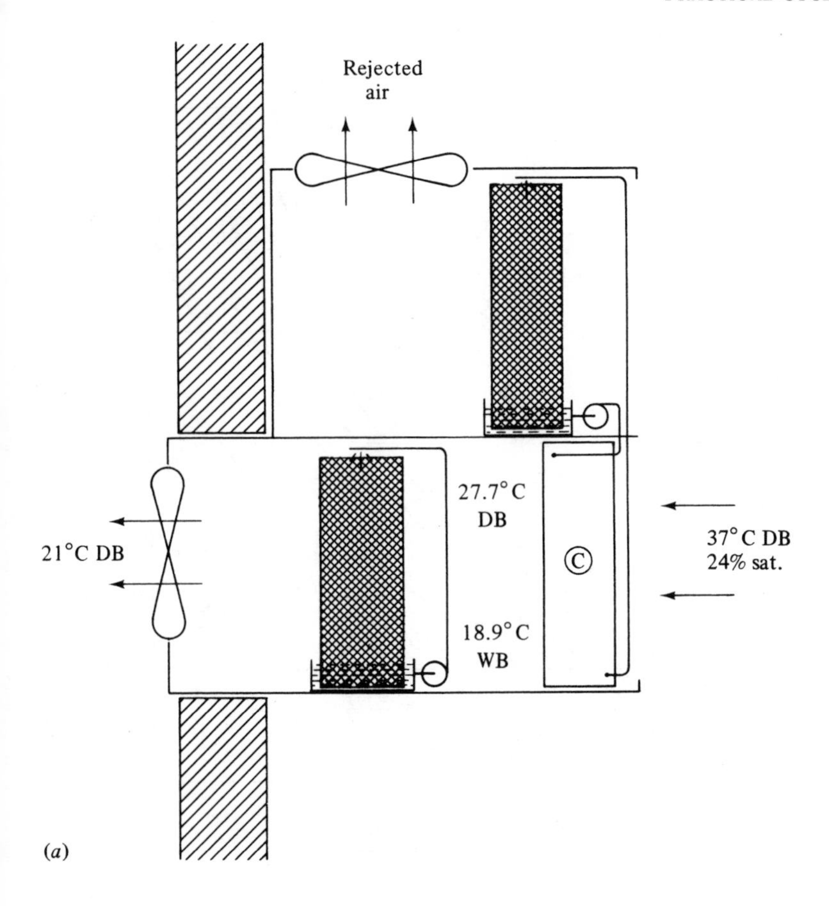

Figure 18-4 Two-stage desert cooler: (*a*) apparatus.

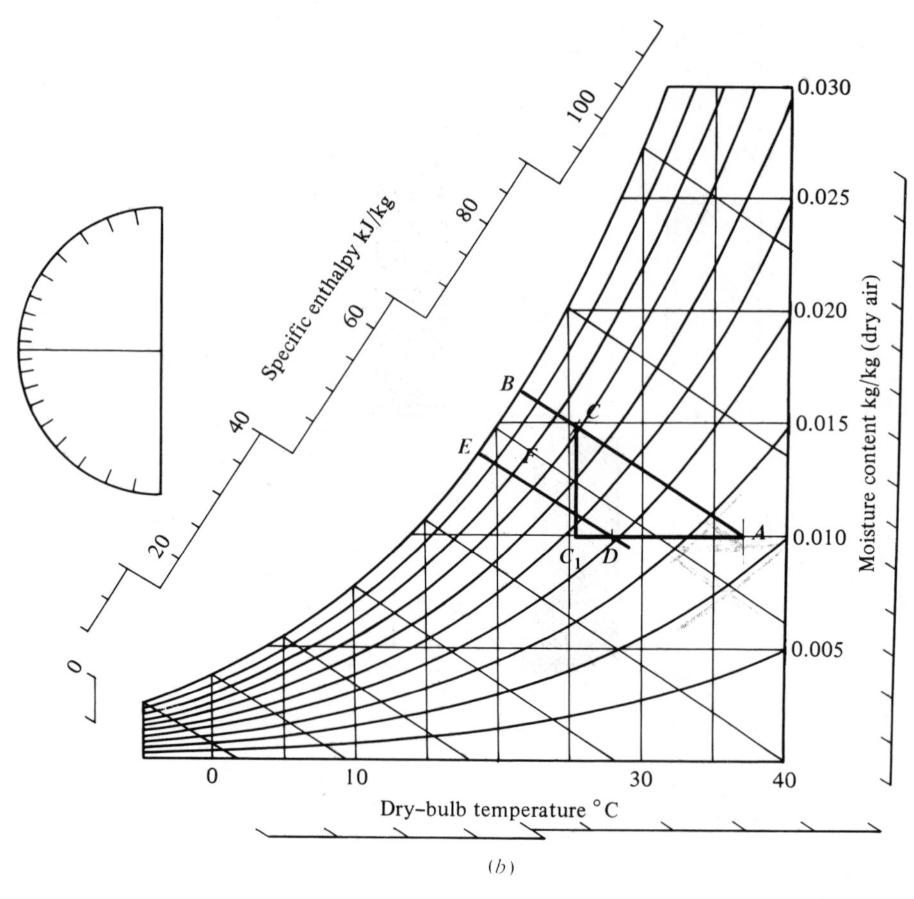

Figure 18-4 (*cont.*) (*b*) process lines.

18-6 COOLING TOWERS

The cooling tower is an evaporative cooler with the prime purpose of cooling the water, rather than the air. The normal forms have the water and air in counter flow or cross flow, although towers are built with parallel flow (see Fig. 18-5).

The process is complex and cannot be simplified into an ultimate balance condition, from which to work back to a supposed operating point, and factors are used for design and application which are based on similar apparatus.

The heat from the water is transferred to the air, so the available heat gain by the air will depend on its initial *enthalpy*. This is usually expressed in terms of ambient *wet bulb* temperature, since the two are almost synonymous and the wet bulb is more easily recognized. This is used as a yardstick to describe performance in terms of the *approach* of the leaving water temperature to the ambient wet bulb. The two could only meet ultimately in a tower of infinite size having an air flow infinitely larger than the water flow, so the term is descriptive rather than a clear indication of tower efficiency.

Assuming the mass flows of air and water to be equal, an approximate balance can be found.

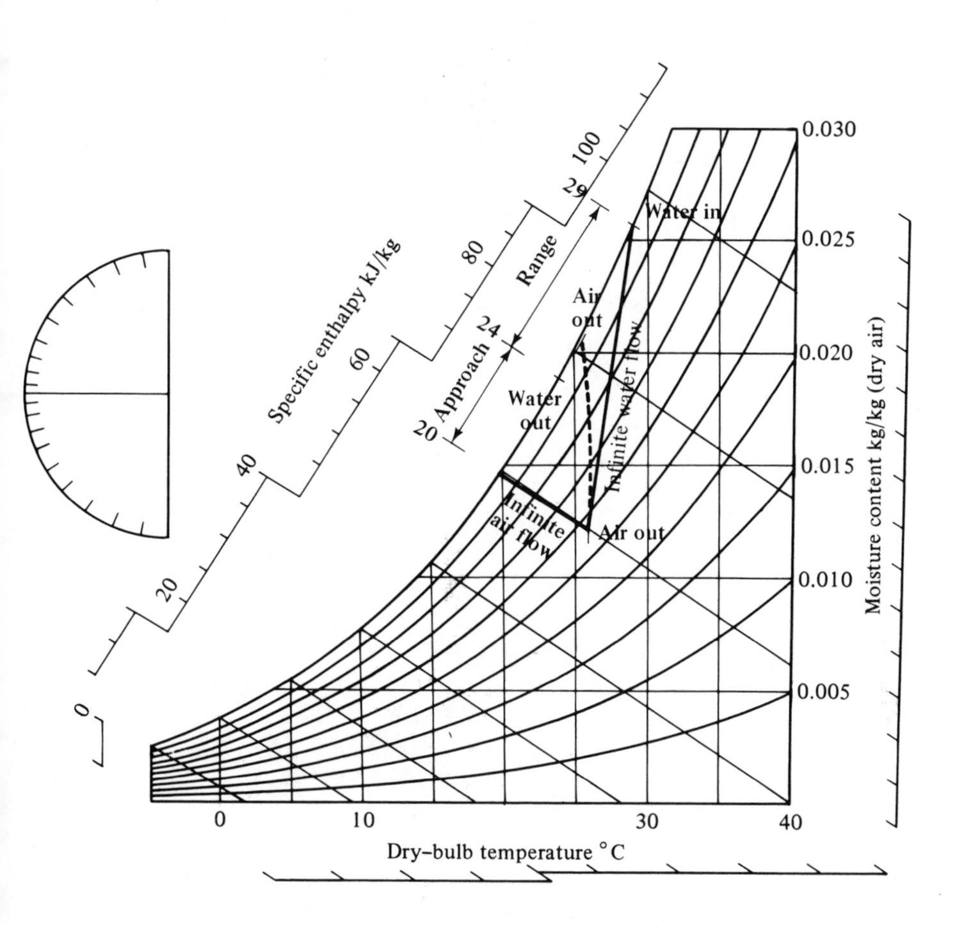

Figure 18-5 Water tower—possible process lines under one set of atmospheric conditions.

Example 18-3 Air enters a cooling tower at 26°C dry bulb and 20°C wet bulb. Water at the same mass flow enters at 29°C and leaves at 24°C. If the air leaves the tower at 98 per cent saturation, what is its final condition?

$$\text{Heat from water} = 4.187(29 - 24) = 21 \text{ kJ/kg}$$

$$\text{Enthalpy of entering air} = 57.1 \text{ kJ/kg}$$

$$\text{Enthalpy of leaving air} = 78.1 \text{ kJ/kg}$$

From the chart, the air leaves at about 25.7°C dry bulb.

Calculations of this sort are only of importance to the tower designer. Manufacturers' application data will give the cooling range or capacity in terms of wet bulb, inlet water temperature, and mass flow.[15, 19]

CHAPTER

NINETEEN

AIR-CONDITIONING LOAD ESTIMATION

19-1 COMPONENTS OF LOAD

The cooling load to maintain steady temperature and humidity in a conditioned space will have four components:

1. Heat leakage through the fabric by conduction from warmer surroundings
2. Heat gain by radiation through transparent surfaces—usually solar but occasionally by other means (radiant heat from a process, such as furnaces)
3. Heat gain by forced or natural convection—air infiltration and fresh air supply—sensible and latent heat
4. Internal heat sources—lights, people, machines, etc.—sensible and latent heat

19-2 CONDUCTION HEAT GAINS

Conduction of heat through plain surfaces under steady-state conditions is given by the product of the area, temperature difference, and the overall conductance of the surface (see Sec. 1-8):

$$Q = A \times \Delta T \times U$$

where
$$U = \frac{1}{R_{si} + R_1 + R_2 + R_3 + \cdots + R_{so}}$$

and R_{si} is the inside surface thermal resistance, R_{so} is the outside surface thermal resistance, and R_1, R_2, etc., are the thermal resistances of the composite layers of the fabric.

Example 19-1 A building wall is made up of pre-cast concrete panels 40 mm thick, lined with 50 mm insulation and 12 mm plasterboard. The inside resistance is 0.3 m^2 K/W and the outside resistance 0.07 m^2 K/W. What is the U factor?

$$U = \frac{1}{0.3 + 0.040/0.09 + 0.050/0.037 + 0.012/0.16 + 0.07}$$

$$= \frac{1}{2.24}$$

$$= 0.45 \text{ W/m}^2 \text{ K}$$

The conductivity figures 0.09, 0.037, and 0.16 can be found in Section A3 of the CIBS Guide.[2]

Figures for the conductivity of all building materials, of the surface coefficients, and many overall conductances can be found in standard reference books.[1, 2, 37]

The dominant factor in building surface conduction is the absence of steady-state conditions, since the ambient temperature, wind speed, and solar radiation are not constant. It will be readily seen that the ambient will be cold in the morning, will rise during the day, and will fall again at night. As heat starts to pass inwards through the surface, some will be absorbed in warming the surface itself and there will be a time lag before the effect reaches the inner face, depending on the mass, conductivity, and specific heat capacity of the materials. Some of the absorbed heat will be retained in the surface material and then lost to ambient at night. The effect of thermal time lag can be expressed mathematically (CIBS Guide, A3, A5).

The rate of heat conduction is further complicated by the effect of sunshine onto the outside. Solar radiation reaches the earth's surface at a maximum intensity of about 0.9 kW/m^2. The amount of this absorbed by a plane surface will depend on the absorption coefficient and the angle at which the radiation strikes. The angle of the sun's rays to a surface (see Fig. 19-1) is always changing, so this must be estimated on an hour-to-hour basis. Various methods of reaching an estimate of heat flow are used, and the sol-air temperature (see CIBS Guide, A5) provides a simplification of the factors involved. This, also, is subject to time lag as the heat passes through the surface.

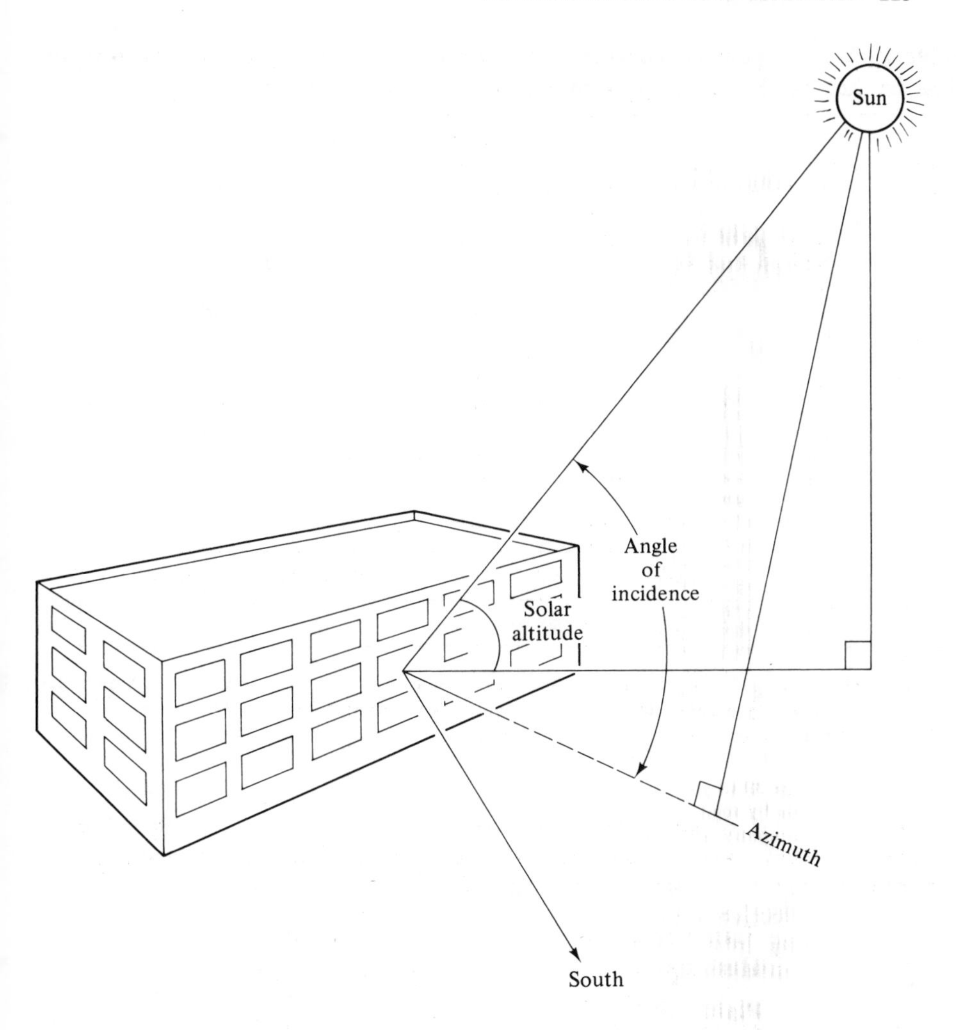

Figure 19-1 Angle of incidence of sun's rays on window.

19-3 SOLAR HEAT

Solar radiation through windows has no time lag and must be estimated by finite elements (i.e., on an hour-to-hour basis), using calculated or published data for angles of incidence and taking into account the type of window glass (see Table 19-1).

Since solar gain can be a large part of the building load, special glasses and window constructions have been developed, having two or more layers

Table 19-1 Heat gain by convection and radiation from single common window glass for 22 March and 22 September† (Btu h/ft^2 of masonry opening)

Time of year	Sun time	Direction for North latitude (read down)									
		N	NE	E	SE	S	SW	W	NW	Horizontal	
22 March and 22 Sept. North latitude	6 am	−4.4	−4.4	−4.4	−4.4	−4.4	−4.4	−4.4	−4.4	−4.4	22 Sept. and 22 March South latitude
	7 am	−3.3	−2.0	−1.0	−2.4	−3.3	−3.3	−3.3	−3.3	−2.8	
	8 am	−1.1	0.2	2.1	1.9	−0.1	−1.1	−1.1	−1.1	0.6	
	9 am	2.2	2.9	5.2	5.6	4.0	2.2	2.2	2.2	4.9	
	10 am	5.5	5.5	7.8	8.8	7.9	5.5	5.5	5.5	9.0	
	11 am	9.9	9.9	11.1	12.8	12.7	10.9	9.9	9.9	13.9	
	12 Noon	13.2	13.2	13.2	15.2	16.2	15.2	13.2	13.2	17.4	
	1 pm	16.5	16.5	16.5	17.5	19.3	19.3	17.6	16.5	20.5	
	2 pm	17.6	17.6	17.6	17.6	20.0	20.9	19.9	17.6	21.1	
	3 pm	18.7	18.7	18.7	18.7	19.6	22.1	21.8	19.4	21.4	
	4 pm	17.6	17.6	17.6	17.6	18.6	19.6	20.8	18.9	18.8	
	5 pm	16.5	16.5	16.5	16.5	16.5	18.4	18.8	17.8	17.0	
	6 pm	14.3	14.3	14.3	14.3	14.3	14.3	14.3	14.3	14.3	
		S	SE	E	NE	N	NW	W	SW	Horizontal	
		Direction for South latitude (read up)									Time of year

† This table is for 30 to 50 degrees North latitude. It can be used for 22 March and 22 September in the South latitude by reading up from the bottom. Room temperature is 78°F.

and with reflective and heat absorbing surfaces. These can reduce the energy passing into the conditioned space by as much as 75 per cent. Typical transmission figures are as follows:

Plain single glass	0.75 transmitted
Heat absorbing glass	0.45 transmitted
Coated glass, single	0.55 transmitted
Metalized reflecting glass	0.25 transmitted

Windows may be shaded, by either internal or external blinds, or by overhangs or projections beyond the building face. The latter is much used in the tropics to reduce solar load (see Fig. 19-2). Windows may also be shaded part of the day by adjacent buildings.

All these factors need to be taken into account, and solar transmission estimates are usually calculated or computed for the hours of daylight through the hotter months, although the amount of calculation can be much reduced if the probable worst conditions can be guessed. For example, the greatest solar gain for a window facing west will obviously be after midday, so no time would be wasted by calculating for the morning. Comprehensive data on solar radiation factors, absorption coefficients, and methods of calculation can be found in reference books.[1, 2, 21, 37, 38]

Figure 19-2 Solar shading by building construction at Gulf Hotel, Bahrain. *(Courtesy of Associated Continental Architects in association with Michael Lyell Associates, London.)*

There are several abbreviated methods of reaching an estimate of these varying conduction and direct solar loads, if computerized help is not readily available. One of these[39] suggests the calculation of loads for five different times in summer, to reach a possible maximum at one of these times. This maximum is used in the rest of the estimate (see Fig. 19-3).

Where cooling loads are required for a large building of many separate rooms, it will be helpful to arrive at total loads for zones, floors, and the complete installation, as a guide to the best method of conditioning and the overall size of plant. In such circumstances, computer programs are available which will provide the extra data as required.

19-4 FRESH AIR

The movement of outside air into a conditioned building will be balanced by the loss of an equal amount at the inside condition, whether by intent (positive fresh air supply or stale air extract) or by accident (infiltration

Air-Conditioning Load Calculation Sheet

Job .... Date ....

Summer cooling load **Outside design condition**
Inside design condition

Table A Solar heat gains glass walls and roof sensible heat

Glass aspect	Glass area, ft ^{2}	Window factor table 6	Shade factor table 8	JUNE 10 am. $F1$ Table 3	Btu/h	4 pm. $F2$ Table 3	Btu/h	SEPTEMBER 10 am. $F3$ Table 3	Btu/h	2 pm. $F4$ Table 3	Btu/h	4 pm. $F5$ Table 3	Btu/h
Wall aspect	Wall, ft ^{2}	U		$F6$ Table 4	Btu/h	$F7$ Table 4	Btu/h	$F8$ Table 4	Btu/h	$F9$ Table 4	Btu/h	$F10$ Table 4	Btu/h
Roof	Roof, ft ^{2}	U		$F11$ Table 5	Btu/h	$F12$ Table 5	Btu/h	$F13$ Table 5	Btu/h	$F14$ Table 5	Btu/h	$F15$ Table 5	Btu/h
Total for each Time of Day				—		—		—		—		—	

Table B Transmission sensible gains

Surface	A, ft ^{2}	U	Δt Table 9	Btu/h

Table C Internal sensible gains

Type	No.	Unit Btu/h	Btu/h

Figure 19-3 Air-conditioning load calculation sheet (part). *(Courtesy Electricity Council.)*

though window and door gaps, and door openings). Since a building for human occupation must have some fresh air supply and some mechanical extract from toilets and service areas, it is usual to arrange an excess of supply over extract, to maintain an internal slight pressure and so reduce accidental air movement and ingress of dirt.

The amount of heat to be removed (or supplied in winter) to treat the fresh air supply can be calculated, knowing the inside and ambient states. It must be broken into sensible and latent loads, since this affects the coil selection.

Example 19-2 A building is to be maintained at 21°C dry bulb and 45 per cent saturation in an ambient of 27°C dry bulb, 20°C wet bulb. What are the sensible and latent air cooling loads for a fresh air flow of 1.35 kg/s?

There are three possible calculations, which cross-check.

(*a*) Total heat:

$$\begin{aligned} \text{Enthalpy at 27°C DB, 20°C WB} &= 57.00 \text{ kJ/kg} \\ \text{Enthalpy at 21°C DB, 45\% sat.} &= \underline{39.08} \text{ kJ/kg} \\ \text{Heat to be removed} &= 17.92 \\ Q_t = 17.92 \times 1.35 &= 24.2 \text{ kW} \end{aligned}$$

(*b*) Latent heat:

$$\begin{aligned} \text{Moisture at 27°C DB, 20°C WB} &= 0.0117 \text{ kg/kg} \\ \text{Moisture at 21°C DB, 45\% sat.} &= \underline{0.0070} \text{ kg/kg} \\ \text{Moisture to be removed} &= 0.0047 \\ Q_l = 0.0047 \times 1.35 \times 2440 &= 15.5 \text{ kW} \end{aligned}$$

(*c*) Sensible heat:

$$Q_s = [1.006 + (4.187 \times 0.0117)](27 - 21) \times 1.35 = 8.6 \text{ kW}$$

Where there is no mechanical supply or extract, factors are used to estimate possible natural infiltration rates. Empirical values may be found in several standard references, and the CIBS Guide (Ref. 2, A4) covers this ground adequately.

Where positive extract is provided, and this duct system is close to the supply duct, heat exchange apparatus (see Fig. 19-4) can be used between them to pre-treat the incoming air. For the air flow in Example 19-2, and in Fig. 19-5, it would be possible to save 5.5 kW of energy by apparatus costing some £800 (price as at March 1980). The winter saving is somewhat higher.

(*a*)

Figure 19-4 Air-to-air heat exchangers: (*a*) wheel, (*b*) static plate, (*c*) two coil. *(Courtesy Curwen & Newbery Limited.)*

(b)

(c)

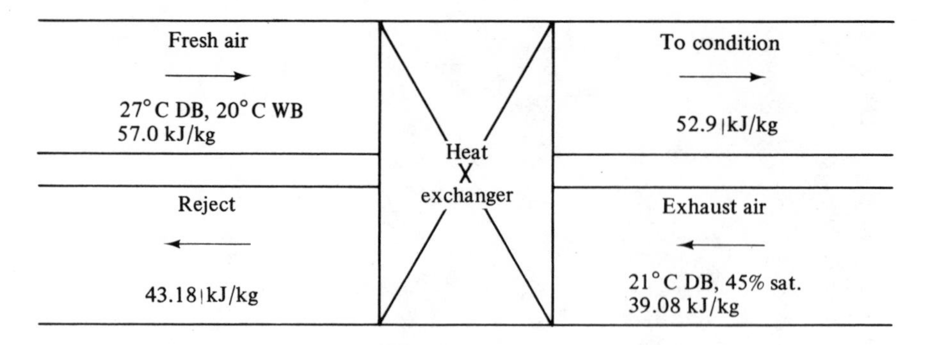

Figure 19-5 Heat recovery to pre-cool summer fresh air.

19-5 INTERNAL HEAT SOURCES

Electric lights, office machines, and other items of a direct energy-consuming nature will liberate all their heat into the conditioned space, and this load may be measured and taken as part of the total cooling load.

Lighting, especially in offices, can consume a great deal of energy and justifies the expertise of an illumination specialist to get the required light levels without wastage, both on new and existing installations. Switching should be arranged so that a minimum of the lights can be used in daylight hours. It should always be borne in mind that lighting energy requires extra capital and running cost to remove again.

Ceiling extract systems are now commonly arranged to take air through the light fittings, and a proportion of this load will be rejected with the exhausted air.

Example 19-3 Return air from an office picks up 90 per cent of the input of 15 kW to the lighting fittings. Of this return air flow 25 per cent is rejected to ambient. What is the resulting heat gain from the lights?

$$\text{Total lighting load} = 15 \text{ kW}$$
$$\text{Picked up by return air, } 15 \times 0.9 = 13.5 \text{ kW}$$
$$\text{Rejected to ambient, } 13.5 \times 0.25 = 3.375 \text{ kW}$$
$$\text{Net room load, } 15.0 - 3.375 = 12.625 \text{ kW}$$

The heat input from human occupants depends on their number (or an estimate of the probable number) and intensity of activity. This must be split into sensible and latent loads. The standard work of reference is CIBS Table A7.1, shown in Table 16-1.

The energy input of part of the plant must be included in the cooling load. In all cases include fan heat, either net motor power or gross motor input, depending on whether the motors are in the conditioned space or not. Also, in the case of packaged units within the space, heat is given off from the compressors and may not be allowed in the manufacturer's rating.

19-6 ASSESSMENT OF TOTAL LOAD ESTIMATES

Examination of the items which comprise the total cooling load may throw up peak loads which can be reduced by localized treatment such as shading, modification of lighting, removal of machines, etc., as may be possible. A detailed analysis of this sort can result in substantial savings in plant size and future running costs.

A careful site survey should be carried out if the building is already erected to verify the given data and search for load factors which may not be apparent from the available information.[21]

It will be seen that the total cooling load at any one time comprises a large number of elements, some of which may be known with a degree of certainty, but also many which are transient and which can only be estimated to a reasonable closeness. Even the most sophisticated and time-consuming of calculations will contain a number of approximations, so short-cuts and empirical methods are very much in use. A simplified calculation method is given by the Electricity Council[39], and abbreviated tables are given in Refs 21, 37, and 38. Full physical data will be found in Ref. 1 and CIBS Guide Book A (Ref. 2).

There are about 37 computer programs available, and a full list of these with an analysis of their relative merits is given by the Design Office Consortium, Cambridge, Evaluation Report No. 5.

Since the estimation will be based on a desired indoor condition at all times, it may not be readily seen how the plant size can be reduced at the expense of some temporary relaxation of the standard specified. Some of the programmes available can be used to indicate possible savings both in capital cost and running energy under such conditions.[40] In a cited case where an inside temperature of 21°C was specified, it was shown that the installed plant power could be reduced by 15 per cent and the operating energy by 8 per cent if short-term rises to 23°C could be accepted. Since these would only occur during the very hottest weather, such transient internal peaks may not materially detract from the comfort or efficiency of the occupants of the building.

CHAPTER

TWENTY

AIR MOVEMENT

20-1 STATIC AND VELOCITY PRESSURE

Atmospheric air at sea level exerts a *static pressure*, due to its weight, of 101.325 kPa. At 20°C the dry air weighs 1.2 kg/m^3. Densities at other conditions of pressure and temperature can be calculated from the Gas Laws:

$$\rho = 1.2\left(\frac{p}{101.325}\right)\left(\frac{273.15 + 20}{273.15 + t}\right)$$

where p is new pressure, in kilopascals, and t is new temperature, in degrees Celsius.

Example 20-1 What is the density of dry air at an altitude of 4500 m (57.5 kPa barometric pressure) and a temperature of −10°C?

$$\rho = 1.2\left(\frac{57.5}{101.325}\right)\left(\frac{293.15}{263.15}\right)$$

$$= 0.76 \text{ kg/m}^3$$

To this figure must be added the proportion of any moisture present. Thus, if this air contains water vapour in the proportion of 0.002 kg/kg dry air, the total density of the moist air will be

$$0.76 \times 1.002 = 0.762 \text{ kg/m}^3$$

If air is in motion, it will have *kinetic energy* of

$$0.5 \times \text{mass} \times (\text{velocity})^2$$

Example 20-2 If 1 m^3 of air at 20°C dry bulb, 60 per cent saturation, and a static pressure of 101.325 kPa is moving at 7 m/s, what is its kinetic energy?

Dry air at this condition weighs 1.2 kg/m^3, to which must be added the water vapour, determined from tables as 0.0089 kg/kg:

$$\text{Density of mixture} = 1.2 \times 1.0089$$
$$= 1.21 \text{ kg/m}^3$$
$$\text{Kinetic energy} = 0.5 \times 1.21 \times 7^2$$
$$= 29.65 \text{ kg/m s}^2$$

The dimensions of this kinetic energy are seen to be kg m^{-3} m^2 s^{-2} or kg/m s^2, which are the dimensions of pascals. This kinetic energy can therefore be expressed as a pressure and is termed the *velocity pressure*.

The *total pressure* of the air at any point in a closed system will be the sum of the static and velocity pressures. Losses of pressure due to friction will occur throughout the system and will show as a loss of total pressure, and this energy must be supplied by the air-moving device, usually a fan. (See Fig. 20-1 and Table 20-1.)

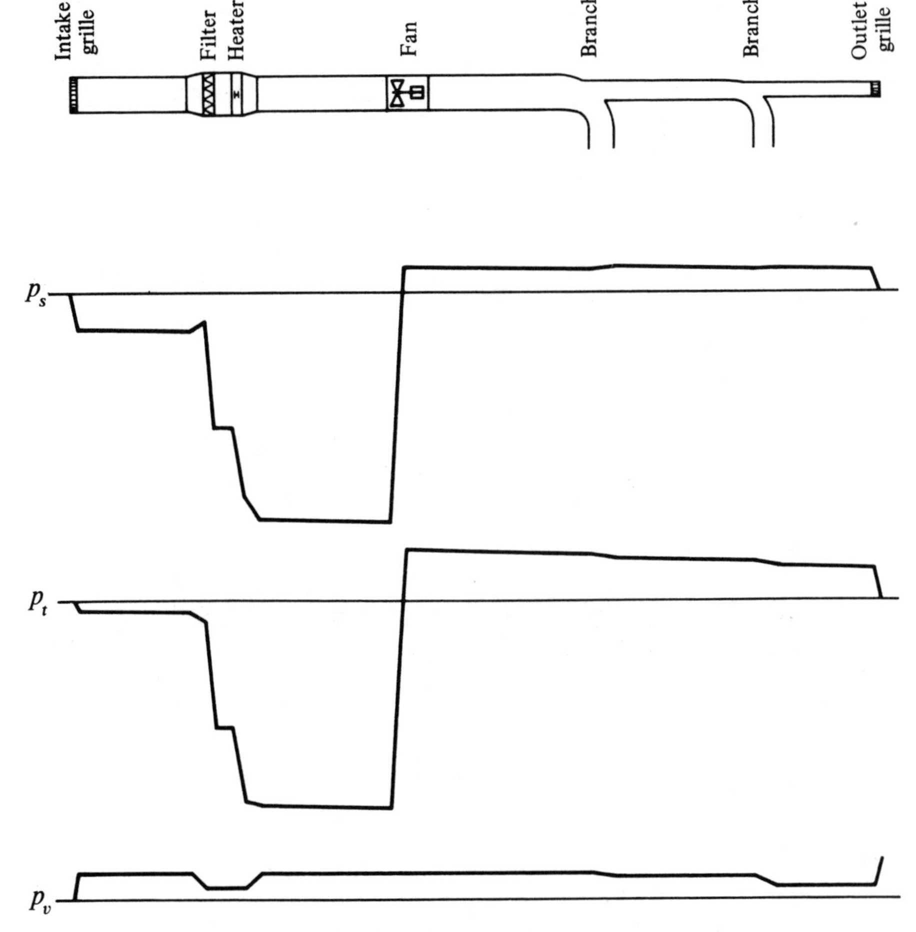

Figure 20-1 Ventilation system showing velocity, static, and total pressures.

Table 20-1 Duct pressure loss

Item	Type	Size, mm	Length, m	Air flow, m^3/s	Velocity, m/s	p_v, Pa	Resistance factor	Pressure loss, Pa	p_t, Pa	p_s, Pa
1	Inlet grille	700 × 400	—	2.0	7.1	30.25	0.40	12.1	−12.1	−42.35
2	Duct	700 × 400	4.5	2.0	7.1	30.25	1.0	4.5	−16.6	−46.85
3	Enlarge	700 × 400			7.1	30.25	0.27	8.2	−24.8	
		to 900 × 600	—	2.0	3.7	8.13				−32.93
4	Filter	900 × 600	—	2.0	3.7	8.13	120 Pa	120	−144.8	−152.93
5	Heater	900 × 600	—	2.0	3.7	8.13	85 Pa	85	−229.8	−237.93
6	Reduce	900 × 600		2.0	3.7	8.13				
		to 700 × 400	—		7.1	30.25	0.07	2.1	−231.9	−262.15
7	Duct	700 × 400	3.5	2.0	7.1	30.25	1.0	3.5	−235.4	−265.65
8	Fan									
9	Duct	700 × 400	3.5	2.0	7.1	30.25	1.0	3.5	55.9	25.65
10	Branch, straight	700 × 400	—	2.0	7.1	30.25	0.09	2.7	52.4	22.15
		to 500 × 400		1.33	6.6	26.1				
11	Duct	500 × 400	3.2	1.33	6.6	26.1	0.9	3.0	49.7	23.6
12	Branch, straight	500 × 400	—	1.33	6.6	26.1	0.09	2.3	46.7	20.6
		to 400 × 400		0.67	4.4	11.6				
13	Duct	400 × 400	3.3	0.67	4.4	11.6	0.8	2.6	44.4	32.8
14	Outlet	400 × 400	—	0.67	4.4	11.6	3.6	41.8	41.8	30.2

Required fan pressure = 265.65 + 25.65 = 291.3 Pa.

20-2 MEASURING DEVICES

The static pressure within a duct is too small to be measured by a bourdon tube pressure gauge, and the vertical or *inclined manometer* is usually employed (Fig. 20-2). The pressure tapping into the duct must be normal to the air flow.

Instruments for measuring the velocity as a pressure effectively convert this energy into *pressure*. The transducer used is the *pitot tube* (Fig. 20-3), which faces into the airstream and is connected to a manometer. The outer tube of a standard pitot tube has side tappings which will be normal to the air flow, giving *static pressure*. By connecting the inner and outer tappings to the ends of the manometer, the difference will be the *velocity pressure*.

Sensitive and accurate manometers are required to measure pressures below 15 Pa, equivalent to a duct velocity of 5 m/s, and accuracy of this method falls off below 3.5 m/s. The pitot head diameter should not be larger than 4 per cent of the duct width, and heads down to 2.3 mm diameter can be obtained. The manometer must be carefully levelled.

Air speed can be measured with mechanical devices, the best known of which is the vane anemometer (Fig. 20-4). In this instrument, the air turns the fan-like vanes of the meter, and the rotation is counted through a gear train on indicating dials, the number of turns being taken over a finite time. Alternatively, the rotation may be detected electronically and converted to velocity on a galvanometer. The rotating vanes are subject to small frictional errors and such instruments need to be specifically

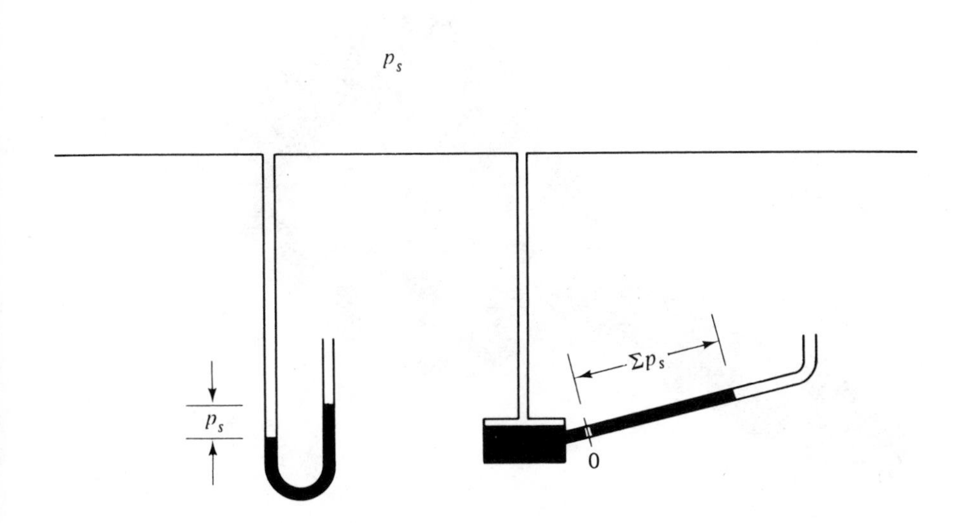

Figure 20-2 Vertical and inclined manometers.

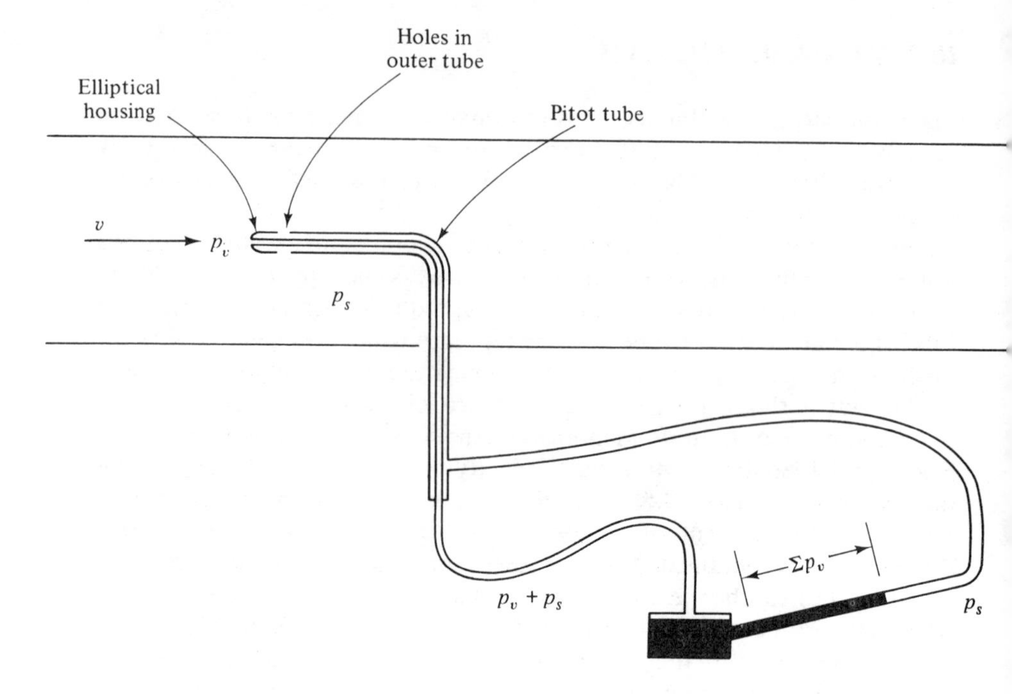

Figure 20-3 Pitot tube.

Figure 20-4 Vane anemometer. *(Courtesy Airflow Developments Limited.)*

calibrated if close accuracy is required. Accuracies of 3 per cent are claimed. Moving air can be made to deflect a spring-loaded blade and so indicate velocity directly.

A further range of instruments detects the cooling effect of the moving air over a heated wire or thermistor, and converts the signal to velocity on a galvanometer. Air velocities down to 1 m/s can be measured with claimed accuracies of 5 per cent, and lower velocities can be indicated.

Air flow will not be uniform across the face of a duct, so that readings must be taken at a number of positions and an average calculated. Methods of testing and positions for measurements are covered in BS.858: Part 1: 1963. In particular, air flow will be very uneven after bends or changes in shape, so measurements should be taken in a long, straight section of duct, where this can be found.

More accurate measurement of air flow can be achieved with nozzles or orifice plates. In such cases, the measuring device imposes a considerable resistance to the air flow, so that a compensating fan is required. This method is not applicable to an installed system and is used mainly as a development tool for factory-built packages, or for fan testing. Details of these test methods will be found in BS.858, BS.2852, and ASHRAE 16-61.

20-3 AIR-MOVING DEVICES

Total pressures required for air-conditioning systems and apparatus are rarely in excess of 2 kPa, so can be obtained with dynamic air-moving machinery rather than positive-displacement pumps. The *centrifugal fan* (Fig. 20-5) imparts a rotation to the entering air and the resulting centrifugal force is converted to pressure and velocity in a suitable outlet scroll. Air leaving the tips of the blades will have both radial and tangential velocities, so the shape of the blade will determine the fan characteristics.

The forward-curve fan blade increases the tangential velocity considerably (see Fig. 20-5*b*). As a result, the power required will increase with mass flow, although the external resistance pressure is low, and oversize drive motors are required if the system resistance can change in operation. The backward-curve fan runs faster and has a flatter power curve, since the air leaves the blade at less than the tip speed (see Fig. 20-5*c*).

Since centrifugal force varies as the square of the speed, it can be expected that the centrifugal fans, within certain limits, will have the same characteristics. These can be summed up in the General Fan Laws:

Volume varies as speed.
Pressure varies as $(\text{speed})^2$.
Power varies as $(\text{speed})^3$.

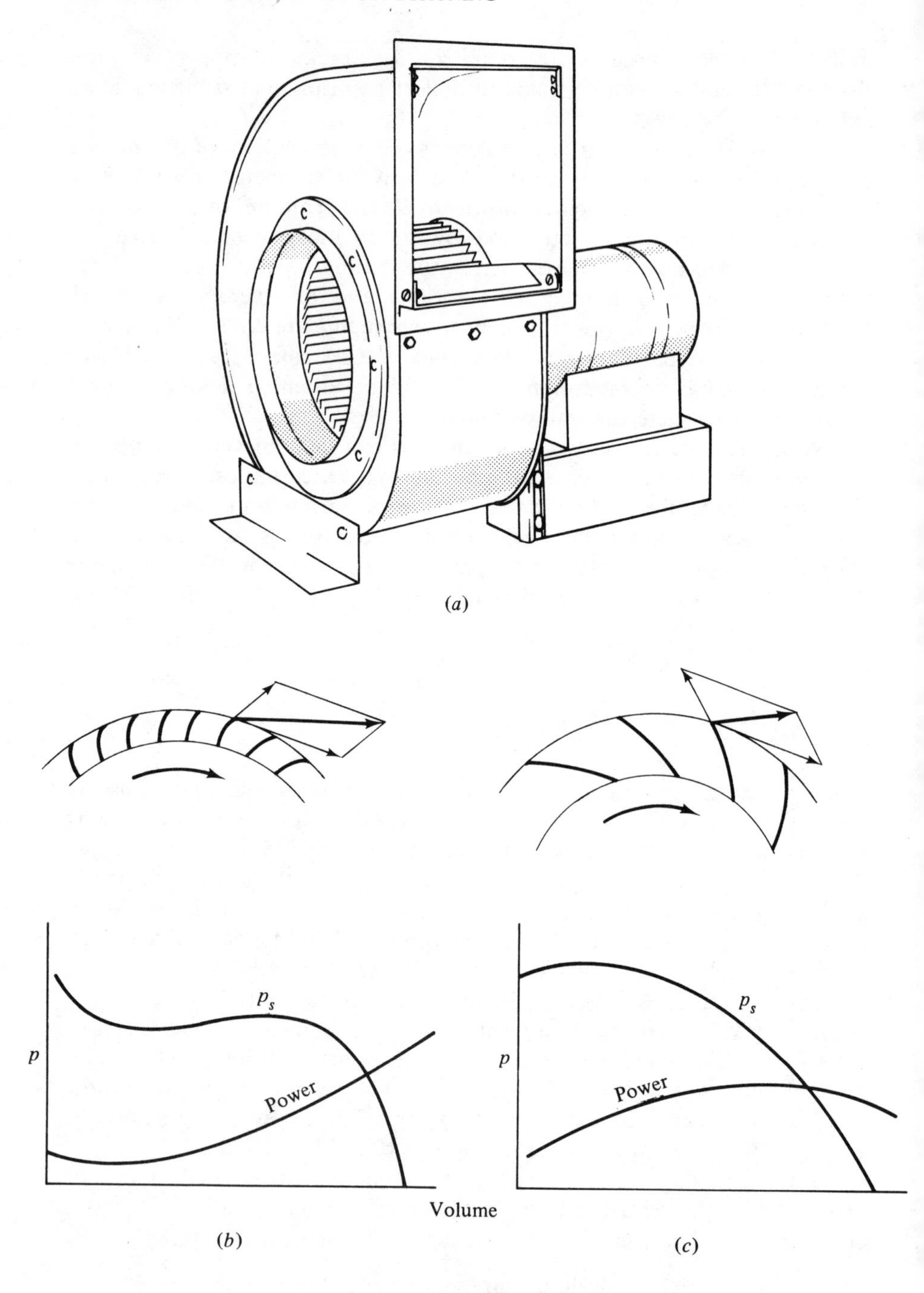

Figure 20-5 Centrifugal fan: (*a*) construction, (*b*) forward-curved blades and typical performance curves, (*c*) backward-curved blades and typical performance curves.

Where a centrifugal fan is belt driven and some modification of performance may be required, these laws may be applied to determine a revised speed and the resulting power for the new duty. Since the resistance to air flow will also vary as the square of the speed of the air within the duct (see Sec. 20-4), it follows that a change of fan speed proportional to the required change in volume should give a close approximation for the new duty.

At no-flow (stall) conditions, these fans will not generate any velocity pressure and the absorbed power will be a minimum, used only in internal turbulence.

Large volumes of air at low pressures can be moved by the *propeller fan* (Fig. 20-6). The imparted energy is mainly in an axial direction and any large external resistance will cause a high proportion of slip over the blades.

The working pressure limits of the propeller fan, depending on its diameter, are of the order of 150 Pa. The characteristic curve has a pronounced 'trough', which should be avoided in application if at all possible, since wide variations in air flow can occur for a small change in pressure. These fans are rarely used on duct systems, which would have a stabilizing effect on this volume variation. Performance varies with aperture shape, clearance, and position.

Peak efficiency and pressure capability can be achieved with axial-flow fans by using blades of correct aerofoil shape and ensuring a low tip clearance. Such fans are termed aerofoil, *axial flow* (Fig. 20-7), or tube axial, to differentiate from propeller fans. The pitch angle of the blades will determine the working characteristics and best working efficiency. Commercially available fans are commonly made so that the angle of pitch can be selected for its application and pre-set at the factory or on installation. Some large axial-flow fans can be obtained with blades which can be varied in pitch while running, similar to variable-pitch aircraft propellers, so that the fan performance can be varied as required by the system load.

Air leaves the blades of an axial-flow fan with some turning motion, and the provision of straightening vanes after the rotor will recover some of this energy, adding to the performance and efficiency of the fan; pre-rotational vanes also help slightly.

Higher pressures can be obtained by putting two axial-flow fans in series. If placed close together and contrarotated, the spin imparted by the first can be recovered by the second, and more than twice the pressure capability can be gained.

The best efficiency of the axial-flow fan is to the right of the trough seen in the pressure curve, and the optimum band of performance will be indicated by the manufacturer. In particular, the air flow should not be less than the given minimum figure, since the fan motor relies on the air mass flow for cooling.

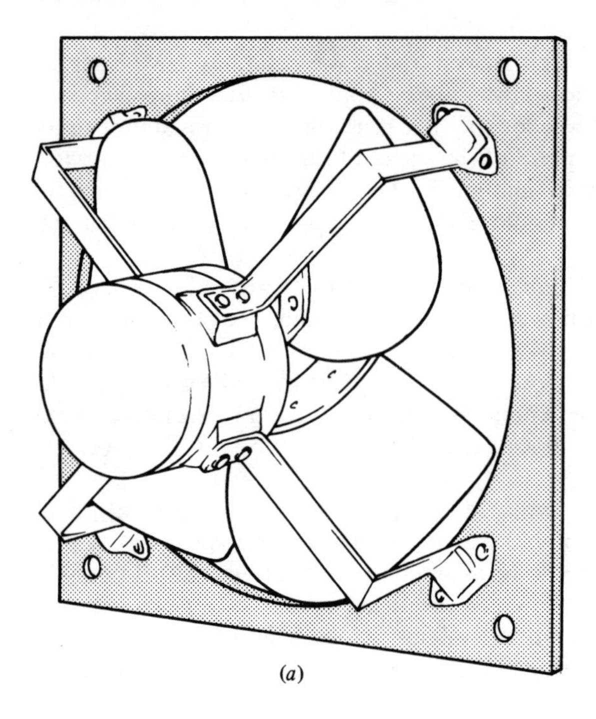

(*a*)

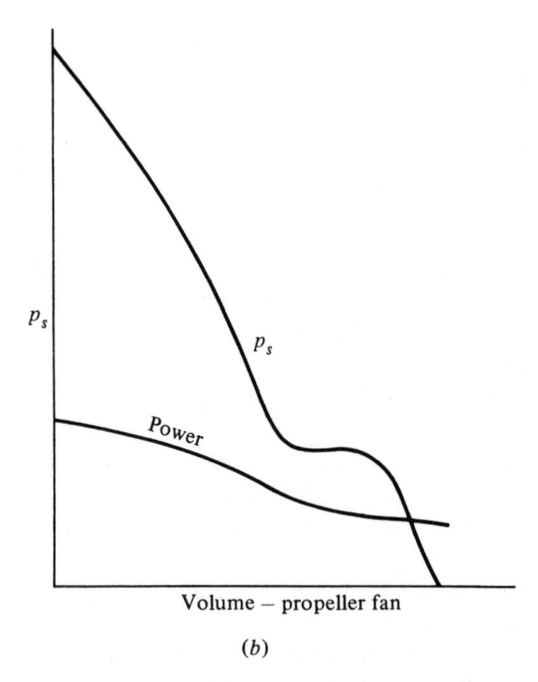

(*b*)

Figure 20-6 Propeller fan: (*a*) construction, (*b*) typical performance curves.

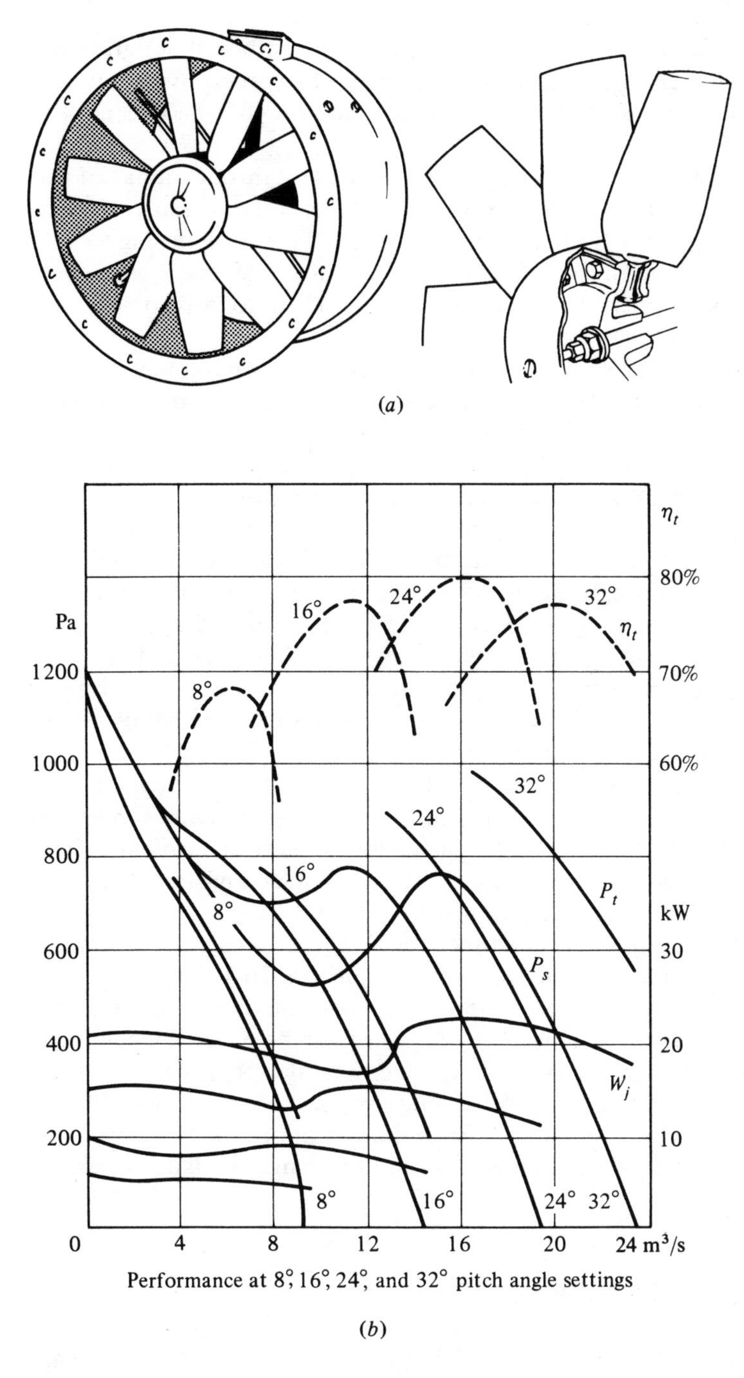

Figure 20-7 Axial flow fans: (*a*) construction, (*b*) typical performance curves. *(Reproduced, with permission, from Woods Practical Guide to Fan Engineering.*[42]*)*

It is possible to readjust the blade angles on site but, if so, great care must be taken to get them all at the same angle. The procedure is not to be recommended. Most such fans are direct drive, so the speed cannot be changed.

It will be seen that there is no change in velocity through an axial-flow fan, and the blade energy is used in increasing the static pressure of the air flow. Since the velocity through the fan casing will probably be higher than adjacent duct velocities, these fans commonly have inlet and outlet cones, which must be properly designed and constructed to minimize energy losses.

The *mixed-flow* fan combines the geometry of the axial-flow and centrifugal fans and can give a very high efficiency at a predetermined operating load, but is less flexible in operation outside that point in its curve. It requires an accurately fitting housing, and is not in general use on commercial applications because of the close working tolerances.

The *cross-flow* or tangential fan sets up an eccentric vortex within the fan runner, the air coming inwards through the blades on one side and leaving outwards through the blades on the other. It can, within mechanical limits, be made as long as necessary for the particular duty.

The cross-flow fan generates only very slight pressure and its use is limited to appliances where the air pressure drop is low and predetermined. Its particular shape is very suitable for many kinds of air-handling devices such as fan coil units and fan convectors.

The fans used in air-conditioning duct systems are centrifugal or axial flow. Since both types are available in a wide range of sizes, speeds, and manufacture, the final choice for a particular application is often reduced to a suitable shape—the centrifugal having its inlet and outlet ducts at 90°C, while the axial flow can be in-line.

The centrifugal fan may be direct-coupled, i.e., having the fan runner on an extension of the motor shaft, or belt driven. In the latter case the motor must be mounted with the fan, to withstand belt tension. This arrangement has the advantage that the speed can be selected for the exact load, and can be changed if required. The axial-flow fan usually has the motor integral, so is restricted to induction motor speeds of 2900, 1450, or 960 r/min, and cannot be altered. Precise application and possible future duty changes may be accommodated within the range of blade angles.

20-4 NOISE AND VIBRATION

All manufacturers now publish sound pressure levels for their products and such figures should be scrutinized and compared as part of a fan-selection decision. Fans are statically, and sometimes dynamically, balanced by the manufacturer. If it is necessary to dismantle a fan for transport, it should be rebalanced on commissioning, imposing a load close to that ultimately required.

Fans are balanced in a clean condition, but will tend to collect dirt in operation, which will adhere unevenly to the blades. It is therefore essential to provide antivibration mountings for all fan assemblies, including their drive motors. Since the fan will then be free to move relative to the ductwork, which is fixed, flexible connections will be needed to allow for this movement. With belt-driven fans, care must be taken that the anti-vibration mountings are suitable for the rotational speeds of both fan and motor. Fans with high tip speeds will generate noise levels which may need attenuation. The normal treatment of this problem is to fit an acoustically lined section of ductwork on the outlet or on both sides of the fan. Such treatment needs to be selected for the particular application regarding frequency of the generated noise and the degree of attenuation required, and compotent suppliers will have this information. The attenuators will be fixed, and located after the flexible connectors, so these latter will also need acoustic insulation to prevent noise breaking out here.[41]

Most large fans need to be cleaned thoroughly every year to remove deposits of dirt and so limit vibration.

20-5 FLOW OF AIR IN DUCTS

General laws for the flow of fluids were determined by Reynolds, who recognized two flow patterns, laminar and turbulent. In laminar flow the liquid can be considered as a series of parallel strata, each moving at its own speed, and not mixing. Strata adjacent to walls of the duct will be slowed by friction and will move slowest, while those remote from the walls will move fastest. In turbulent flow there is a general forward movement together with irregular transfer between strata.

In air-conditioning systems, all flow is turbulent, and formulas and charts show the resistance to air flow of ducting of various materials, together with fittings and changes of shape to be met in practice. The reader is referred to the tables and charts in CIBS Guide C4 (Ref. 4) and in Ref. 42 (chap. 6), and the fuller theoretical analysis in Ref. 38 (chap. 15).

High duct velocities show an economy in duct cost, but require more power which will generate more noise. Velocities in common use are as follows:

High-velocity system, main ducts	20 m/s
High-velocity system, branch ducts	15 m/s
Low-velocity system, main ducts	10 m/s
Low-velocity system, branch ducts	6 m/s
Ducts in quiet areas	3–4 m/s

Ducting construction must be stiff enough to retain its shape, be free from air-induced vibration (panting) and strong enough to allow air-tight joints along its length. Such construction is adequately covered by HVCA[43] Specification No. DW.141 for sheet metal, No. DW.151 for plastics, and No. DW.181 for g.r.p.

The frictional resistance to air flow within a duct system follows the general law

$$H = a\frac{v^2}{d}$$

where a is a coefficient based on the roughness of the duct surface and the density of the air. Where square or rectangular ducts are to be calculated, their dimensions are reduced to an equivalent diameter.

Within the limits of operation of normal air-conditioning systems, tables or charts can be drawn up, based on this law.

Example 20-3 What is the resistance pressure drop in a duct measuring 700 × 400 mm, if the air flow through it is 2 m^3/s? What is the velocity?

From the chart (Table 20-1), reading down the 700 × 400 line until it meets the horizontal line through 2 m/s gives

$$\text{Pressure drop} = 1.0 \text{ Pa/m}$$

$$\text{Velocity} = 7.1 \text{ m/s}$$

It should be noted that the energy for this pressure drop must come from static pressure, since the velocity, and hence the velocity pressure, remains constant.

Frictional resistance to air flow of fittings such as bends, branches, and other changes of shape or direction will depend on the shape of the fitting and the velocity, and such figures are tabulated with factors to be multiplied by the velocity pressure. Tables of such factors will be found in standard works of reference.[1, 4, 42]

Example 20-4 The duct specified above has in it two bends, for which a pressure loss factor of 0.28 is shown in the tables. What is the total pressure loss?

$$\text{Pressure loss per bend} = p_v \times 0.28$$

$$p_v = 0.5 \times 1.2 \times v^2$$

where $v = 7.1$

$$p_v = 30.25 \text{ Pa}$$

$$\text{Pressure loss} = 2 \times 0.28 \times 30.25$$
$$= 16.94 \text{ Pa}$$

The sizing of ductwork for a system will commence with an assumption of an average pressure-loss figure, based on a working compromise between small ducts with a high pressure drop and large ducts with a small pressure drop. An initial figure for a commercial air-conditioning plant will be 0.8 to 1.0 Pa/m. This will permit higher velocities in the larger ducts with lower velocity in the branches within the conditioned spaces, where noise may be more noticeable.

Pressure drops for proprietary items such as grilles and filters will be obtained from manufacturers.

An approximate total system resistance can be estimated from the design average duct loss and the maximum duct length, adding the major fittings. However, this may lead to errors outside the fan power and it is safer to calculate each item and tabulate as shown in Table 20-1. Only the longest branch need be taken for fan pressure.

It will be seen that, where there are a number of branches from a main duct, there will be an excess of available pressure in these branches. In order to adjust the air flows on commissioning, dampers will be required in the branch ducts or, as is more usually provided, in the necks of the outlet grilles. The latter arrangement may be noisy, if some of these dampers have to be closed very far to balance the air flow, with a resulting high velocity over the grille blades.

20-6 FLOW OF AIR UNDER KINETIC ENERGY

Any static pressure at the outlet of a duct will be lost as the air expands to atmospheric pressure. This expansion, which is very small, will be in all directions, with no perceptible gain of forward velocity. (See Table 20-1.) Static pressure can be converted to velocity at the outlet by means of a converging nozzle or by a grille. In both cases the air outlet area is less than the duct area, and extra forward velocity is generated from the static pressure. The leaving air will form a jet, the centre of which will continue to move at its original velocity, the edges being slowed by friction and by entrainment of the surrounding air. (See Fig. 20-8.) The effect is to form a cone, the edges of which will form an included angle of 20 to 25°, depending on the initial velocity and the shape of the outlet. Since the total energy of the moving air cannot increase, the velocity will fall as the mass is increased by entrained air, and the jet will lose all appreciable forward velocity when this has fallen to 0.25 to 0.5 m/s.

If the air in a horizontal jet is warmer or cooler than the surrounding air, it will tend to rise or fall. This effect will lessen as the jet entrains air, but may be important if wide temperature differences have to be used or in large rooms.[44, 45]

If an air jet is released close to a plane surface (ceiling or wall usually), the layer of air closest to the surface will be retarded by friction and the jet

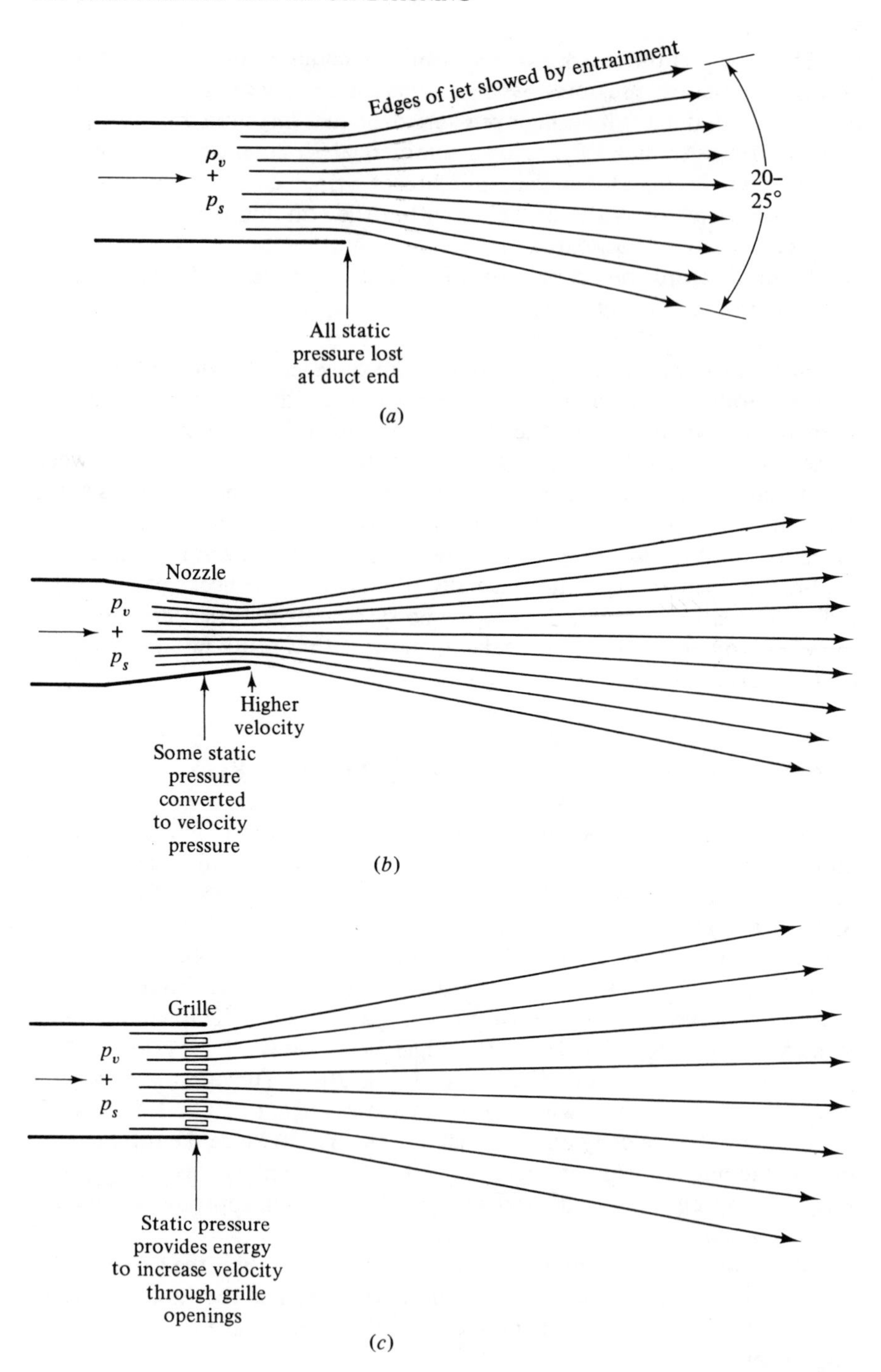

Figure 20-8 (*a*) Air leaving open-ended duct. (*b*) Air leaving nozzle. (*c*) Air leaving grille.

will tend to cling to the surface. Use of this effect is made to distribute air across a ceiling from ceiling slots or from grilles high on the walls. (See Fig. 20-9.) Air is entrained on one side only and the cone angle is about half of that with a free jet. This produces a more coherent flow of input air with a longer throw.

If the air jet is held within a duct expansion having an included angle less than 20°, only duct friction losses will occur. Since there is no entrained air to take up some of the kinetic energy of the jet, a large proportion of the drop in kinetic energy will be regained as static pressure. (See Fig. 20-10.)

The optimum angle for such a duct expansion will depend on the air velocity, since the air must flow smoothly through the transition and not 'break away' from the duct side with consequent turbulence and loss of energy. This included angle is about 14°. With such an expansion, between 50 and 90 per cent of the loss of velocity pressure will be regained as static pressure.[38, 42]

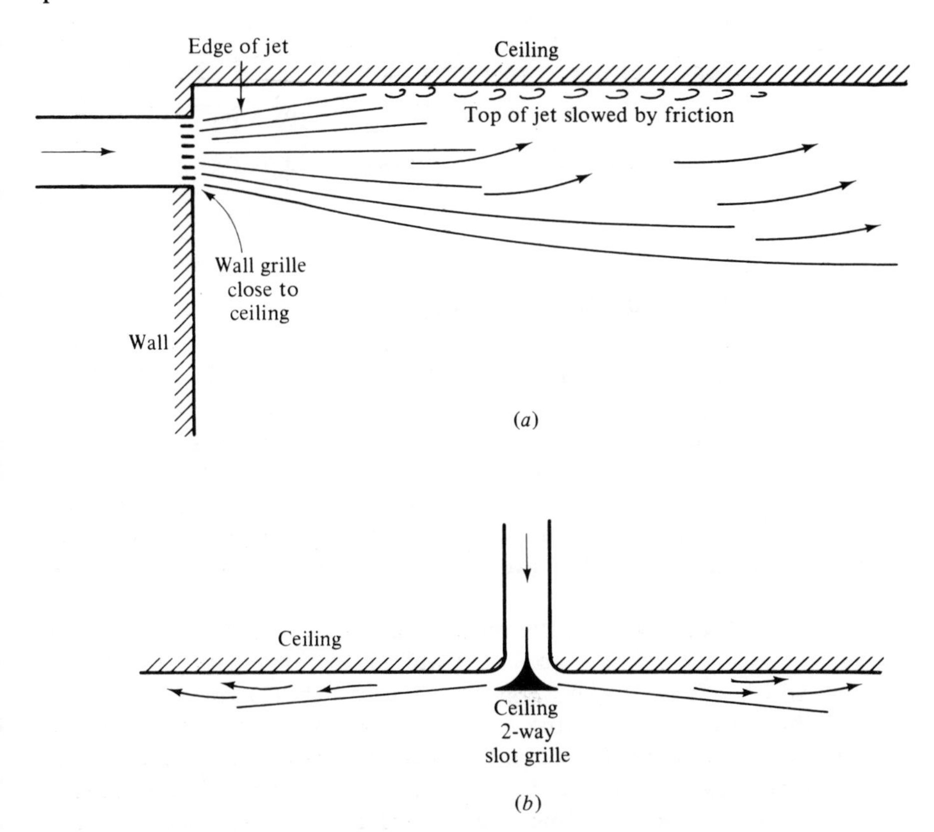

Figure 20-9 Restriction of jet angle by adjacent surface: (*a*) wall grille close to ceiling, (*b*) ceiling slots.

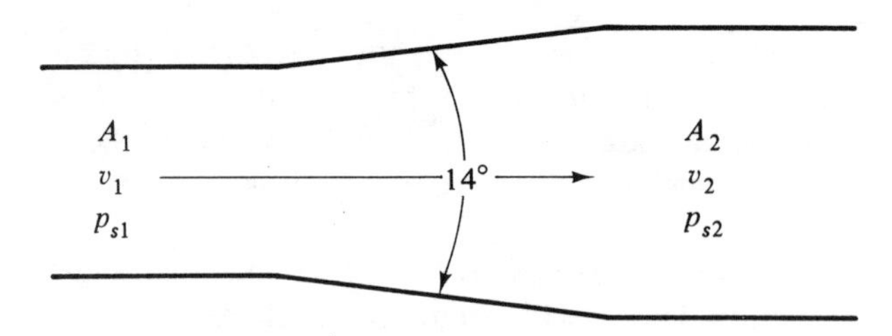

Figure 20-10 Duct expansion with static pressure regain.

Example 20-5 Air moving in a duct at 8 m/s is gently expanded to a velocity of 5.5 m/s. If the friction losses are 20 per cent of the available velocity pressure change, what is the amount of *static regain*?

Velocity pressure entering expansion $= 0.5 \times 1.2 \times 8^2$

$= 38.4$ Pa

Velocity pressure leaving expansion $= 0.5 \times 1.2 \times 5.5^2$

$= 18.15$ Pa

Friction losses $= 0.2\ (38.4 - 18.15) = 4.05$ Pa

Static regain $= 0.8\ (38.4 - 18.15) = 16.2$ Pa

20-7 FLOW OF AIR IN A ROOM

Since incoming air may be as much as 11 K colder or 25 K warmer than the conditioned space, the object of the duct and grille system must be to distribute this air and mix it with the room air with the least discomfort to the occupants. The subjective feeling of discomfort will depend on the final temperature difference, the velocity, and the degree of activity, cold air being less acceptable than warm. Velocities at head level should be between 0.1 and 0.45 m/s and comprehensive factors will be found in the CIBS Guide, Table B3.1.

Figure 20-11*a* shows a typical office or hotel room with supply duct in the central corridor ceiling space and a wall grille blowing air towards the window, which will usually be the greatest source of heat gain or loss. High-level discharges of this type work best when cooling, since the incoming air jet will fall as it crosses the room. On heating it will tend to rise, so must have enough velocity to set up a forced circulation in the pattern shown.

Figure 20-11*b* shows perimeter units under the window and discharging upwards to absorb the heat flow through the window. The angle and velocity of discharge should be enough for the air then to set up a circula-

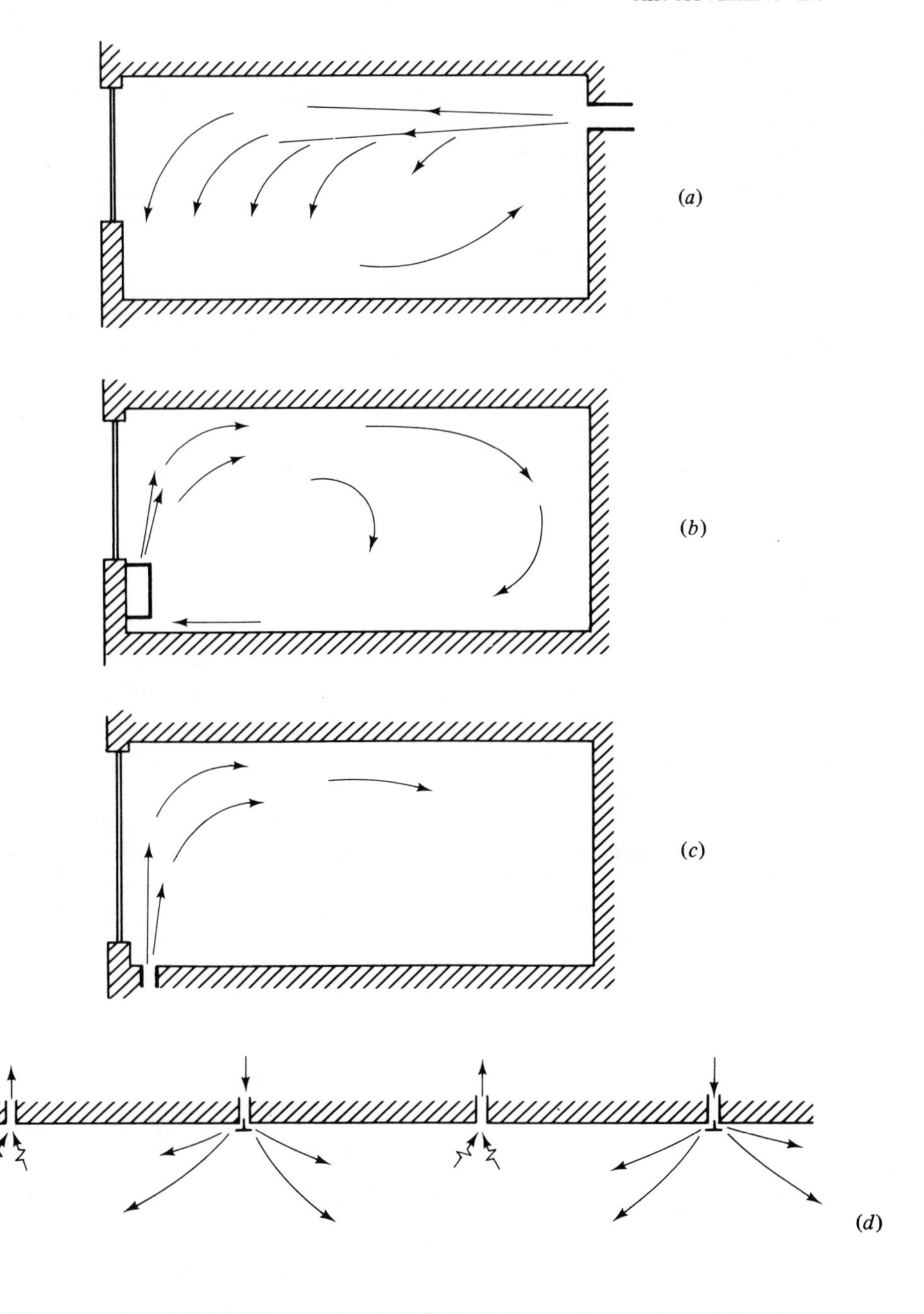

Figure 20-11 Room air circulation patterns: (*a*) grilles on corridor wall, (*b*) sill outlets, (*c*) floor outlets, (*d*) ceiling slots.

tion within the room to reach the far wall or, in an open plan room, to the area covered by an adjacent grille. In such units, air returns to the face or underside of the unit.

Figure 20-11*c* shows floor slots, setting up a pattern similar to the perimeter units. This arrangement has been adopted in some buildings having all glass walls. The position of the return grille varies with room layout and stagnant zones can occur.

Figure 20-11*d* shows ceiling grilles or slots, requiring all ducting to be within the ceiling void. This system is generally adopted for open plan rooms, since the area can be divided into strips with alternate supply and extract slots, or into squares (or near-squares) for supply and extract by grilles.

It will be seen that all of these patterns require some consideration and planning, since the best equipment cannot cope with an impossible air circuit.[45]

20-8 GRILLES

The air inlet grille will be recognized as a device for converting static pressure to outlet velocity, having the required speed and direction to take the conditioned air across the room and entrain the surrounding air so as to reach the occupants at a suitable temperature.

Wall grilles will have directional vanes in one or both planes, which can be set on commissioning. These need to be set by a competent person who is aware of the required room flow pattern. It is advisable that such a setting adjustment should be operated by special tools to prevent subsequent tampering.

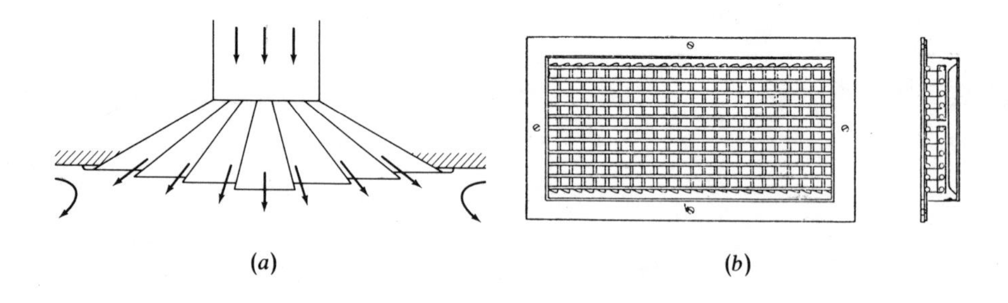

Figure 20-12 Air discharge grilles: (*a*) ceiling (*Courtesy of Anemostat Corporation of America.*), (*b*) wall.

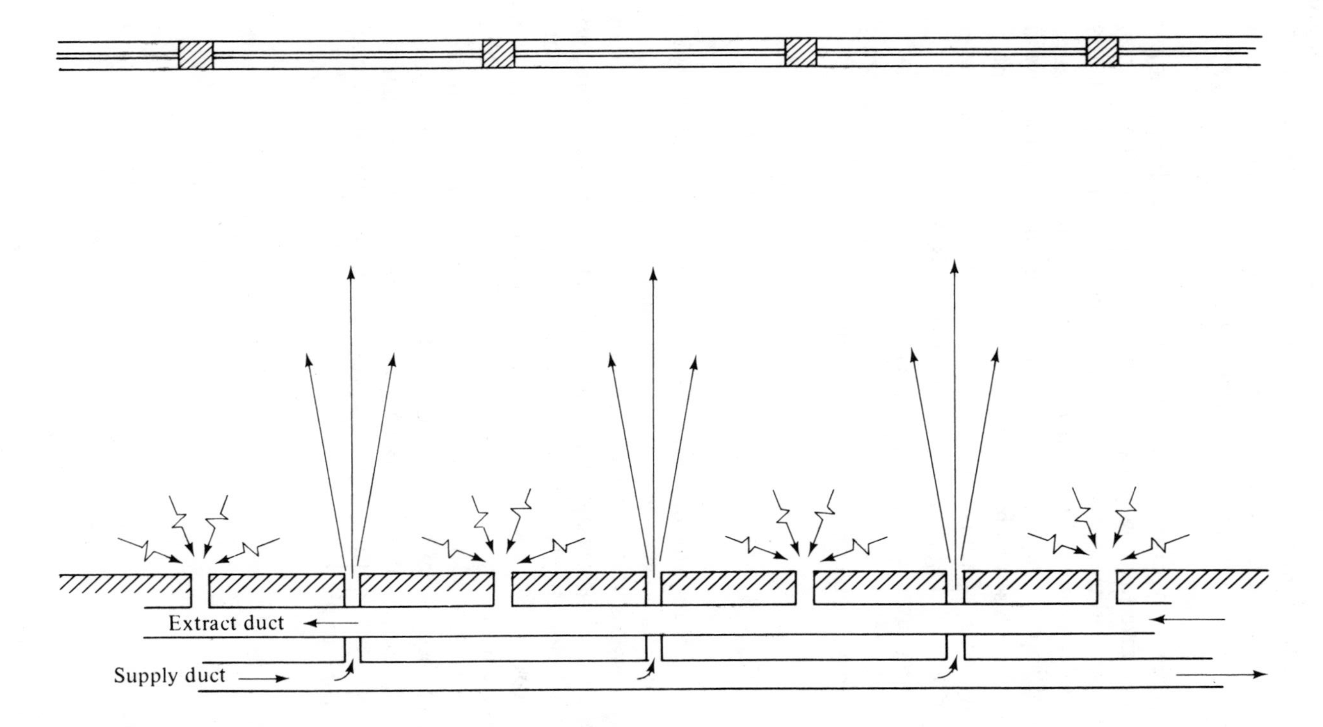

Figure 20-13 Discharge and return grilles on same wall.

Perimeter unit grilles must direct the air upwards and slightly away from the window to start the circulatory pattern. Fixed angles are preferable on these units, rather than the adjustable segments supplied with many packaged products. A common fault with such an installation is obstruction of inlet and outlet grilles by office equipment. It will be a definite advantage if the upper surface is sloped, to discourage its use as a shelf.

The geometry of ceiling grilles and slots can be fixed or adjustable. In the former case the flow pattern is set by the spacing and volume (=velocity), so site adjustment cannot always compensate for a faulty layout. Slots have a limited throw, of the order of 5 m at full volume, and the layout of supply and return slots must take into account any operating variation in this volume. The setting on commissioning of all these, especially those of adjustable geometry, should be left to competent hands and then locked. Figure 20-12 shows two kinds of grille.

For very large areas, such as assembly halls or sports arenas, jets of air will be required to obtain the large throw distances. Localized draughts may be unavoidable in such installations.

20-9 RETURN AIR

Air entering a return duct will be moved by the difference in pressure, the duct being at a lower static head than the room. Such movement will be radially towards the inlet and *non-directional*. At a distance of only 1 m from the grille this pressure gradient will be quite low, so return grilles can be located close to supply grilles, providing the overall circulation pattern ensures coverage of the space. In Fig. 20-11*a*, the return air grilles can also be in the corridor wall, if far enough from the inlets. (See Fig. 20-13.)

With ceiling inlet and extract systems, the opportunity is presented of removing heat from light troughs. This can reject a proportion of the cooling load, possibly as high as 20 per cent, in the exhaust air. The recirculated air is also warmer, improving heat transfer at the cooling coil. (See also Example 19-3.)

CHAPTER

TWENTY-ONE

AIR-CONDITIONING METHODS

21-1 REQUIREMENT

The cooling load of an air-conditioned space comprises estimates of the sensible and latent heat gains, and is $Q_S + Q_L$. This rate of heat flow is to be removed by a cooling medium which may be air, water, brine, or refrigerant, or a combination of two of these. (See Fig. 21-1.)

Example 21-1: All air A space is to be held at 21°C dry bulb and 50 per cent saturation, and has an internal load of 14 kW sensible and 1.5 kW latent heat gain. The inlet grille system is suitable for an inlet air temperature of 12°C. What are the inlet air conditions and the mass air flow?

$$\text{Inlet air temperature} = 12.0^\circ\text{C}$$

$$\text{Air temperature rise through room, } 21\text{–}12 = 9.0 \text{ K}$$

$$\text{Air flow for sensible heat, } \frac{14}{9 \times 1.02} = 1.525 \text{ kg/s}$$

$$\text{Moisture content of room air, } 21^\circ\text{C}, 50\% = 0.007\,857 \text{ kg/kg}$$

$$\text{Moisture to pick up, } \frac{1.5}{2440 \times 1.525} = 0.000\,403$$

$$\text{Moisture content of entering air} = 0.007\,454$$

From tables,[4] this gives about 85 per cent saturation.

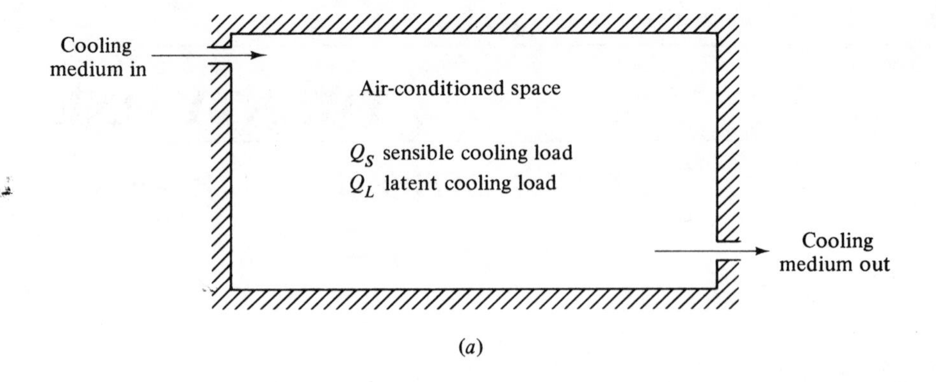

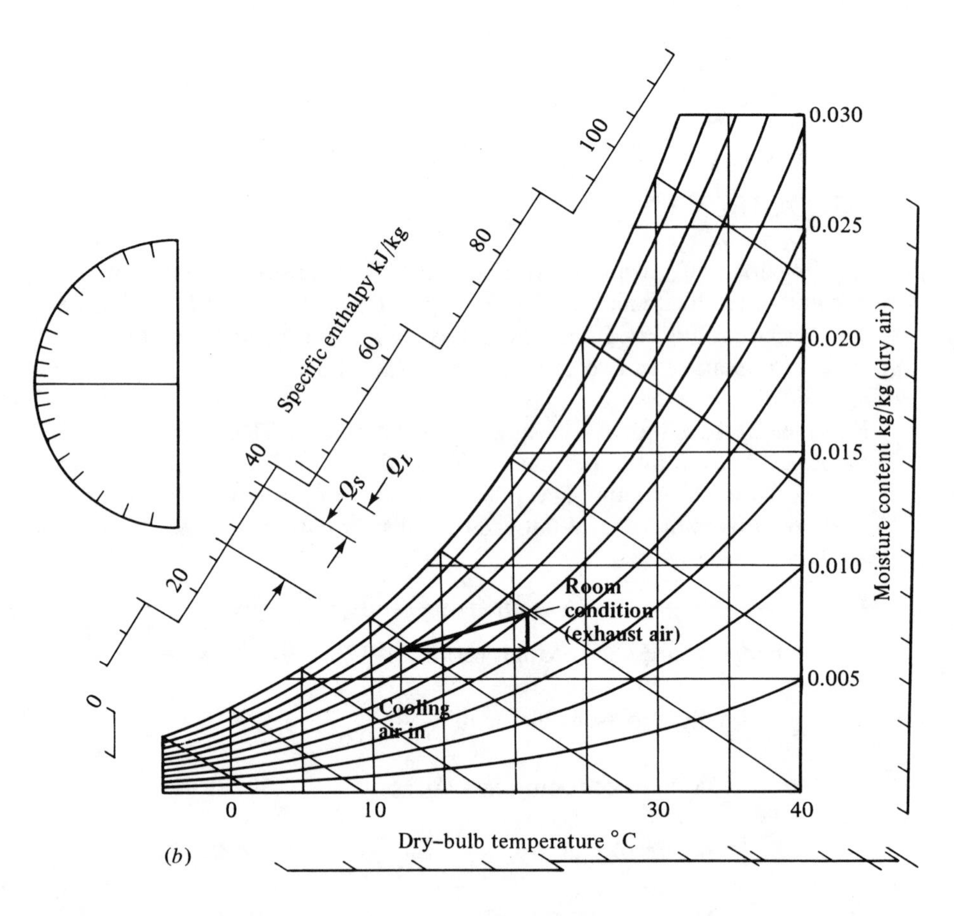

Figure 21-1 Removal of sensible and latent heat from conditioned space: (*a*) flow of cooling medium, (*b*) process line.

Note that the figure of 1.02 in the third line is a general figure for the specific heat capacity of moist air, commonly used in these calculations. (The true figure for this particular example is slightly higher.) The figure of 2440 for the latent heat is, again, a general quantity in common use, and is near enough for these calculations.

Example 21-2: Chilled water For the same duty, a chilled water fan coil unit is fitted within the space. Water enters at 5°C and leaves at 10.5°C. The fan motor takes 0.9 kW. What water flow is required?

$$\text{Total cooling load, } 14.0 + 1.5 + 0.9 = 16.4 \text{ kW}$$

$$\text{Mass water flow, } \frac{16.4}{4.187 \times (10.5 - 5)} = 0.71 \text{ kg/s}$$

Example 21-3: Refrigerant For the same duty, liquid R.22 enters the expansion valve at 33°C, evaporates at 5°C, and leaves the cooler at 9°C. Fan power is 0.9 kW. What mass flow of refrigerant is required?

$$\text{Total load, as Example 21-2} = 16.4 \text{ kW}$$

Enthalpy of R.22, evaporated at 5°C,
super heated to 9°C = 309.39 kJ/kg

Enthalpy of liquid R.22 at 33°C = 139.84 kJ/kg

Refrigerating effect = 169.55 kJ/kg

$$\text{Required refrigerant mass flow, } \frac{16.4}{169.55} = 0.097 \text{ kg/s}$$

Example 21-4: Primary air and chilled water For the same application, primary air reaches an induction unit at the rate of 0.4 kg/s and at conditions of 13°C dry bulb and 72 per cent saturation. Chilled water enters the coil at 12°C and leaves at 16°C. What will be the room condition and how much water will be used?

The chilled water enters higher than the room dew point temperature, so any latent heat must be removed by the primary air, and this may result in a higher indoor condition to remove the design latent load:

Moisture in primary air, 13°C DB, 72% sat. = 0.006 744 kg/kg

$$\text{Moisture removed, } \frac{1.5}{2440 \times 0.4} = 0.001\,537 \text{ kg/kg}$$

Moisture in room air must rise to = 0.008 281 kg/kg

which corresponds to a room condition of 21°C dry bulb, 53 per cent saturation.

Sensible heat removed by primary air,

$$0.4 \times 1.02 \times (21 - 13) = 3.26 \text{ kW}$$

Heat to be removed by water, $14.9 - 3.26 = 11.64$ kW

$$\text{Mass water flow, } \frac{11.64}{4.187 \times (16 - 12)} = 0.7 \text{ kg/s}$$

21-2 CENTRAL STATION SYSTEM. ALL AIR

The centralization of all plant away from the conditioned space, originating from considerations of safety, also ensures the best access for operation and maintenance and the least transmission of noise. Since all air passes through the plantroom, it is possible to arrange for any proportion of outside air up to 100 per cent. This may be required for some applications, but the option of more outside air for other duties will reduce the refrigeration load in cold weather. For example, in the systems considered in Sec. 21-1, there may still be a cooling load required when the ambient is down to 12°C dry bulb, but this is the design supply air temperature, so all cooling can be done with ambient air and no mechanical refrigeration.

The distribution of air over a zone presupposes that the sensible and latent heat loads are reasonably constant over the zone (see Fig. 21-2). As soon as large variations exist, it is necessary to provide air cold enough to satisfy the greatest load, and reheat the air for other areas. Where a central plant serves a number of separate rooms and floors, this resolves into a system with reheat coils in each zone branch duct (see Fig. 21-3). It will be recognized that this is wasteful of energy and can, in the extreme, require a reheat load almost as high as the cooling load.

To make the central air system more economical for multizone installations, the quantity of cooled air to the individual zones can be made variable, and reduced when the cooling load is less. This will also reduce the amount of reheat needed. This reheat can be by means of a coil, as before, or by blending with a variable quantity of warmed air, supplied through a second duct system (see Fig. 21-4).

In the first of these methods, the reduction in air mass flow is limited by considerations of distribution velocities within the rooms, so at light load more air may need to be used, together with more reheat, to keep air speeds up. Within this constraint, any proportion of sensible and latent heat can be satisfied, to attain correct room conditions. However, full humidity control would be very wasteful in energy and a simple thermostatic control is preferred.

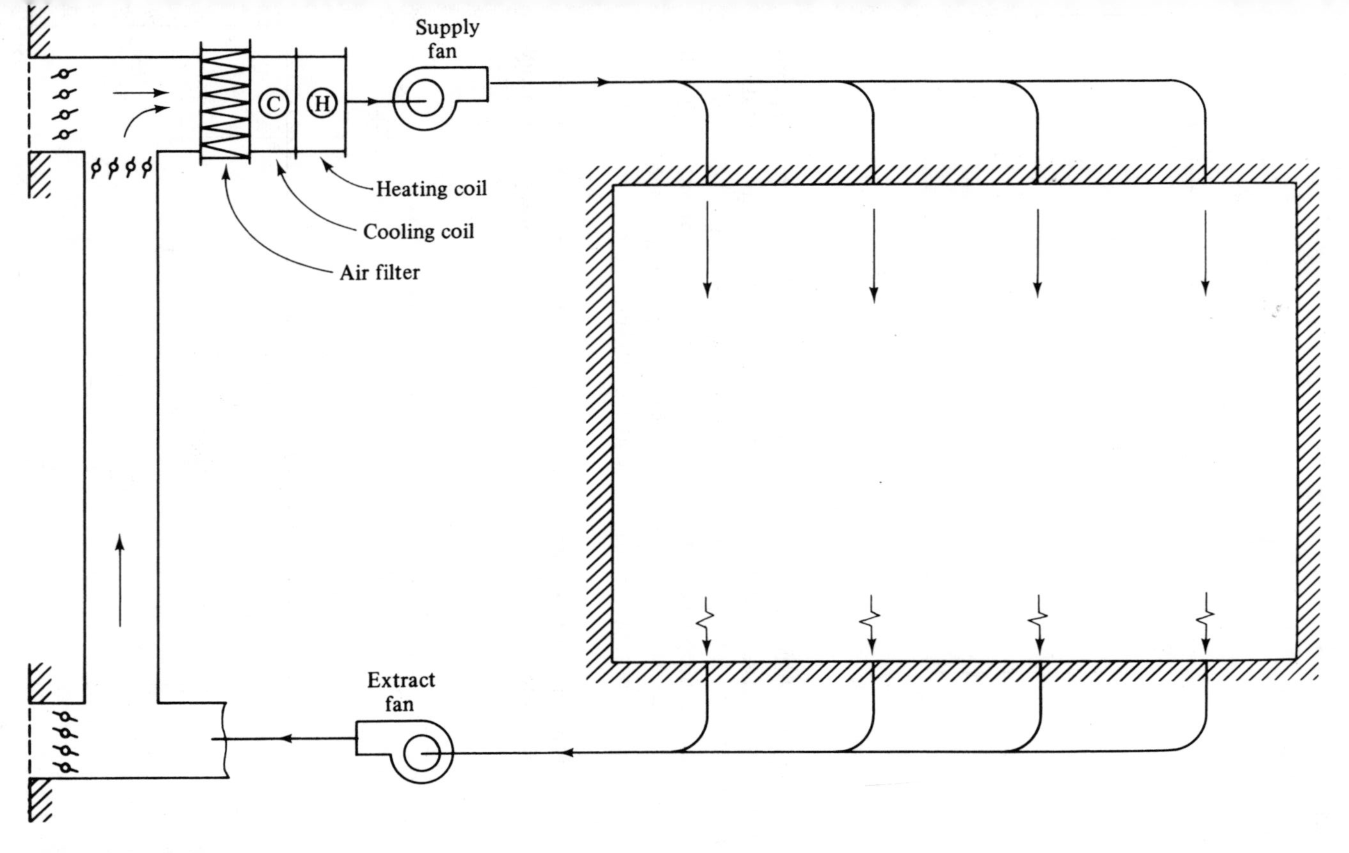

Figure 21-2 All-air system.

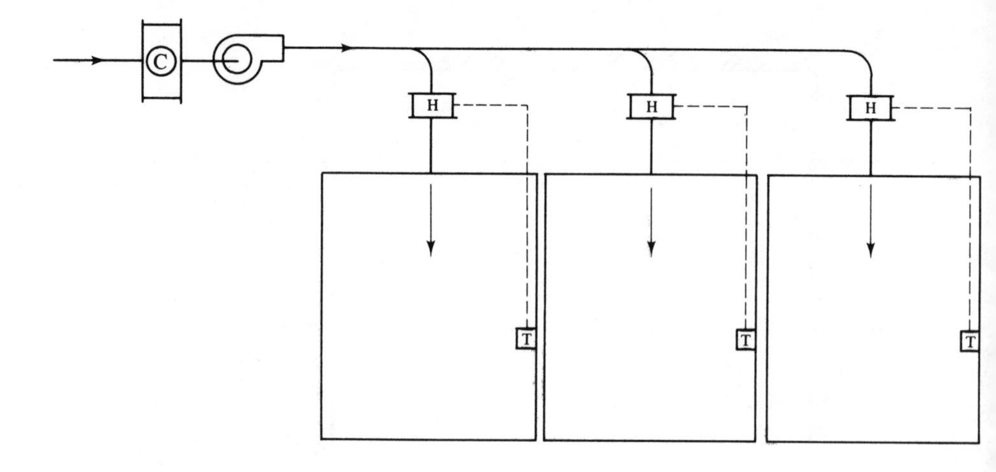

Figure 21-3 Reheat for individual zones.

Example 21-5 A room is to be maintained at 21°C, with a preferred 50 per cent saturation, using air at 13°C dry bulb, 78 per cent saturation and reheat. The load is 0.7 sensible/total ratio. (See Fig. 21-5.)

Air at the supply condition can be reheated to about 18°C and will rise from 18 to 21°C in the room, picking up the quantity of heat 'B' as shown. The final condition will be 50 per cent saturation, as required.

Alternatively, supply air is used directly, without reheat. It now picks up the quantity of heat 'A' (about three times as much) and only

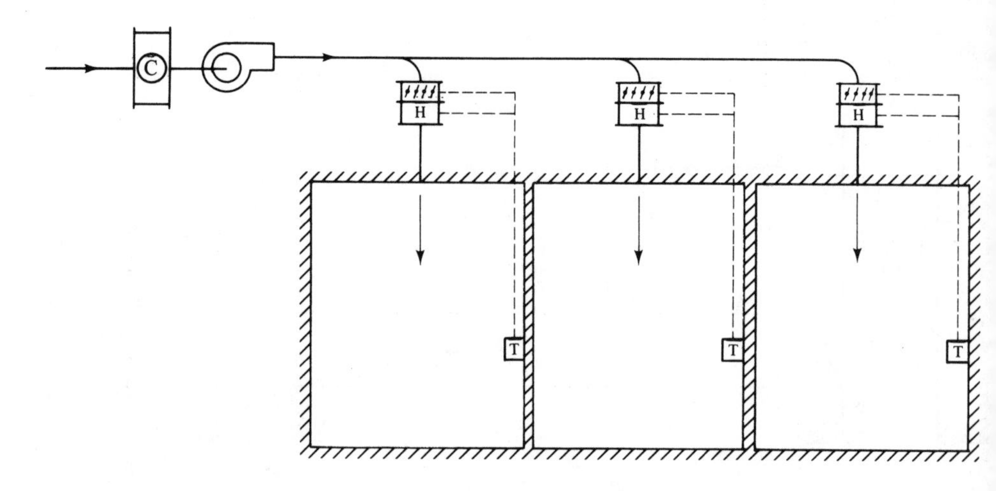

Figure 21-4 Variable air flow with reheat to individual zones.

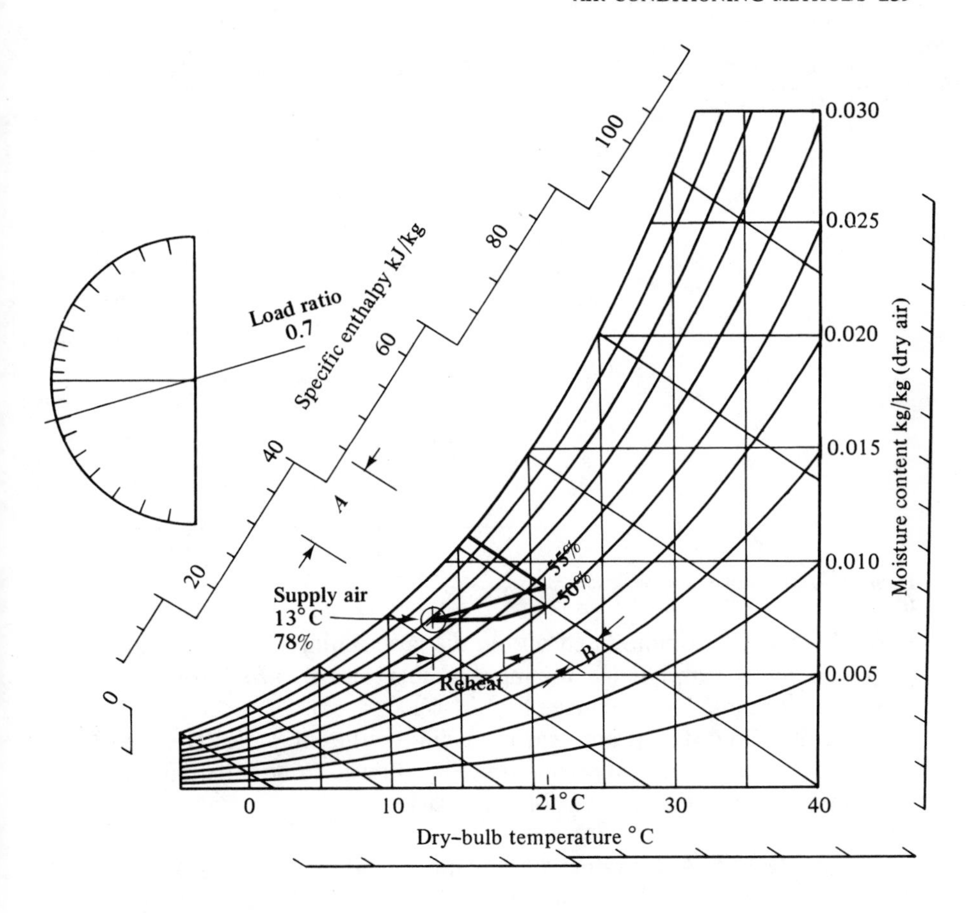

Figure 21-5 Zone differences.

one-third the amount of air is needed. The final condition will be about 55 per cent saturation. This is still well within comfort conditions, and should be acceptable.

With this variable volume method, the cold-air supply system will be required to deliver less air into the building during colder weather and must be capable of this degree of 'turn-down'. Lower than 30 per cent plant design mass flow it may be necessary to spill air back to the return duct, with loss of energy and, in some types, cold air in the ceiling void when trying to heat the room. If the final throttling is at the inlet grille, the reduction in grille area will give a higher outlet velocity, which will help to keep up the room circulation, even at lower mass flow. One type releases the room air in pulses, to stimulate room circulation.

The *dual-duct* system, having the second method of heating by blending cold and warm air, has reached a considerable degree of sophistication,

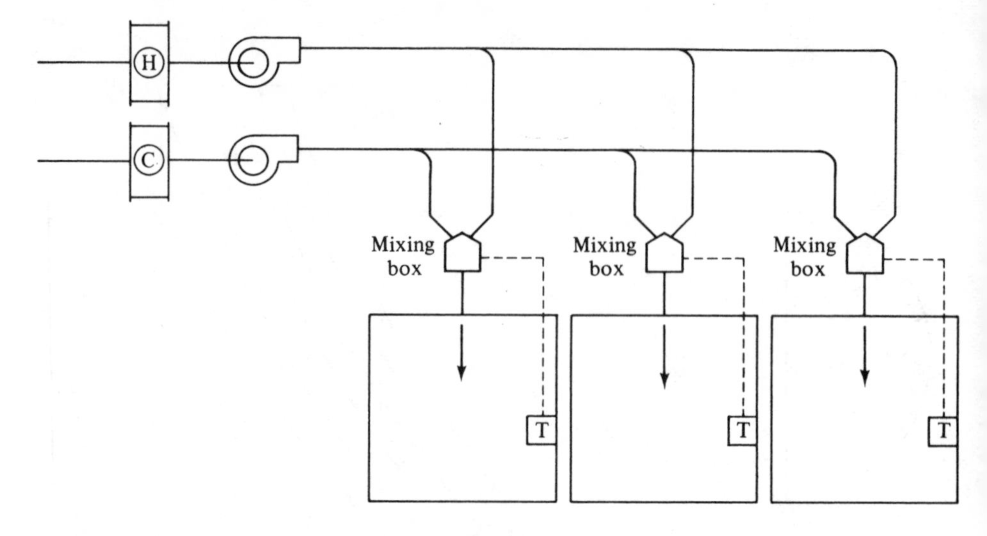

Figure 21-6 Dual duct supplying separate zones.

normally being accommodated within the false ceiling and having cold and warm air ducts supplying a mixing chamber and thence through ceiling grilles or slots into the zone (see Fig. 21-6).

The blending of cold and warm air will be thermostatically controlled, so that the humidity in each zone must be allowed to float, being lowest in the zones with the highest sensible heat ratio.

Example 21-6 A dual-duct system supplies air at 14°C dry bulb, 75 per cent saturation through one duct and at 25°C dry bulb, 45 per cent saturation, through the other. Two zones are to be maintained at 21°C and in both cases air leaves the mixing boxes at 17°C. Room A has no latent load. Room B has a sensible/total heat ratio of 0.7. What room conditions will result? (See Fig. 21-7.)

Air leaving the mixing boxes will lie along the line *HC*. For these two zones it will be at *M* (17°C dry bulb). For room A, air will enter at *M* and leave at *A*, the process line being horizontal, since there is no latent heat load. The final condition is about 50 per cent saturation. For room B, air enters at *M* and the slope of the line *MB* is from the sensible/total angle indicator. Condition B falls at about 56 per cent saturation.

This example gives an indication of the small and usually acceptable variations found with a well-designed dual-duct system. Since a constant total flow is required with the basic dual-duct circuit, a single fan may be used, blowing into cooling and heating branches. Where variation of volume is employed, one or two fans may be used, as convenient for the

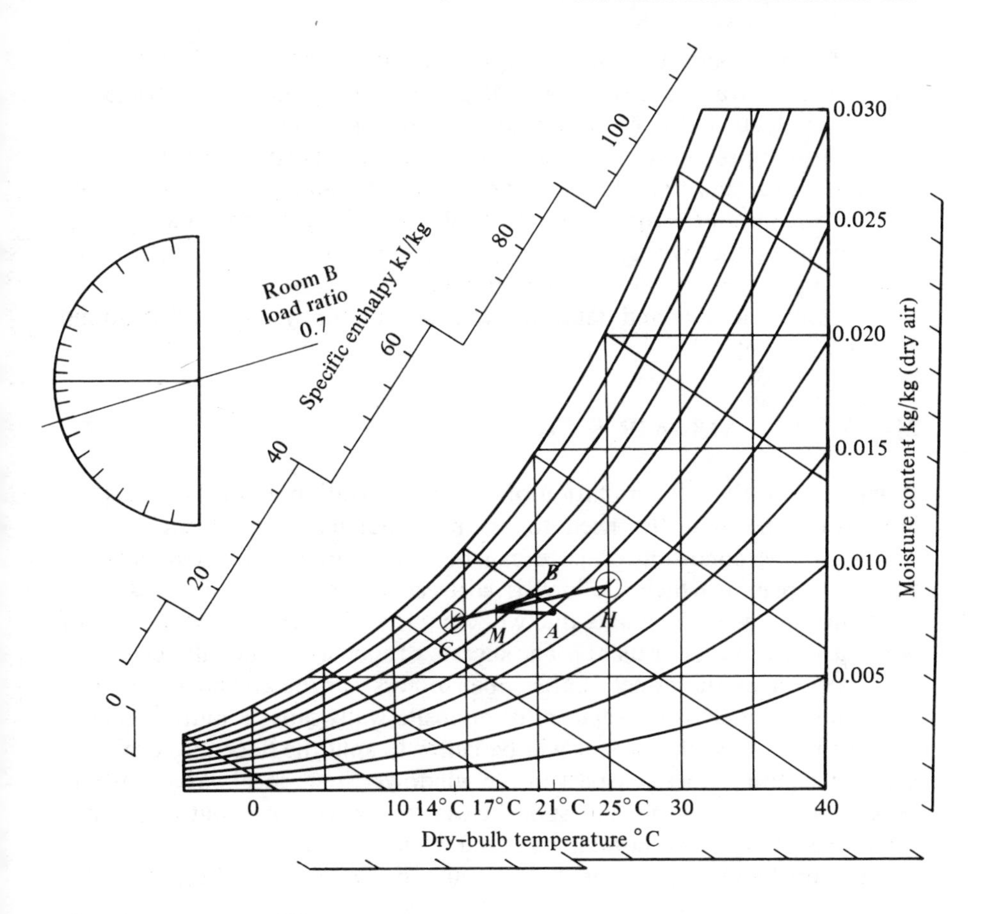

Figure 21-7 Dual-duct differences.

circuit. In all cases an independant extract fan and duct system will be required, so that the proportion of outside/recirculated air can be controlled.

Since about 0.1 m^3 of air flow is required for each kilowatt of cooling, the mass air flow for a large central station system will be large and the ductwork to take this very bulky. This represents a loss of available building space, both in terms of vertical feed ducts and the extra ceiling space to accommodate branches on each floor. For a tall building, it may be necessary to have a number of plantrooms for air handling equipment (fans, coils, filters) with the refrigeration machinery central. Instances will be seen in major cities of tall buildings having 'blank' floors to accommodate air handling plant.

Reduction of duct size can be achieved by increasing the velocity from a *low velocity* of 3 to 6.5 m/s to a *high velocity* of 12 to 30 m/s. Such velocities cause much higher pressure losses, requiring pressures in excess of 1 kPa, for which ductwork must be carefully designed and installed, to conserve energy and avoid leakage. The use of high velocity is restricted to the supply ducts and is not practical for return air ducting.

With a supply system pressure of 1 kPa and another 250 Pa for the return air duct, the total fan energy of a central all-air system will amount to 12.5 per cent of the maximum installed cooling load, and a much greater proportion of the average operating load. This power loss can only be reduced by careful attention to design factors.

A comprehensive and detailed analysis of all-air systems will be found in Ref. 17 (Chap. 3).

21-3 ZONE, ALL-AIR SYSTEMS

It will be seen that the limitation of the central station all-air system is the large ductwork and the need to arrange dual ducts or reheat to each branch. If the conditioned space can be broken down into a number of zones or areas in which the load is fairly constant, then a single-zone air handling unit with localized ductwork may be able to satisfy conditions without reheat in its branches. The success of such a system will depend on the selection of the zones. Large open offices can be considered as one zone, unless windows on adjacent or opposite walls cause a diurnal change in solar load. In such cases, it will be better to split the floor into arbitrary areas, depending on the aspect of the windows. Some local variations will occur and there may be 'hot spots' close to the windows, but conditions should generally be acceptable by comfort standards.

The air handling unit for the zone may be one of several types:

1. Direct expansion, supplied with refrigerant from the central plantroom
2. Chilled water air handling unit taking chilled water from a package or the central plantroom
3. Water-cooled packaged direct expansion unit, using condenser water from an external tower
4. Remote condenser (split) air-cooled direct expansion unit; condenser remote, possibly on roof
5. Air-cooled direct expansion unit, mounted adjacent to an outside wall, or through the roof

21-4 CENTRAL STATION, COMBINED AIR AND CHILLED WATER

The chilled air of the central station system serves the purpose of providing the proportion of fresh air needed, and carrying heat energy away from the space. These functions can be separated, using a more convenient fluid for the latter. Since the heat is at a temperature well above 0°C, the obvious choice of fluid is water, although brines are used for some applications.

The central plant is now required to supply chilled water through flow and return pipes, plus a much smaller quantity of fresh air. No air return duct may be required.

The chilled water will be fed to a number of air handling units, each sized for a suitable zone, where the conditions throughout the zone can be satisfied by the outlet air from the unit. This constraint has led to an increasing tendency to reduce the size of the zones in order to offer the widest range of comfort conditions within the space, until the units now serve a single room, or part of a room. Such units are made in wall-mounting form for perimeters or ceiling-mounted form to cover open areas. (See Fig. 21-8.) Larger units may be free standing.

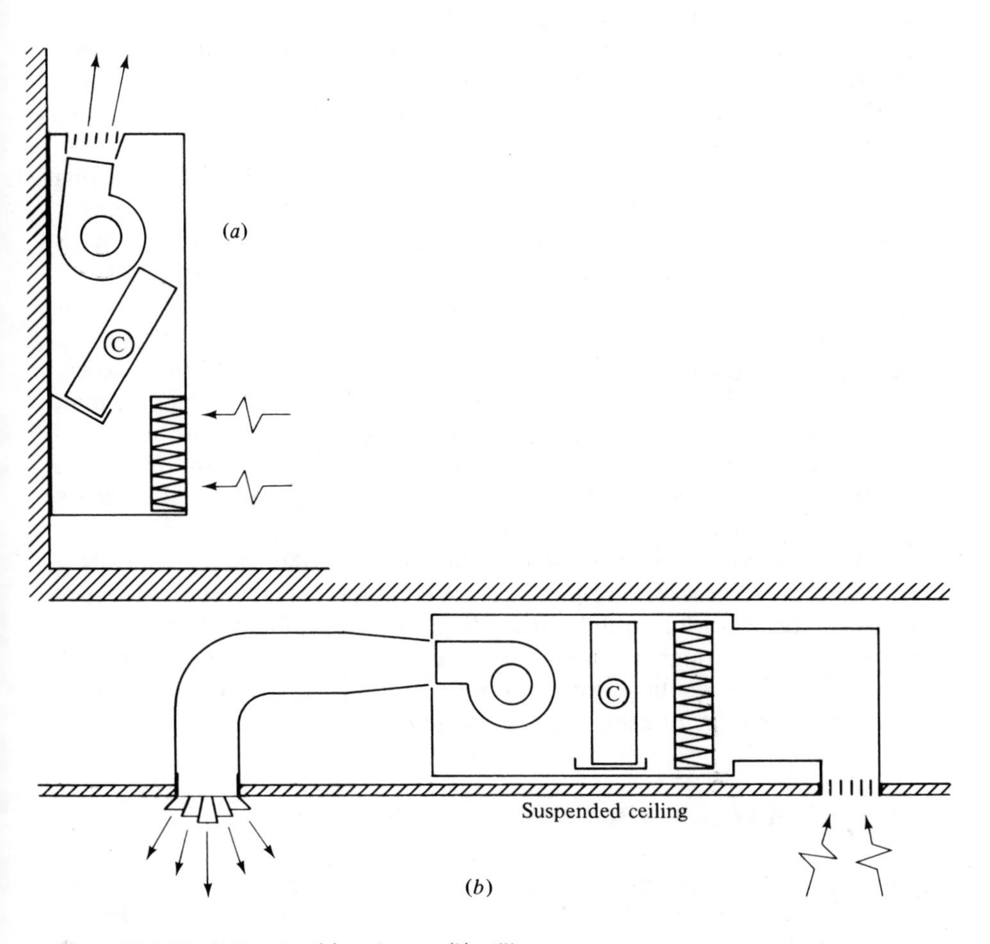

Figure 21-8 Fan coil units: (*a*) perimeter, (*b*) ceiling.

Two methods are used to circulate the room air over the chilled water coil. In the first, an electric fan draws in the air, through a filter, and then passes it over the coil before returning it to the space. The fan may be before or after the coil. The fresh air from the plantroom may be introduced through this unit, or elsewhere. The coil is normally operated with a fin temperature (ADP) below room dew point, so that some latent heat is removed by the coil, which requires a condensate drain. Multispeed fans are usual, so that the noise level can be reduced at times of light load.

The second method makes use of the pressure energy of the *primary* (fresh) air supply to induce room (*secondary*) air circulation. This air, at a pressure of 150 to 500 Pa is released through nozzles within the coil assembly and the resulting outlet velocity of 16 to 30 m/s entrains or *induces* room air to give a total circulation four or five times as much as the primary supply. This extra air passes over the chilled water coil. Most *induction units* are wall mounting for perimeter cooling, but they have been adapted for ceiling mounting. With the induction system, latent heat extraction can usually be handled by the primary air and they run with dry coils. Some systems have been installed, having high latent loads, which remove condensate at the coil.

In climates which have a well-defined summer and winter, heating when required can be obtained with fan coil or induction units, by supplying warm water to the coil instead of cold. Some variation of this is possible with induction systems which can, at times, have cold primary air with warm water, or vice versa, giving a degree of heating–cooling selection.

Most climates, however, have mid-seasons of uncertain weather so that heating and cooling may be required on the same day, and this is accentuated by buildings with large windows which may need cooling on winter days. For these applications, units need to have a continuous supply of both chilled and warm water and a suitable control to choose one or the other without wastage. This usually implies two separate coils and four pipes, with separate chilled and warm water circuits. (See Fig. 21-9.)

An alternative system, lower in first cost, is the *three-pipe* system. Chilled and warm water are piped to the coil unit and chosen by the room thermostatic valve for cooling or heating duty as required. Water leaving the coil passes through a common third return pipe back to the plantroom. At times of peak cooling load, very little warm water is used and there is little or no wastage of energy in this mixing of the water streams.

21-5 PACKAGED AIR-COOLING UNITS

This is no clear demarcation between a zone, served by a unit package, and a single room or part of a room. A zone is an arbitrary selection by the design engineer and air handling packages are available in a very

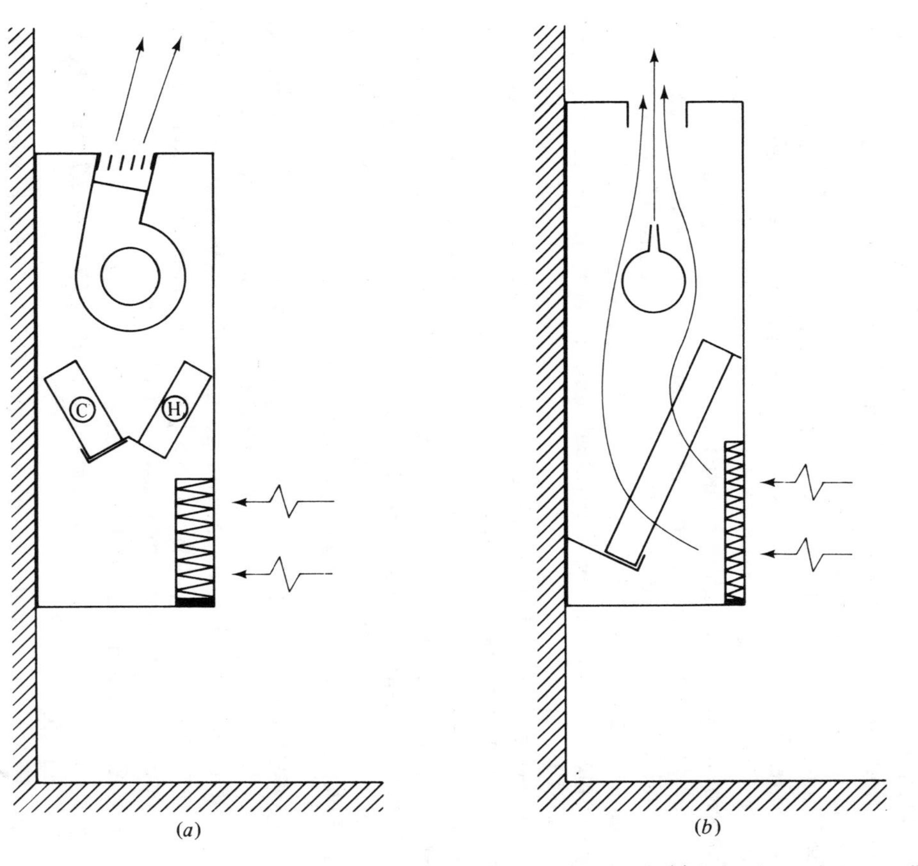

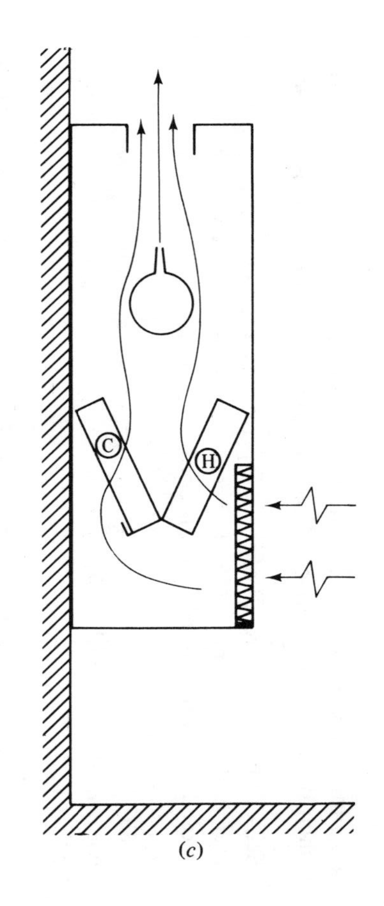

Figure 21-9 (*a*) Fan coil unit, two coil. (*b*) Induction unit, single coil. (*c*) Induction unit, two coil.

wide range of sizes to cope with such a range of loads. By definition, all such units are room air-conditioners, and fall into three classifications:

1. Self-contained, where all of the refrigeration circuit components are integral parts of the unit. If not specifically stated, it is assumed to be air cooled, i.e., with an integral air-cooled condenser.
2. Water cooled, having an integral water-cooled condenser.
3. Split, having the condenser remote from the air handling section, and connected to it with refrigeration piping. The compressor may be in either section.

Control of the indoor cooled condition will be by thermostat in the return airstream, and thus based on room dry bulb temperature. The resulting moisture content will depend on coil characteristics and air flow. Packaged air-conditioners for tropical applications commonly have a design coil sensible/total ratio in the order of 0.7 with entering air at 50 per cent saturation, and will give indoor conditions nearer 45 per cent saturation if used in temperature climates with less latent load (see Chapter 25).

Winter heating methods fitted within room air-conditioners may be electric resistance elements, hot water or steam coils, or reverse cycle (heat pump). One model of water-cooled unit operates with a condenser water temperature warm enough to be used also in the heating coil.

The *heat reclaim* packaged unit system comprises water-cooled room units with reverse cycle valves in the refrigeration circuits. The water circuit is maintained at 21 to 26°C, and may be used as a heat source or sink, depending on whether the individual unit is heating or cooling. (See Fig. 21-10.)

If the water circuit temperature rises above about 26°C, the cooling tower comes into operation to reject the surplus. If the circuit drops below 21°C, heat is taken from a boiler or other heat source to make up the deficiency. During mid-season operation within a large installation, many units may be cooling and many heating, so that energy rejected by the former can be used by the latter. With correct system adjustment, use of the boiler and tower can be minimized.

Example 21-7 A large office building is to be fitted with a packaged unit in each room. During mid-season, it is estimated that 350 rooms will require cooling at an average rate of 3.5 kW and another 120 rooms will require 2 kW of heating. Three alternative systems are proposed. Calculate running costs at this time.

(*a*) Air cooled units, COP 2.8, with electric heaters
(*b*) Air/air heat pump units, having a cooling COP of 2.7 and a heating COP of 2.2
(*c*) Heat reclaim units, having a cooling COP of 3.1, a heating COP of 2.6, and requiring an average of 28 kW for pumps and the tower fan

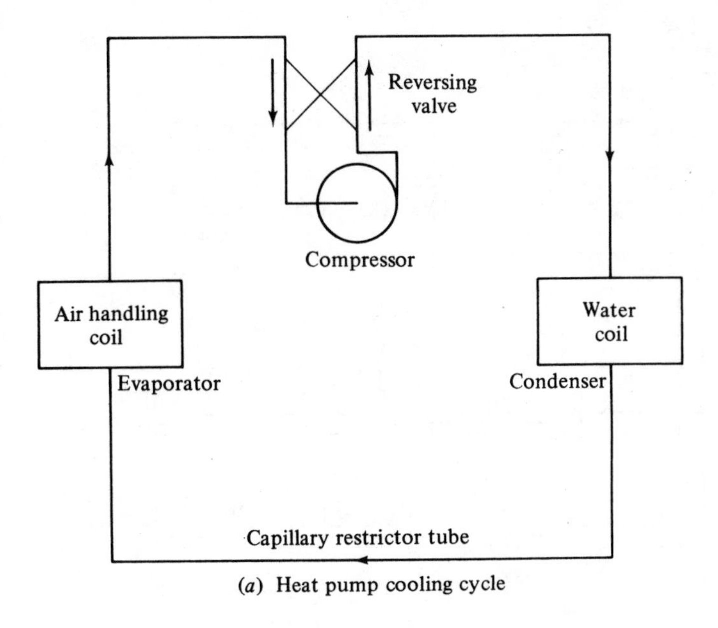

(*a*) Heat pump cooling cycle

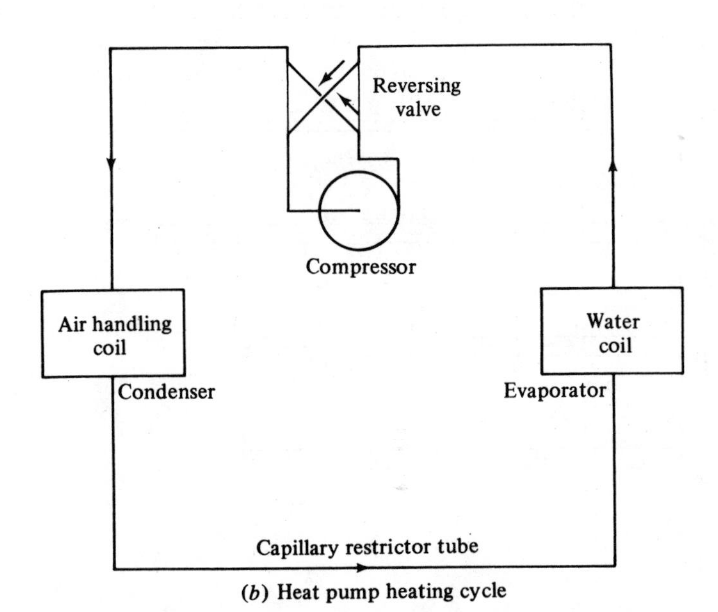

(*b*) Heat pump heating cycle

Figure 21-10 Heat reclaim system: (*a*) unit cooling, (*b*) unit heating.

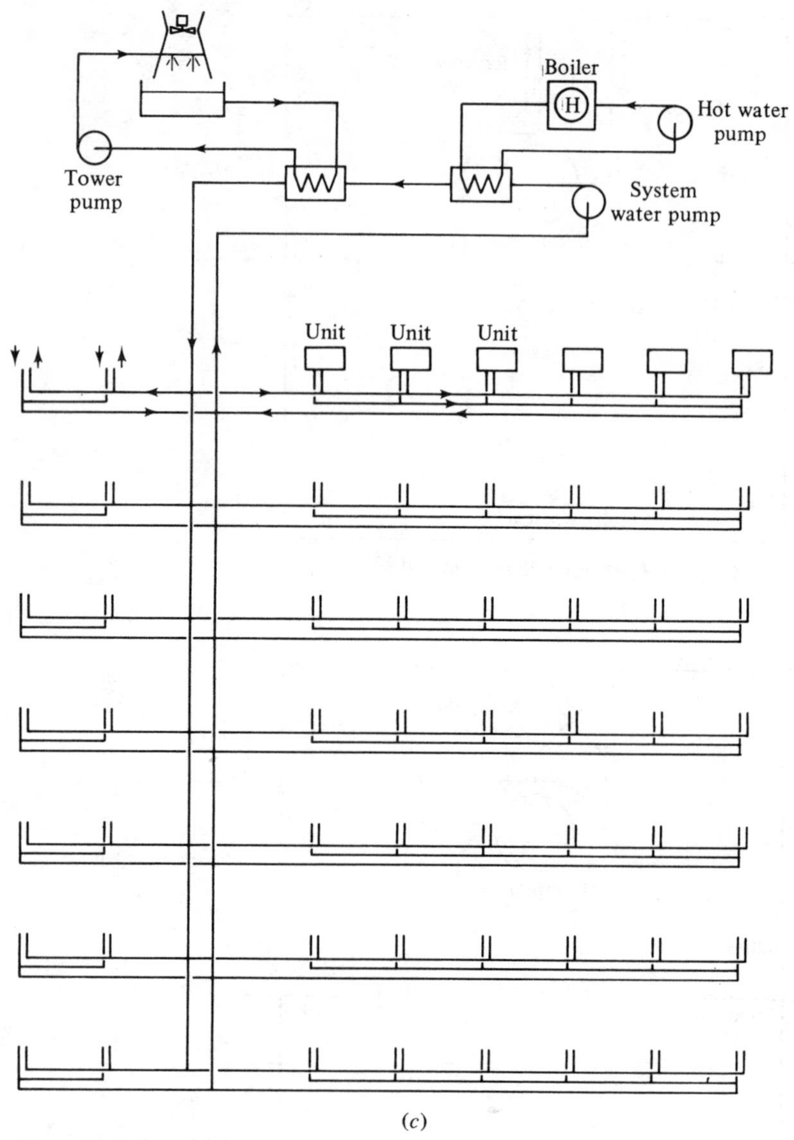

Figure 21-10 (*cont.*) (*c*) system.

Electricity costs 3.45p/kW h and gas for the boiler in system (*c*) costs 18p/therm and is burnt at an efficiency of 78 per cent, giving an overall cost of 0.79p/kW h.

	Hourly running costs, £		
	(*a*)	(*b*)	(*c*)
$350 \times \frac{3.5}{2.8} \times 0.0345$	15.1		
$120 \times 2 \times 0.0345$	8.3		
$350 \times \frac{3.5}{2.7} \times 0.0345$		15.6	
$120 \times \frac{2.0}{2.2} \times 0.0345$		3.8	
$350 \times \frac{3.5}{3.1} \times 0.0345$			13.6
$120 \times \frac{2.0}{2.6} \times 0.0345$			2.8
28×0.0345			1.0
$120\left(2 - \frac{2.0}{2.6}\right) \times 0.0079$			1.2
	23.4	19.4	18.6

It should be noted that this example is general, and indicates the type and method of cost analysis which should be made before the selection of an air-conditioning system for any building.

21-6 NOISE LEVELS

All-air systems have a noise level made up of the following:

1. Noise of central station machinery transmitted by air, building conduction, and duct borne
2. Noise from air flow within ducts
3. Grille outlet noise

The first of these can be reduced by suitable siting of the plantroom, by antivibration mounting and possible enclosure of the machinery. Air flow noise is a function of velocity and smooth flow. High velocity ducts usually need some acoustic treatment.

Grille noise will only be serious if long throws are used, or if poor duct design requires severe throttling on outlet dampers.

Apart from machinery noise, these noises are mostly 'white', i.e., with no discrete frequencies and they are comparatively easy to attenuate.

Where machinery of any type is mounted within or close to the conditioned area, discrete frequencies will be set up and will require some knowledge of their pattern before acoustic treatment can be specified. Manufacturers are now well aware of problems to the user and should be able to supply this basic data and offer technical assistance towards a solution.

Where several units of the same type are mounted within a space, discrete frequencies will be amplified and 'beat' notes will be apparent. Special treatment is usually called for, in the way of indirect air paths and mass-loaded panels.[3, 17, 41, 46]

Useful practical guidance can be gained by visiting existing installations before taking major decisions on new plant.

CHAPTER

TWENTY-TWO

CONTROL SYSTEMS

22-1 FUNCTION

The purpose of a control system on a refrigeration or air-conditioning plant is to:

1. Provide automatic operation, i.e., avoid the cost of an attendant labour force.
2. Maintain the controlled conditions closer than could be achieved by manual operation.
3. Provide maximum efficiency and economy of operation.
4. Ensure safe operation at all times.

The control system will consist of a loop, with *detector* (sensor), *controller*, and *controlled device*. The communication between these parts of the loop will be electric, pneumatic, or mechanical. (See Fig. 22-1.)

22-2 DETECTORS (SENSORS)

Types of detector are *two-position* (on–off) and *proportional*. The *two-position* will be set to actuate at upper and lower limits, and will respond when its sensitive element reaches these set limits. Since all devices have some time lag in operation, the controlled condition will overshoot to some extent, depending on the time lag of the detector and the extent to

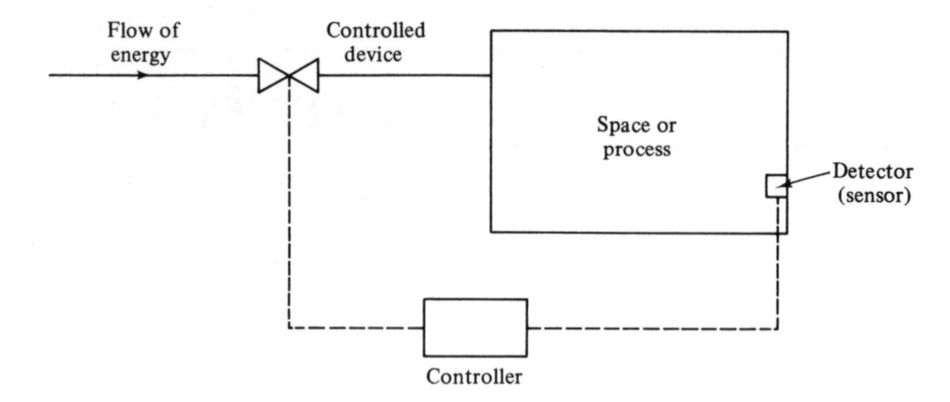

Figure 22-1 Basic control loop.

which the rate of supply of energy to the process exceeds the load. (See Fig. 22-2.) The range of the control will therefore be the differential of the detector plus the upper and lower overshoots under load.

Two-position detectors such as thermostats can be fitted with an anticipatory bias to reduce the amount of overshoot. In such instruments, a small bias heater accelerates the action of the control. An alternative method to reduce overshoot is to introduce a time delay in the controller, so that it acts slower or intermittently.

A two-position detector can be used to operate a *floating control*. At the upper limit it will operate the control in one direction and if it reaches the lower limit it will operate the control in the other direction. Between the two limits the control is not actuated. (See also Sec. 22-4.)

Two-position detectors can be classified according to the purpose:

Thermostatic	bimetal liquid expansion solid expansion vapour pressure
Pressure	diaphragm bellows bourdon tube
Fluid flow	moving vane
Time	clock bimetal and heater dashpot
Humidity	dimensional change of hygroscopic element
Level	float

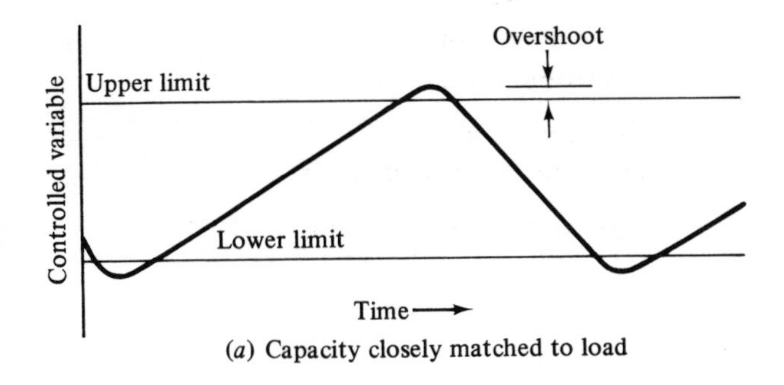

(a) Capacity closely matched to load

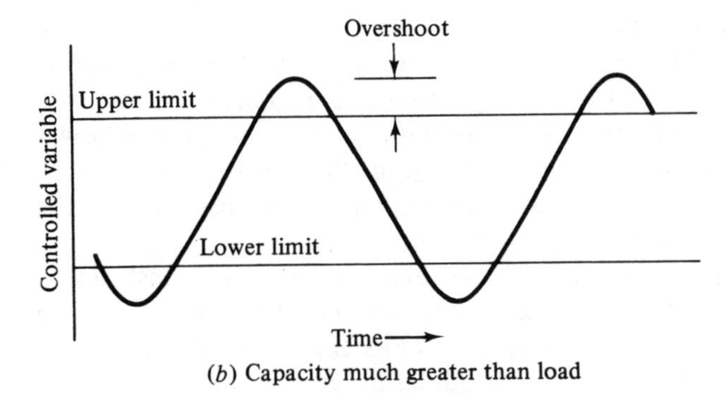

(b) Capacity much greater than load

Figure 22-2 Limits of controlled variable with two-position control: (a) capacity closely matched to load, (b) capacity much greater than load.

Many of these devices are direct acting on the controlled device and do not require a controller to process the signal.

Proportional detectors measure the process condition, which can then be compared by the controller with the required value. They are not direct acting, and need a controller to convert the signal to a working instruction to the controlled device. Proportional detectors include:

Temperature	those above, plus electrical resistance of a metal or a semi-conductor thermocouple infrared radiation
Pressure	those above, plus piezo-electric

Time	those above, plus electronic timing devices
Humidity	that above, plus resistance of a hygroscopic salt
Level	float with impedance coil

22-3 CONTROLLERS

If a controller is used with an on–off detector, it functions only as an *amplifier* to transmit the detector signal to the controlled device. It can modify the speed of this action by a bias or by a slow-speed operating motor, as in the floating control.

The *floating control* normally takes the form of a slowly rotating reversible motor moving a valve or operating a sequence of cams which control, in turn, steps of plant capacity. As the detector reaches its upper or lower limit it energizes the motor to advance or reduce the valve opening or the steps of plant capacity. When the condition has been satisfied and the detector moves away from the limit, the motor stops.

Some proportional detectors are combined in the same instrument with a suitable transducer which can perform some of the functions of a controller. For example, for pneumatic systems the primary sensing element actuates a variable air jet, thus modulating an air pressure which is transmitted to a further controller or direct to the controlled device. Electric and electronic detectors such as the infrared detector include the sensing and amplifying circuits of the instrument.

Controllers generally for use with proportional detectors will measure the displacement of the signal from a pre-set value and transmit a *proportional* signal to the controlling device. They may also be able to measure the rate of change of that signal (*derivative*) or be able to modify the rate of change of the output signal (*integral*). The effect of these capabilities is to anticipate the deviation and so give better response to changes of load. A controller having proportional, derivative, and integral actions is known as a *three-term* controller.

A controller may be arranged to accept input signals from more than one detector, e.g., the flow temperature of a hot water heating system may be raised at the request of an outdoor detector if the ambient falls, or may start the heating earlier in the morning to pre-heat the building before it is occupied; a servo back-pressure regulation valve (Fig. 8-4) can respond both to evaporator pressure and load temperature. With the recent advent of microcomputer devices almost any combination of signals can be processed by an electronic controller, providing the output signals can be made coherent and not conflicting.

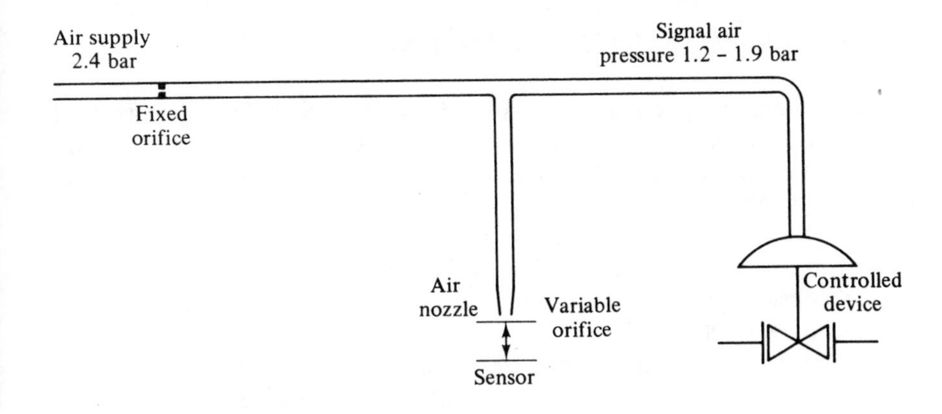

Figure 22-3 Pneumatic operation of controlled device.

Pneumatic controllers, which may include part of the sensing instrument, are supplied with compressed air at 2 bar (1 bar 'gauge'), which is allowed to escape from an orifice controlled by a detector. The resulting pressure modulates about 1.4 bar and is used in a servo piston, diaphragm, or bellows to actuate the controlled device. (See Fig. 22-3.)

22-4 CONTROLLED DEVICES

Controlled devices commonly consist of an actuator, which accepts the signal from the controller and works the final element. Typical examples are as follows:

1. Electric relay	operating	contactor motor motorized valve dampers
2. Electric solenoid	operating	solenoid valve
3. Modulated electronic signal	operating	magnetically positioned valve thyristor power control
4. Pneumatic pressure (and hydraulic)	operating	pneumatic relay valve positioner damper positioner

The effect of a controlled device may not be proportional to its movement. In particular, the shape of valve plugs and the angle of opening of dampers will not give a linear result, and the signal from the controller must take this into account.[3, 47]

22-5 CONTROL SYSTEM PLANNING

Control systems can quickly escalate into unmanageable complexity, and the initial approach to the design of a suitable control system should examine the purpose of each item, and the effect on others, to eliminate those which are not essential.

The action of a control may combine two or more of the purposes, as set out in Sec. 22-1, which may then be interdependent. It is more informative to consider the action of a control and examine what purpose it may serve in the circuit.

Controls for economic operation should ensure that functions are shut down when not needed (the boiler in summer and the chiller in winter). Optimum start controls now complement the starting clock, to advance or retard the starting time according to the ambient.

In planning a control system, a flow diagram is needed to indicate what may influence each item of plant. In many diagrams it will be seen that complexity arises and two items work in conflict. A typical instance is the cooling and dehumidifying of air, to a room condition lower than design, with concurrent operation of a humidifier.

Since most controls will be electrical and largely of the two-position type, it is a convenient notation to set out the initial control scheme as an electrical circuit and in 'book page' form, i.e., from left to right and line by line, to indicate the sequence of operation, with the controlled device always in the right-hand column. This analysis should indicate the different items which might act to produce a final effect and bring errors to light. Figure 22-4 is a simplified control circuit for a small air-conditioning system. Non-electric items can be shown on the same initial scheme, possibly with dotted lines to indicate a non-electric part of the system. The possibilities of abnormal operation should be examined, and grouped as *system not working*, *system unsafe*, and *system dangerous*, and protected accordingly. The last category requires two independent safety controls or one control and an alarm.

22-6 COMMISSIONING OF CONTROL SYSTEMS

The setting up, testing, and recording of all control functions of a refrigeration or air-conditioning system must be seen as part of the commissioning procedure. It requires that all items of equipment within the system are in working order and that the function of each item of control is checked, initially set at the design value (if this is known), readjusted as necessary during the testing stages, and finally placed on record as part of the commissioning documentation.

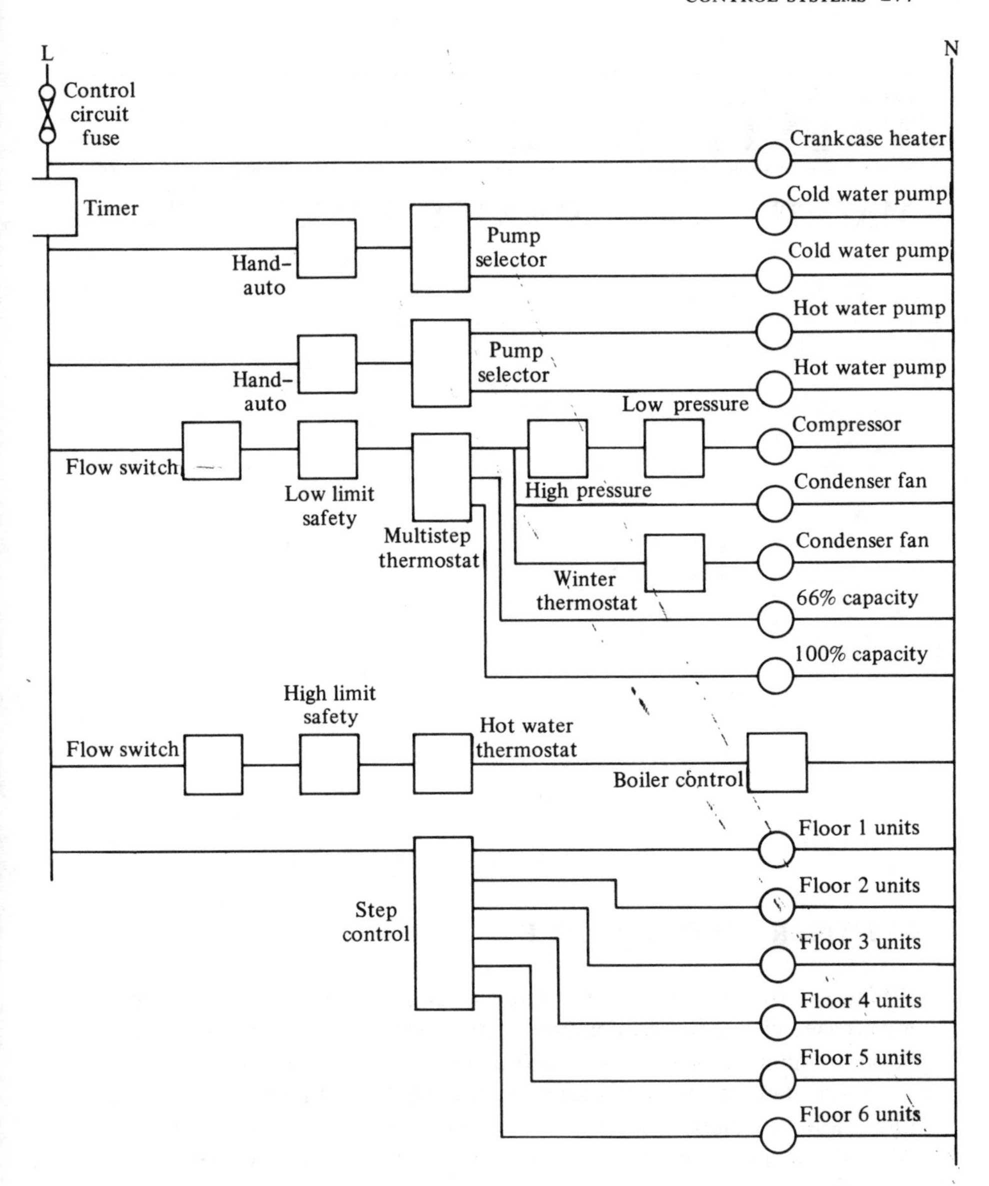

Figure 22-4 Electrical control diagram for small air-conditioning system.

Most controllers have adjustments, not only to the set points but to differentials, time delays, and response rates. It is of paramount importance that these are set up by an engineer who completely understands their function. Such settings should be marked on the instrument itself and recorded separately, since unauthorized persons may later upset these adjustments.[48]

CHAPTER

TWENTY-THREE

MAINTENANCE. SERVICE. FAULT-FINDING

23-1 USER MAINTENANCE

Where the user undertakes the day-to-day running of the plant, including most cases where the equipment is fully automatic in ordinary operation, basic maintenance will be assumed as part of this responsibility.

User maintenance includes operation where not automatic, cleaning of filters and strainers, attention to oil and lubricant levels, belt tensioning, general cleaning, running standby equipment, and verification of control operation.

Accumulated dirt on air filters will increase the resistance and lead to reduced air flow. This is by far the most frequent cause of malfunction of air-conditioning equipment.

Example 23-1 An R.22 direct expansion coil evaporates at 3°C when cooling air from 20 to 11°C. Condensing is at 35°C. If the air flow is reduced by 15 per cent because of a dirty filter, what is the approximate increase in running cost?

Ignoring second-order corrections:

$$\text{Air entering coil} = 20°\text{C}$$

$$\text{Air off coil at full air flow} = 11°\text{C}$$

$$\text{Evaporating temperature at full air flow} = 3°\text{C}$$

$$\text{ln MTD at full air flow, } \frac{17 - 8}{\ln(17/8)} = 11.94 \text{ K}$$

$$\text{Air off coil at 85\% air flow, } 20 - \frac{20 - 11}{0.85} = 9.41°\text{C}$$

$$\text{Coil performance at 85\% air flow, } (0.85)^{0.8} = 0.88$$

$$\text{ln MTD at 85\% air flow, } \frac{11.94}{0.88} = 13.6 \text{ K}$$

See the cooling curves in Fig. 23-1. The evaporating temperature will now fall to about 0.2°C. Compressor manufacturers' tables show 10.3 per cent loss in duty for 1.5 per cent less power at the new condition—an overall power increase of 9 per cent. Second-order corrections refine this to about 8 per cent extra power.

It is the responsibility of the supplying contractor to ensure that the user is aware of the need to clean or replace *air filters* and knows how to carry this out and when. Spare filters should be available so that the change of

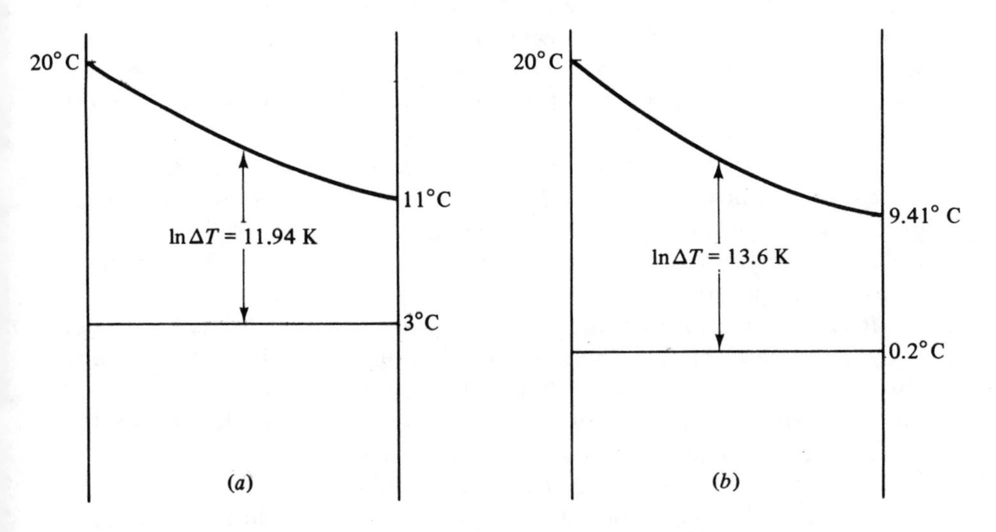

Figure 23-1 Effect of air flow reduction: (*a*) clean filters, (*b*) dirty filters.

clean for dirty can be made in the one operation and the dirty filters taken away in closed bags for cleaning or disposal, to prevent release of dirt in the conditioned area. It is an advantage if the person changing the filters has a hand vacuum cleaner to pick up dirt which may become dislodged, and to clean the filter frames.

Changing of large filters will need to be left until the plant can be shut down for the time required to carry out the work. Under no circumstances should fans run without filters in place, or dirt will be deposited in inaccessible parts of the plant.

The provision of a manometer across the filter to indicate the pressure drop will give a positive indication of the need to clean or replace. Such resistances can be estimated from the filter manufacturer's data and should be recorded at the time of commissioning and also marked at the filter.

Filters of the automatic roller type need to have an independent manometer, which will give warning in the event of malfunction of the winder.

Water strainers are of the cleanable type, either a single-mesh basket, which must be removed after isolating the water flow, or a twin construction which permits the cleaning of one while the other is working. Indication of a dirty strainer will be an increase in pump pressure, and it is essential to have a pressure gauge on the pump discharge.

Strainers should be located where they can be cleaned easily, from the point of view of accessibility, isolation of the water pipe, and where the small amount of escaping water can be tolerated. Strainers in closed water systems will need cleaning soon after first starting up the circuit, but little attention once the pipe dirt is flushed out. Open systems, such as water cooling towers, continuously wash dirt from the air and the frequency of cleaning must be judged from operating conditions, with a tendency to do so often rather than too seldom. Water tower strainers will not remove all the dirt. The larger particles will fall to the bottom of the sump and must be flushed out, possibly twice a year.

In many plants, the day-to-day operation is manually controlled and this requires a knowledge of, and familiarity with, the system which must be given by the installing contractor. A great deal of malfunction and inefficiency, many errors, and a few serious accidents arise from operation by untrained persons.

It is not sufficient that one person only has this knowledge. A clear set of operating instructions should be posted in the main plantroom, enabling any authorized person to start, run, and shut down the system in a correct, safe, and efficient manner. All staff who may be required to operate the plant need to be instructed and have some practice.

It is usual practice to mark the grade of lubricant on each item which might need periodic attention. Most equipment is designed to run for long periods without addition of lubricant and the dangers of adding too much should be noted.

The user will not normally add oil to a refrigeration system, apart from an industrial R.717 plant which will have a routine for the draining of parts of the circuit and replenishing the compressor sump.

Drive-belt tensioning and the replacement of broken or worn belts is a normal maintenance procedure, but may be missed if equipment is out of sight. A routine check will find these out.

The general cleanliness of plant is an indication of the care and interest taken by the maintenance staff, and is an encouragement to others working on it. There is no reason or excuse for accumulations of dirt and refuse on or around any system.

Standby plant needs to be run frequently, both to ensure that it is in working order, if required, and also to keep items such as shaft seals oiled, run-in, and thus gas-tight. The location and function of any changeover valves which must be operated in conjunction with standby plant should be clearly marked.

A great deal of malfunction, and some dangerous situations, arise from incorrect setting of control and safety instruments. It is assumed that these are all set and the correct settings recorded at the time of commissioning, but such settings may afterwards be tampered with by uninformed or unauthorized persons. The correct adjustment of any instruments normally set by the user remains his or her responsibility as a matter of routine operation. It is good practice to arrange that instruments are locked, sealed, or otherwise guarded from tampering—even to the extent of putting dummy controls in a conspicuous place.

The function of safety controls should be checked at least once a year.

Water treatment and corrosion inhibition systems require periodic attention, and full instructions should be left on site by the supplier or installer of the apparatus, whether or not they will be responsible for later attention.

Where water is evaporated from a circuit, such as a cooling tower, evaporative condenser, or humidifier, it must be remembered that there is no way of avoiding a steady bleed-off or frequent flushing, to restrict the concentration of dissolved solids. Much trouble arises from the efforts of well-wishing but misguided persons who stop the flow of bleed-off to 'save water'.

Many systems are shut down for periods of the year, either for process closure or if not required in winter. The advice of the supplier should be sought as to the correct procedure. In the case of refrigerant circuits, it is advisable to pump-down into the receiver or condenser to minimize leakage losses. Water towers should be drained in winter in this climate, if not in use, and the tank heater disconnected.

23-2 MAJOR MAINTENANCE WORK

The average user, unless of an industrial nature, will tackle only the simpler of these maintenance jobs and will entrust the major work to a specialist firm. Large concerns will have their own trained and skilled personnel and will do all work themselves.

The services of an outside maintenance contractor are usually on an annual contract basis and should clearly define the work which is to be carried out by the two parties. Day-to-day operation and the simpler tasks such as air filter cleaning are nearly always excluded from a contract of this sort.

Various types of contract are offered and it is recommended that the original supplier is approached during the commissioning period (or before) in order to find his suggestions. In particular, there may be friction if a rival firm undertakes maintenance while the plant is under guarantee and it is usual to let the first year's contract to the installer.

Most ammonia systems, even those with the most sophisticated separators, have continuous oil migration and require attention every week or, in some cases, every day to ensure maximum working efficiency. Plant operators need to be well trained and practised in this simple maintenance task, to minimize loss of refrigerant. In larger plants, the oil removed can be filtered and used again. It is useful to enter in the running log the quantities of oil removed and put in, since it has been known for large amounts of oil to accumulate in an evaporator without operators being aware of it.

Migration of oil over a long period in a dry expansion circuit should be treated as a design fault, and some action taken to put it right.

Air filters and water strainers have been mentioned in Sec. 23-1. Major work on those components and associated systems will be the cleaning and refilling of the oil sump of a viscous air filter and the periodic draining, cleaning, and flushing of water tanks.

Leaks of R.717 usually make themselves apparent and motivate staff to search out the leak and repair it. In the case of the halocarbons, a regular leak test should be part of the general maintenance schedule.

Under no circumstances should refrigerant be added to a leaking circuit without first making a repair. The one exception to this rule may be a continuous process plant, where the cost of a shut-down may override the cost and inherent danger of a small continuous leak.

Where gas is detected at the shaft gland of an open compressor which is not turning, the compressor should be run for a short time to relubricate the gland. The leak may then cease.

Staff should be forbidden to smoke while leak testing or repairing. Many operatives are ignorant of the danger to their health if smoking in the presence of traces of the halocarbons.

Moisture in halocarbon circuits will be indicated by the colour trace on the sight glass, where this is fitted. Immediate action is required, especially with an hermetic or semi-hermetic compressor, before damage is caused. The drier should be changed and the sight glass watched for reversal of the colour to 'dry'. Bad cases of contamination may need a second change of drier. If the liquid line leaving the drier or strainer (if separate) is colder than the inlet, there is a severe pressure drop within, indicating dirt. A new drier or cleaning of the strainer will cure this.

Heat exchanger surfaces need to be kept clean. Aqueous circuits (evaporator or condenser) can be cleaned with a chemical such as sulphamic acid, brushed or subjected to high-pressure water jets. In each case, all traces of dirt and chemical need to be removed from the circuit before it is put back to work. In cases of doubt, the manufacturer's advice should be sought. A layer of scale 2 mm thick on a condenser tube can cause a power increase of 16 per cent, and the need to clean a condenser can usually be deduced from the condensing pressure.

The checking and readjustment as necessary of all safety controls is an essential part of periodic maintenance—possibly annually. A time should be chosen when temporary stoppage of the plant will not cause inconvenience. Unsafe conditions can be set up by throttling valves, stopping pumps, or removing the load. In each case the relevant safety control should function at the pre-set conditions. Safety checks on specialized items such as fire dampers may be required from time to time by local authorities, and these checks, together with the expert advice available from the testing officers, should be welcomed as proof of the inherent safety of the installation.

It is essential that all major maintenance work and findings are recorded in the plant running log as a guide to the reliability of components, the need for cleaning, and other indicators to future work.

Refrigerant compressors, air compressors, and some other items of mechanical equipment should be subjected to a periodic part strip-down inspection and overhaul, as may be recommended by the manufacturer and indicated by running time meters or estimated running hours from the plant log sheet. Such planned maintenance entails, as its name suggests, some planning. Manuals, diagrams, and drawings should be obtained beforehand, and sources of possible spare parts located. Special tools or instruments may be required. The manufacturer should be able and willing to advise and guide in this work. Failed, worn, and other replaced parts should be retained for later examination and a post-mortem held in cases of doubt; records should be kept. In the case of a shut-down of a process plant, the major components should be tackled in rotation, lest a serious fault or shortage of spares prevent the process re-starting at the end of the closure.

Planned or preventative maintenance is not necessarily the best for all installations. If the service is not essential, and spares are known to be available within a reasonable time, nothing is done beyond obvious routine servicing work. Then, when a breakdown occurs, it is repaired. This approach, although not widely accepted, is an option which should be considered.

23-3 GUARANTEE PERIOD

Most equipment will be guaranteed by the supplier or contractor for 12 months from the date of supply, installation, or commissioning, and these precise details should be agreed and noted to avoid arguments later. In particular, some items may have left the factory several months before commissioning and, if the manufacturer is advised, the guarantee may be extended from that date. Such a guarantee will probably cover the cost of repair or replacement of the item, but not labour charges in removal or refitting. Again, these details should be noted.

Many disputes arise in this first year if the installation is not maintained to the satisfaction of the supplier, and a split responsibility of this sort is to be avoided. As already stated, where possible, maintenance for the first 12 months should be by the original contractors or a firm recommended by them.

23-4 FAULT-FINDING

System faults fall into two general classes: the sudden catastrophe of a mechanical breakdown and the slow fall-off of performance which can be detected as a malfunction in its early stages but will also lead to a breakdown if not rectified. Identification of the first will be obvious. To track down the cause of a malfunction will be more complicated.

Fault-tracing is seen as a multistep process of deduction, ending in normal operation again and a record of the incident to inform other operatives. The steps are as follows:

1. *Detection*, i.e., detection of abnormal operation
2. *Knowledge* of the system to track down the cause
3. *Observation* of exact operating conditions
4. *Identification* of the fault
5. *Decision*: what to do? how? when? can it be left?
6. *Action* to rectify the fault
7. *Test*: is it now normal?
8. *Record* note in log, for future information

A lot of help in fault-tracing may be had from charts for specific pieces of apparatus, prepared by the manufacturer.

Detailed examination of a sophisticated item may be beyond the skills of the plant operators and require the assistance of a specialist, such as an electronics engineer. Where such complications are part of the system, it is an advantage to know beforehand where such specialist help can be reached.

It should be accepted that fault analysis can be a slow process and that it usually defies prior estimates of the time it will take, regardless of the pressures of persons who are affected by the interruption of the service. In any case, hasty decisions and random efforts to get the plant working again are to be shunned, since more damage may result.

Training courses are available in analytical methods of fault-tracing. Minicomputers are also in use which will monitor a number of parameters and draw attention to any observed abnormality. The operator then has to interpret the data presented and make his own judgement of the cause. Further refinements of such aids are expected within the next few years.

23-5 SPARE PARTS

Except in the case of a planned overhaul, spare parts will only be wanted in an emergency, and then in a hurry. Most manufacturers can guarantee supplies from their own stocks and will dispatch quickly, providing they are given enough information to correctly identify the parts required.

There are two predominant problems. First, there is the necessity to keep the plant in operation, which may vary from a non-essential service such as the comfort cooling of an infrequently used room to the precise temperature control of an expensive or potentially dangerous process. The second problem is the time taken in transit, which might be as long as a year in parts of the world subject to excessive docks and customs delays.

The scale of spares to be held under these varying conditions must be judged by the user, taking into account the problems which might be met and seeking advice from the supplier. For remote installations, the latter may be asked to recommend 'Spares for one year's operation'—a classification which is subject to the vagaries of mechanical breakdown known colloquially as 'Murphy's Law'. A suggested scale of spares is given in Table 23-1.

23-6 TRAINING

The nature of refrigeration and air-conditioning equipment requires specialized training for operating and service personnel. The basic skills of those entering the industry should be the ability to read an engineering

Table 23-1

	Availability of spares at site			
Type of installation	Very good	Fair	Poor	Very bad
Air-conditioning				
comfort, small	Nil	Nil	A	B
comfort, large	Nil	R	A	B
essential process	D	D	D + A	D + B
Cold stores				
small, above 0°C	Nil	R	D + A	D + B
large, above 0°C	R	D	D + A	D + B
freezer, small	R	D	D + A	D + B
freezer, large	D	D	D + A	D + B
Industrial process	D	D	D + A	D + B

Code: R refrigerant
A refrigerant, drier, solenoid valve coil, compressor suction and discharge valves and gaskets, electrical controls, solenoids and overloads, any other items specifically recommended by the manufacturer
B as A, together with expansion valve
D duplicate equipment throughout plant to comply with Lloyd's Rules[24] which says, in effect, that the process shall not be at risk in the event of the failure of any one component

drawing and to read and understand flow, circuit, and electrical diagrams. When recruiting labour it will be an advantage if the candidate already has some knowledge of electrical circuits.

A contractor supplying new equipment should be required to instruct the staff who will operate it, and to advise where further instruction may be had. If the senior operator is allocated to work alongside the contractor during the final erection and commissioning stages, he should pick up enough knowledge of the system to give him the confidence and skill to run it.

Contractors generally select school-leavers, who then alternate sandwich technical courses with workshop or field training and experience under supervision. The more able students progress through the firm and can reach technician or engineer status. One of the larger contractors in the United Kingdom[49] runs short practical courses in various aspects of operation and maintenance, which are open to outside students. At least one major manufacturer[50] also runs five-day courses on operation and maintenance.

Courses with and without practical content, part- and full-time, and of various standards are run by technical colleges and colleges of higher education.[51] Two training organizations[52] set up short courses, mainly for plant engineers who wish to have a working knowledge of refrigeration and air-conditioning without becoming too deeply involved.

Degree and higher academic courses are held for full-time and part-time students at some of the universities and the Polytechnic of the South Bank, London. Details of these can be obtained direct or through the education authorities.

23-7 THE RUNNING LOG

The detection of abnormal operation can only occur if normal operation is monitored. Since refrigeration is a thermal cycle, the obvious readings to be taken will be temperature and the related refrigerant pressure.

The skilled operator or the visiting service mechanic will have a working knowledge of the pressures and temperatures to be expected, but will not be able to make an accurate assessment of the actual conditions without plant measurements for comparison. The commissioning log (see Sec. 24-4) will show readings taken at that time, but only at one set of running conditions.

It is therefore essential on a plant of any size to maintain some kind of running record, so that performance can be monitored with a view to detecting inefficiency and incipient troubles. The degree of complexity of this running log must be a matter of judgement, and a small amount of useful information is to be preferred to a mass of data which would be confusing. The following would seem to be basic:

1. Compressor suction and discharge pressures and corresponding temperatures.
2. Oil pressure gauge. It would be helpful to add a column so that true oil pressure can be entered (i.e., oil – suction).
3. The load temperature (room, water, brine, etc.).
4. Load flow rate or pump pressure.
5. Ambient temperature, dry bulb *and* wet bulb if possible.
6. Condenser water flow rate or pump pressure.
7. Any motor currents where ammeters are fitted.

These, together with space for comments, date, and time, should be set out as shown below.

Sarsaparilla Brewing & Bottling Co. Ltd. Running log.
Plant *Borough Road* Line No. *3* Date *8 Sept. 80*

	Pressures				Temperatures		Brine			Ambient				
Time	S	D	OG	O	S	D	In	Out	P	DB	WB	A	Comments	C
08.00	3.8	11	5.7	1.9	−8	27	+4	−1	3.1	11	9	190	—	*H*
10.00	3.1	11.5	4.9	1.8	−14	29	−3	−8	3.1	15	12	175	—	*H*
12.00	3.1	11.5	4.9	1.8	−14	29	−2	−7.5	3.1	19	13	175	—	*H*
14.00	3.1	12	4.9	1.8	−14	30	−3	−8	3.1	20	14	170	—	*H*

23-8 EXERCISES

Exercise 23-1 The motor driving an open compressor is switched on. When entering the plantroom 10 min later the compressor is not turning. List six possible reasons.

Answer

No mains electric supply
Fuse blown
One phase blown, out on single-phasing trip
Belts broken or slipping
Out on high-pressure cut-out (various)
Short of oil
Out on thermostat
Flow switch open, in load or condenser water

Exercise 23-2 A discharge pressure gauge reads 0.6 bar higher than the reading a week ago. List four possible reasons.

Answer

Higher ambient, dry or wet bulb
Higher load temperature or more flow
Dirty condenser
Condenser fan stopped
Non-condensible gas in system
Pump strainer dirty (condenser water)

Exercise 23-3 An automatic plant on a frozen food store uses R.717 and has three compressors working on five flooded evaporators. Draft a brief job specification for the senior operating mechanic.

Answer

Experience on ammonia
Knowledge of electrical controls and interlocks
House close by, prepared to work 'on call'
Reliable
Safety conscious
Able to instruct staff

CHAPTER

TWENTY-FOUR

COMMISSIONING

24-1 SPECIFICATION

The commissioning of a refrigeration or air-conditioning plant starts from the stage of static completion and progresses through the setting-to-work procedure and regulation to a state of full working order to specified requirements.

Since the final object of commissioning is to ensure that the equipment meets with a specified set of conditions, this specification must be clearly stated and, hopefully, would have been clearly stated when the contract was placed.

A contract should state at least the following:

1. The medium or product to be cooled, or the area to be cooled
2. The total required cooling capacity, or mass throughput of product with ingoing and outgoing temperatures
3. The required limits of control
4. A realistic ambient condition for condenser water or air, and for fresh air supply

Example 24-1 The equipment is to maintain a temperature of -10°C in a cold room measuring $X \times Y \times Z$ and insulated with 100 mm expanded polystyrene, and freeze 20 t/d of chilled beef entering at 0°C, assuming an ambient air temperature of 26°C.

Example 24-2 The plant is to cool water at the rate of 120 litre/s from 18 to 4°C. The ambient wet bulb temperature is 19°C.

Example 24-3 The plant is to have a capacity of 325 kW when cooling a 30 per cent aqueous solution of propylene glycol from −4 to −7.2°C. Water is available from the main cooling tower at 23°C.

Example 24-4 The direct expansion coil is to cool 6.7 m^3/s air from 21°C dry bulb, 50 per cent saturation, to 10°C dry bulb, 85 per cent saturation, when evaporating at 4°C.

Example 24-5 The air-conditioner is to maintain 23°C dry bulb plus or minus 0.75°C and 50 per cent saturation plus or minus 4 per cent, in the room shown on drawing *XYZ*, assuming an internal load of 28 kW, including the four occupants. The maximum ambient conditions are 28°C dry bulb, 19.5°C wet bulb.

If no such specification exists at the time of commissioning, some basis of acceptance must be agreed between the parties concerned.

Basic flow diagrams should be available and, if not, the commissioning engineer must draw one up, against which actual plant performance can be checked. (See Fig. 24-1.)

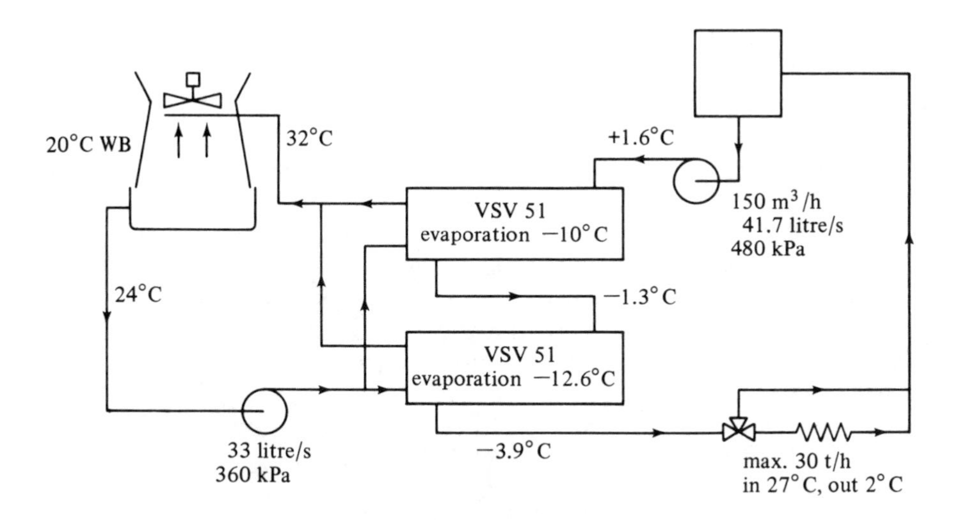

Figure 24-1 Basic flow diagram for liquid chilling process plant.

24-2 AUTHORITY

The work of commissioning must be under the control of a single competent authority, whether it be the main contractor, a consultant, or the user. Since this authority must accept the installation, it should be so stated in the original contract. Other specialists may be required during the course of the work and they submit their test figures and other data to this central authority.

24-3 SETTING TO WORK

The setting-to-work procedure needs to be carried out in a logical sequence, since subsystems are interdependent. The following order will be typical:

1. Check all wiring and electrical controls.[48]
2. Check action of all controls as far as may be possible without running any item.[48]
3. Check all water systems filled. Start pumps and check rotation, flows, and pressures.[53]
4. Start fans; check rotation, flows, and pressures.[54]
5. Balance duct and grille flows.[54]
6. Start main refrigeration system.[28] Allow to run on load until steady conditions are reached.
6. Set automatic controls to their approximate values, so the system will run without attention.

The services of specialist personnel and plant mechanics will be required during this period to operate the equipment and carry out any day-to-day attention. Care should be taken that this work does not come into conflict with the recommendations of suppliers, or invalidate their warranties. Where major items have not yet been accepted from suppliers, it will be advisable to retain their own commissioning engineer or other attendant until the project is complete. All necessary maintenance must be carried out, since any premature failure of a component may be blamed on such an omission.

The whole system is now left to run for a shake-down period, which may be from a few hours to several days, depending on the size and complexity. During this time, all components will be checked for vibration, leaks, or other malfunction, and remedial action taken.

Low-temperature systems and cold stores should be brought down slowly, to allow for shrinkage in the structure. A fall of 5 K/d is reasonable, moving more slowly through the band +2 to −2°C.

At the end of the shake-down period all strainers and filters are cleaned ready for the final test. If compressor oil is seen to be contaminated, this should be changed (see Sec. 9-11).

24-4 CALIBRATION AND FINAL COMMISSIONING

Final adjustments should now be made to the following, and any other items of this sort:

1. Air flows, by setting of dampers. This entails measurement at various points and comparison with the design figures.
2. Concentration of any brines present.
3. Water and other liquid flows.
4. Starter overloads and the settings of safety controls such as pressure cut-outs and safety thermostats.

In the final commissioning stage, readings are taken and recorded of all measurable quantities in the system, and compared with the specification and design figures. The following, as applicable, should be considered as the absolute minimum to be taken and recorded:

1. Ambient conditions, dry and wet bulbs
2. All fluid flows, temperatures, and pump, fan, and filter pressures
3. Refrigerant pressures and temperatures at expansion valve inlet, evaporator outlet, and compressor suction and discharge
4. Settings of all adjustable controls
5. Electric motor currents

It is probable that a full load cannot be obtained during the final test, for reasons of low ambient or lack of completion of other equipment for the process. In such circumstances, the commissioning engineer must establish the load which prevailed at the time and make an estimate of the system performance, on the basis of time run, or otherwise interpret the figures obtained. In such cases it may be advisable to agree to a tentative acceptance of the plant and carry out a full-load test at a later date.

Errors may come to light during this work and be possible to correct. If not, the acceptance or otherwise will be a matter for negotiation between contractor and customer. If no agreement can be reached, it may be necessary to refer the commissioning procedure to arbitration.

24-5 COMMISSIONING RECORDS

The commissioning log, which is the basis of the handover and acceptance of the system, should have as much of the original design data as may be available, cross-checked against readings taken during the final tests. Line flow diagrams, if not already supplied, should be prepared and kept with the final drawings and wiring and other control diagrams. Details of the information which should be gathered at this time are in the Commissioning Codes of the CIBS.[28, 48, 53, 54]

On acceptance, the following should be handed over to the customer:

1. A copy of the commissioning log.
2. Flow, control, electrical, and layout diagrams and drawings.
3. Operating instructions. It is usual to instruct the user's operators as part of these final works.
4. Copies of instructions and manuals for all proprietary items of equipment.
5. Maintenance instructions.

It must be particularly noted that details should be entered at this time of ambient and load conditions, and any other factors which have an interface outside the plant itself. In this way, the relationship between ambient, load, and plant can be established as a guide to future seasonal and load variations. It is helpful to the user if some indication can be given of operating conditions under light load, since the plant may well work most of its life at less than full load and the operators might not be able to interpret the readings taken.

Once this initial data has been recorded, a running log will indicate the performance under service conditions. As the load and ambient conditions change, the plant operators will be able to monitor the day-to-day conditions. This establishes normal running. Only by a clear understanding of what is normal can the abnormal be detected.

CHAPTER

TWENTY-FIVE

CATALOGUE SELECTION

25-1 GENERAL

Manufacturers will publish rating and application data for their products, based on standard test conditions and for the more usual range of uses. They cannot be expected to have accurate figures for every possible combination of conditions for an individual purpose, although most will produce estimates if asked.

The widespread use of packaged units of all sizes requires interpretation of catalogue data by applications engineers, sales engineers, and others, and by the end user.

The first step is to be certain of the basis of the published data and consider in what ways this will be affected by different conditions. Revised figures can then usually be determined. For extensive interpretation work, simple mathematical models of performance can be constructed.

25-2 COMPRESSORS

Refrigeration compressors which will probably be used on flooded evaporators (R.717 and the larger machines generally) will be rated with the suction at saturated conditions, since there will be little or no superheat in practice. Compressors for dry expansion systems may be rated at a stated amount of superheat, commonly 8 K.

Example 25-1 An ammonia compressor is rated at 312 kW with saturated suction at −15°C. It is installed with a very long suction line, causing a pressure drop of 0.4 bar, and picks up 6 K superheat from its evaporator condition. Estimate the capacity loss.

$$\begin{aligned} \text{Evaporator pressure at } -15°\text{C} &= \quad 2.36 \text{ bar} \\ \text{Suction pressure, } 2.36 - 0.4 &= \quad 1.96 \text{ bar} \\ \text{Rating suction temperature} &= -15.0°\text{C} \\ \text{Actual suction temperature, } -15 + 6 \text{ K} &= -\ \ 9.0°\text{C} \end{aligned}$$

The volume pumped by the compressor will remain about the same, but the density of the gas is reduced, and thence the mass flow.

Using the General Gas Laws:

$$\frac{m_2}{m_1} = \frac{p_2 T_1}{p_1 T_2} = \frac{1.96 \times 258.15}{2.36 \times 264.15} = 0.81$$

So the capacity loss is of the order of 19 per cent, or 59 kW. There may also be a slight drop due to the higher compression ratio, ignored here as the condensing pressure is not known.

Example 25-2 An R.22 compressor is rated at 15.9 kW when evaporating at −5°C, with 8 K superheat. Estimate the gain in capacity if it can be run safely with half the superheat.

$$\begin{aligned} \text{Rating suction temperature, } -5 + 8 &= \quad 3°\text{C} \\ &= 276.15 \text{ K} \\ \text{Working suction temperature, } -5 + 4 &= -1°\text{C} \\ &= 272.15 \text{ K} \end{aligned}$$

$$\text{Ratio of mass pumped} = \frac{m_2}{m_1} = \frac{T_1}{T_2} = \frac{276.15}{272.15} = 1.015$$

This gives a gain in capacity of about 1.5 per cent, or 0.24 kW.

Example 25-3 An hermetic compressor is rated at 18.2 kW for R.22 when evaporating at 7°C, suction superheated to 35°C, condensing at 54°C, and with 8 K sub-cooling of the liquid. Assuming that the inlet gas picks up another 30 K as it passes over the compressor motor, estimate the change in capacity if the suction is superheated to 12°C.

(*a*) Change in mass flow:

Compressor inlet temperature, rating, 35 + 30 = 65°C

= 338.15 K

actual, 12 + 30 = 42°C

= 315.15 K

$$\frac{m_2}{m_1} = \frac{T_1}{T_2} = \frac{338.15}{315.15} = 1.073$$

(*b*) Change in enthalpy (kJ/kg):

Enthalpy of suction gas at 35°C =	329.8	
Enthalpy of suction gas at 12°C =		311.7
Enthalpy of liquid at (54 − 8) 46°C =	157.0	157.0
Refrigerating effect (kJ/kg) =	172.8	154.7

$$\text{Reduction in overall working capacity} = 1.073 \times \frac{154.7}{172.8}$$

$$= 0.96$$

Working capacity = 17.5 kW

25-3 CONDENSING UNITS

Rating curves for condensing units (see also Sec. 10-2) will be for stated entering temperatures of the condensing medium—air or water. These may not go as high as the particular application may demand, and figures must be extrapolated.

Example 25-4 An air-cooled condensing unit is rated at 13.7 kW on R.22 when evaporating at 5°C and with ambient air at 43°C. Estimate the duty with ambient air at 52°C.

Some assumptions must be made regarding the condenser coil performance, and this may have a ΔT of 14 K between the entering air and condensing refrigerant and subcooling the liquid 5 K, with suction gas entering the compressor with 6 K superheat.

Rating condensing temperature, $43 + 14 = 57°C$

Working condensing temperature, $52 + 14 = 66°C$

Enthalpy of suction gas at (5 + 6) 11°C =	312.1	312.1
Enthalpy of liquid at (57 − 5) 52°C =	165.3	
Enthalpy of liquid at (66 − 5) 61°C =		178.5
Refrigerating effect (kJ/kg) =	146.8	133.6

In addition, the compression ratio has increased considerably, and there must be a correction for loss of volumetric efficiency.

Suction pressure at 5°C =	5.82 bar	5.82 bar
Discharge pressure at 57°C =	22.84 bar	
Discharge pressure at 66°C =		27.76 bar
Compression ratio =	3.92	4.77

From Fig. 2-6, the volumetric efficiency will fall, possibly from 0.75 to 0.68:

$$\text{Estimated new duty} = 13.7 \times \frac{133.6}{146.8} \times \frac{0.68}{0.75} = 11.3 \text{ kW}$$

25-4 EVAPORATORS

The rating of an evaporator will be proportional to the temperature difference between the refrigerant and the cooled medium. Since the latter is changing in temperature as it passes over the cooler surface (see Sec. 1-8), an accurate calculation for a particular load is tedious and subject to error.

To simplify the matching of air cooling evaporators to condensing units, evaporator duties are commonly expressed in *basic ratings* (see Fig. 25-1), in units of kilowatts per kelvin (formerly in British thermal units per hour per degree Fahrenheit). This rating factor is multiplied by the ΔT between the entering air and the refrigerant.

Example 25-5 An air cooling evaporator has a mass air flow of 8.4 kg/s and a published 'rating' of 3.8 kW/K. What will be its rated duty at −15°C coldroom temperature with refrigerant at −21°C? What is the true ln MTD?

Entering air temperature = −15°C

Refrigerant temperature = −21°C

'Rating' temperature difference = 6 K

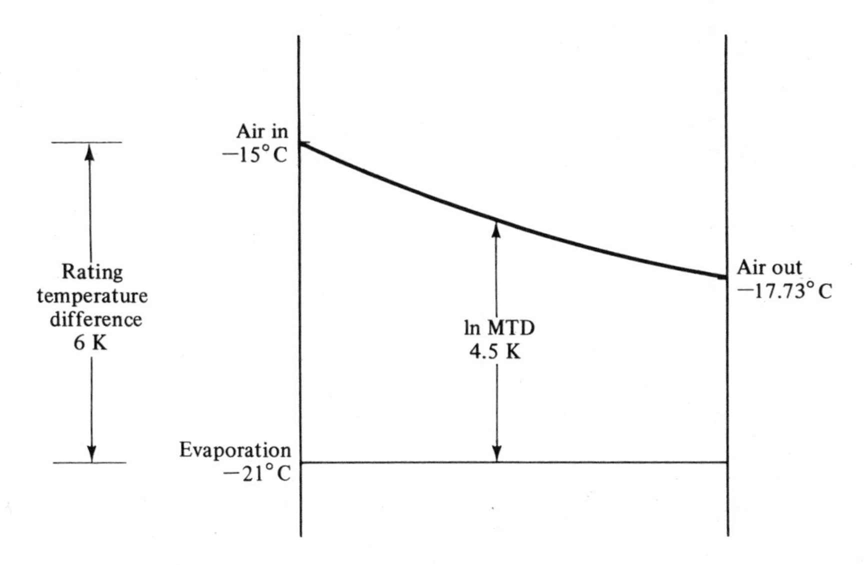

Figure 25-1 Basic rating and ln MTD.

$$\text{Rated duty, } 3.8 \times 6 = 22.8 \text{ kW}$$

$$\text{Reduction in air temperature, } \frac{22.8}{1.006 \times 8.4} = 2.73 \text{ K}$$

$$\text{Air leaving temperature} = -17.73°\text{C}$$

$$\text{True ln MTD, } \frac{6 - 3.27}{\ln(6/3.27)} = 4.5 \text{ K}$$

It follows that there would be an error at other conditions and the *basic rating* is only accurate at one point, so this short-cut factor must only be used within the range specified by the manufacturer.

The method of balancing such an evaporator with a condensing unit is graphical. The condensing unit capacity is shown as cooling duty against evaporator temperature, line *CD* in Fig. 25-2. The coil rating is plotted as the line *AB*, with *A* at the required coldroom (or 'air-on') temperature, and the slope of the line *AB* corresponding to the *basic rating*. The intersection of this line with the condensing unit curve *CD* gives the graphical solution of the system balance point. Similar constructions for higher condenser air conditions (*EF*, *GH*) or different room temperatures ($A_1\,B_1$) will show balance points for these conditions.

The graph also indicates the change in evaporating temperature and coil duty when the ambient is lower or higher than the design figure. This will show if there is any necessity to control the evaporating temperature in order to keep the correct plant operation. (See also Secs 8-8, 8-9, and 8-10.)

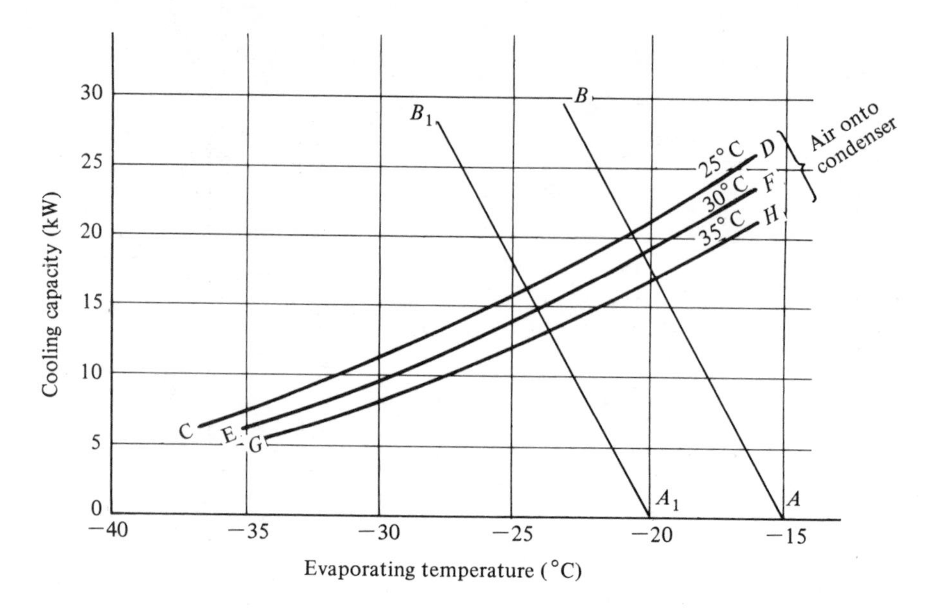

Figure 25-2 Graphical balance of evaporator with condensing unit.

25-5 REDUCTION OF AIR FLOW

Frequently, coil data will be available for a design air flow, but the system resistance reduces this flow to a lower value. There is a double effect: the lowering of the ln MTD and the lower heat transfer from the coil by convection.

The outer surface coefficient is the greatest thermal resistivity (compared with conduction through the coil material and the inside coefficient), and a rough estimate of the total sensible heat flow change can be made on the basis of[6, 7]

$$h = \text{constant} \times (V)^{0.8}$$

Example 25-6 An air cooling coil extracts 45 kW sensible heat with air entering at 24°C and leaving at 18°C, with the refrigerant evaporating at 11°C. Estimate the cooling capacity at 95, 90, and 85 per cent mass air flow.

$$\text{Design mass air flow} = \frac{45}{1.006 \times (24 - 18)} = 7.455 \text{ kg/s}$$

An approximate analysis comes out:

	Mass air flow (%)				
	100	95	90	85	
Mass air flow	7.455	7.08	6.71	6.34	kg/s
Air temperature on coil	24	24	24	24	°C
ΔT for 45 kW	6	6.3	6.7	7.1	K
Air temperature off coil	18	17.7	17.3	16.9	°C
ln MTD, refrigerant at 11°C	9.7	9.5	9.2	9.0	K
h in terms of design figure (from $V^{0.8}$)	100	96	92	88	%
Capacity, $45 \times h \times \dfrac{\ln \text{MTD}}{9.7}$	45	42.3	39.3	36.7	kW

This first estimate is now subject to second-order corrections, but gives a rough indication of the fall-off in coil duty.

25-6 ROOM AIR-CONDITIONERS

The catalogue rated cooling capacity of a room air-conditioner, if not qualified, will be based on ASHRAE Standard 16–69. This specifies test conditions of air onto the evaporator at 80°F dry bulb, 50 per cent RH (26.7°C, 48.1 per cent saturation), and air onto the condenser at 95°F dry bulb, 75°F wet bulb (35°C dry bulb, 23.9°C wet bulb). The original basis of the specification was the conditions prevailing in the mass-market area of the United States.

For these units, British Standard 2852 : 1970 gives three sets of alternative rating conditions, corresponding to ASHRAE, for tropical and temperate markets. They are:

	Room air	Temperature	Outside	Air temperature
	DB	WB	DB	WB
Condition *A*	27	19	35	24
Condition *B*	29	19	46	24
Condition *C*	21	15	27	19

and catalogue ratings quoting BS.2852 will be qualified with the appropriate conditions letter.

The International Document ISO R 859 evolved from existing national standards and does not specify any test conditions, only test methods. Any catalogue ratings quoting this ISO must be qualified with test conditions.

Performance of the average commercial room air-conditioner at BS.2852, condition *C*, will be some 10 to 15 per cent lower than at condition *A*, since it will evaporate some 5 K lower. This reduction factor should be applied to any unqualified unit rating if used under UK ambient conditions.

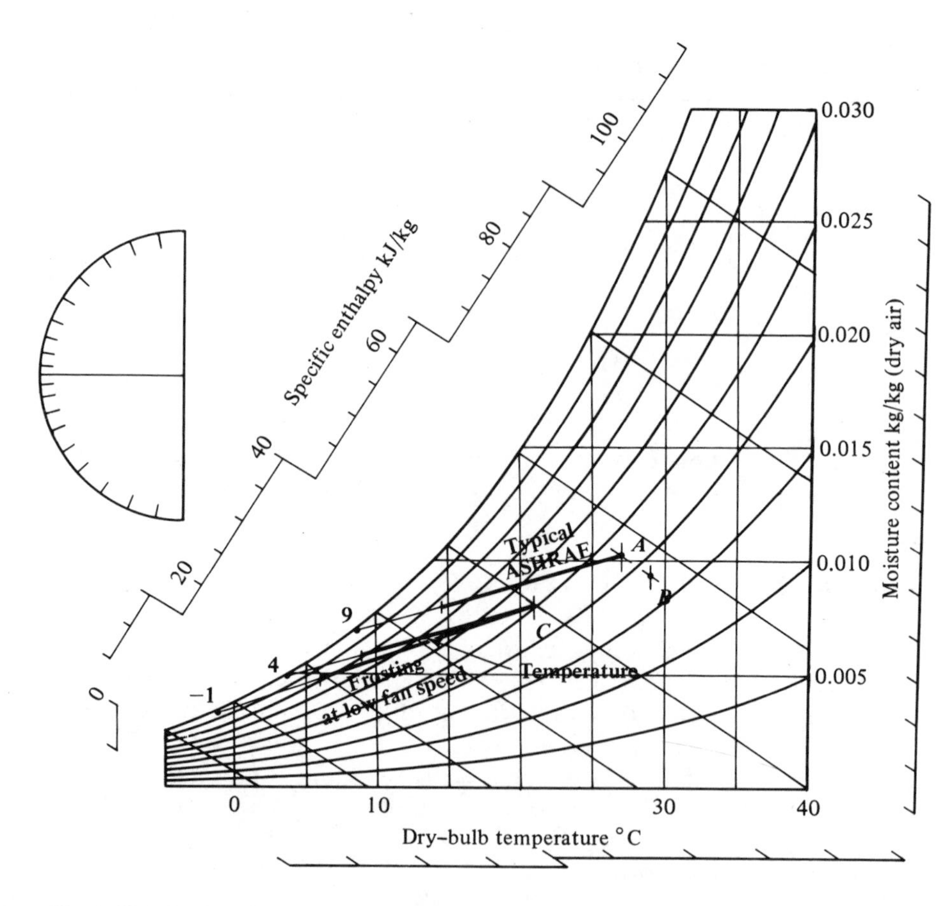

Figure 25-3 Typical process lines for room air-conditioners.

A further complication arises with the application of room air-conditioners which have been designed primarily for the tropical markets. (See Fig. 25-3). The evaporating temperature may fall to a level where the coil frosts (−3°C or thereabouts), and steps have to be taken to defrost the coil. As will be seen from Example 25-6, this will be accentuated at low fan speeds where there is a speed control.

25-7 ANALYTICAL CATALOGUE SELECTION

Since a large proportion of refrigeration and air-conditioning equipment will be bought from catalogue data, an analytical approach should be adopted to ensure correct selection. The principles to be applied are those of *value analysis*—to start with the basic need and no preconceived method, to consider all the different methods of satisfying the need, and to evaluate each of these objectively before moving towards a choice.

The details of such an approach will vary considerably, and the following guidelines should be taken as an indication of the factors to be considered, rather than as an exhaustive list:

1. What is the basic need?

   To cool something, a dry product, in air: temperature
   humidity
   maximum air speed

   other solid product

   a liquid: what liquid?
   temperature range
   viscosity

   To keep something cool, a solid product / an enclosed space } conditions

2. What is the load?
   - Temperature
   - If at ambient, can it be done without mechanical refrigeration?
   - Product cooling load
   - Heat leakage, sensible and latent
   - Convection heat gains, sensible and latent
   - Internal heat gains
   - Time required
3. Constraints
   - Degree of reliability
   - Position of plant
   - Automatic/manned?
   - Refrigerant
   - Same type of equipment as existing?
4. Possible methods
   - Direct expansion
   - Indirect—what medium?
   - Part by tower water or ambient air?
   - Thermal storage?
   - Existing plant spare capacity
5. Location
   - Plantroom
   - Adjacent space
   - Within cooled space
   - Maintenance access

6. Condenser
   - Inbuilt: water
     - air
   - Remote
   - Availability of cooling medium
   - Maintenance access
7. Economy of first cost and running costs
8. Options

If these steps have been carried through in an objective manner, there will be at least three options for most projects, and possibly as many as five.

Enquiries can now go out for equipment to satisfy the need, based on the options presented. No attempt should be made to reach a decision until these have been evaluated.

REFERENCES AND NOTES

1. *Handbook of Fundamentals*, American Society of Heating, Refrigerating and Air-Conditioning Engineers (ASHRAE), New York.
2. *Guide Book A*, Chartered Institution of Building Services (CIBS), London.
3. *Guide Book B*, CIBS, London.
4. *Guide Book C*, CIBS, London.
5. R. M. E. Diamant, *Insulation Deskbook*, Heating & Ventilating Publications, Croydon, 1977.
6. A. J. Ede, *An Introduction to Heat Transfer*, Pergamon Press, Oxford, 1967.
7. M. Knudsen and D. L. Katz, *Fluid Dynamics and Heat Transfer*, McGraw-Hill, 1958.
8. R. J. Dossat, *Principles of Refrigeration*, John Wiley, New York, 1961.
9. W. M. Kays and A. L. London, *Compact Heat Exchangers*, National Press, Palo Alto, 1955.
10. T. K. Sherwood, R. L. Pigford, and C. R. Wilke, *Mass Transfer*, McGraw-Hill, New York, 1975.
11. V. Chlumsky, and R. W. Webb (eds), *Reciprocating and Rotary Compressors*, SNTL, Prague, 1965.
12. *Refrigeration Safety*, BS.4434: 1980.
13. *Arcton Refrigeration Service Engineer's Handbook*, ICI Ltd., 1978.
14. *Refrigerants*, BS 4580 : 1970.
15. *Equipment Handbook*, ASHRAE, New York.
16. D. E. Kvalnes, *The Sealed Tube Test for Refrigeration Oils*, ASHRAE Trans., 1965.
17. *Systems Handbook*, ASHRAE, New York.
18. E. F. Wojtkowski, *System Contamination and Cleanup*, ASHRAE Journal, June 1964.
19. J. D. Gurney and I. A. Cotter, *Cooling Towers*, Maclaren, London, 1966.
20. F. L. Pettman, *Design and Manufacture of Packaged Air Conditioning Units*, Proc. Inst. Refrigeration, London, 1962.
21. C. T. Gosling, *Applied Air Conditioning and Refrigeration*, Applied Science Publishers, London, 1974.

22. *Glossary of Refrigeration, etc. Terms*, BS 5643 : 1979.
23. G. Lorentzen, *Design of Refrigerant Recirculation Systems*, Proc. Inst. Refn., 1976.
24. *Rules for the Survey and Classification of Refrigerated Stores*, Lloyd's Register, London, 1976.
25. *Design and Construction of Systems using Ammonia*, Inst. Refn., 1979.
26. Courses in Contract Management are run by HVTC Management Education, London.
27. *Reciprocating Refrigeration Manual*, The Trane Company, La Crosse, Wisconsin, 1964.
28. *Commissioning Code R (Refrigeration)*, CIBS, London, 1973.
29. *Methods for the Testing of Refrigerant Condensing Units*, BS 1586 : 1966.
30. *Applications Handbook*, ASHRAE.
31. J. C. Fidler, *Controlled Atmosphere Storage of Apples*, Proc. Inst. Refn., 1965.
32. S. Forbes-Pearson, *Performance of a High Efficiency Air Blast Freezer*, Proc. Inst. Refn., 1977.
33. H. W. Miller and T. P. Gordon Brown, *Recent Developments in Ground Freezing*, Proc. Inst. Refn., 1967.
34. C. Bailey and R. P. Cox, *The Chilling of Beef Carcases*, Proc. Inst. Refn., 1976.
35. *Refrigerating Techniques in Developing Countries*, International Institute of Refrigeration, Paris, 1965.
36. East Malling Research Station.
37. *Trane Air Conditioning Manual*, The Trane Company, La Crosse, Wisconsin, 1965.
38. W. P. Jones, *Air Conditioning Engineering*, Edward Arnold, London, 1973.
39. *Packaged Air Conditioning*, Electricity Council, London, 1974.
40. *ATKOOL* and *KOSWING*, W. S. Atkins & Partners, Epsom, Surrey.
41. I. Sharland, *Woods Practical Guide to Noise Control*, Woods Acoustics, Colchester, 1973.
42. B. B. Daly, *Woods Practical Guide to Fan Engineering*, Woods, Colchester, 1979.
43. Heating and Ventilating Contractors Association, London.
44. ASHRAE Research Report 1534.
45. P. J. Jackman, BSIRA Laboratory Reports Nos. 65 and 71.
46. C. M. Harris, *Handbook of Noise Control*, McGraw-Hill, New York, 1957.
47. J. E. Haines, *Automatic Control of Heating and Air-Conditioning*, McGraw-Hill, New York, 1953.
48. *Commissioning Code C, Automatic Controls*, CIBS, London, 1973.
49. Haden Carrier, Croydon, Greater London.
50. Hall-Thermotank Products, Dartford.
51. Notably Grimsby and Willesden Colleges of Technology.
52. PERA of Melton Mowbray and Target Training of Grantham.
53. *Commissioning Code W, Water*, CIBS, London, 1976.
54. *Commissioning Code A, Air Distribution*, CIBS, London, 1971.

INDEX